Science and Reform

Selected works of Charles Babbage

Science and Reform
Selected works of Charles Babbage

Chosen with introduction and discussion by Anthony Hyman

The right of the
University of Cambridge
to print and sell
all manner of books
was granted by
Henry VIII in 1534.
The University has printed
and published continuously
since 1584.

CAMBRIDGE UNIVERSITY PRESS

Cambridge
New York   New Rochelle
Melbourne   Sydney

CAMBRIDGE UNIVERSITY PRESS
Cambridge, New York, Melbourne, Madrid, Cape Town, Singapore, São Paulo

Cambridge University Press
The Edinburgh Building, Cambridge CB2 2RU, UK

Published in the United States of America by Cambridge University Press, New York

www.cambridge.org
Information on this title: www.cambridge.org/9780521343114

First published 1989
This digitally printed first paperback version 2007

A catalogue record for this publication is available from the British Library

Library of Congress Cataloguing in Publication data
Babbage, Charles, 1791–1871
Science and reform: selected works of Charles Babbage / [edited] by Anthony Hyman.
      p.   cm.
Includes index.
ISBN 0 521 34311 9
1. Mathematics.   2. Science.   3. Babbage, Charles, 1791–1871.
I. Hyman, Anthony.   II. Title.
QA3.B2   1989
500--dc19   88-10294 CIP

ISBN-13 978-0-521-34311-4 hardback
ISBN-10 0-521-34311-9 hardback

ISBN-13 978-0-521-03676-4 paperback
ISBN-10 0-521-03676-3 paperback

Contents

Acknowledgements

The author gratefully acknowledges the help and use of facilities of the following: the Director of the Science Museum (London), and the Librarians and staff of the library of the Royal Society of London, the Crawford Library, the Oxford Museum of the History of Science, the University Library (Cambridge), and the Library of the Royal Observatory (Edinburgh). Also, thanks are due to the Trustees of the Science Museum (London) for permission to reproduce Babbage's original drawings, given in the Plate section.

Introduction

CHARLES BABBAGE is known today as the great pioneer of computing. As computers spread through commerce and industry, banking and insurance, national and local government, science and mathematics, school and university education, and through society all round the world, so Babbage has joined his countrymen Faraday, Newton, and Darwin in the pantheon of popularly celebrated men of science.

The ubiquitous application of modern computing makes it singularly fitting that Babbage's own interests were also remarkably wide. Starting as a mathematician he turned to calculating and printing numerical tables: the manufacture of number. The study of mechanical devices for use in his Calculating Engines led Babbage on to an extraordinarily thorough and wide-ranging study of British industry. He generalized this study to include political economy, and his book *On the Economy of Machinery and Manufactures* was an important influence on nineteenth-century political economists, particularly John Stuart Mill and Karl Marx.

In May 1812, while he was at Cambridge, Babbage and some friends founded the Analytical Society to discuss mathematics and encourage introduction of the Leibniz 'd' notation for the calculus to replace Newton's 'dot' notation. It was not only that the Leibniz notation was the more powerful of the two: after the death of Newton mathematics had advanced far more rapidly on the continent, while British mathematicians, often unable even to read foreign mathematics because the notation used was unfamiliar, lagged far behind. The Young Turks of the Analytical Society were successful in their campaign for the Leibniz 'd's, and the Cambridge Philosophical Society was the posthumous offspring of the Analytical Society. However, Babbage and his friends had far more ambitious objectives in view: they sought to develop science throughout Britain and use it to change society. In this grand scheme Charles Babbage was the leading figure.

The liberals of the time were profoundly influenced by the French Revolution. Lazare Carnot, 'Organizer of Victory', and his associates had

1

established the Grandes Ecoles with their emphasis on science, and under French influence technical high schools and other institutions for scientific training were established in continental Europe. Moreover Napoleon fostered science with an eye to its military use. Babbage and his friends sought to learn from France and develop national support for scientific institutions, and particularly for scientific and technical education and training in Britain. The movement came to its climax in the Decline of Science campaign, the scientific counterpart of the political movement for the Great Reform Bill.

Babbage wrote a highly polemical book *Reflections on the Decline of Science in England, and on Some of its Causes*, which sought to reform the Royal Society, scientific education in England, and much else besides. The movement centred on the campaign to elect John Herschel, rather than a royal duke, as president of the Royal Society. Herschel was defeated and the Royal Society did not become a professional body until later in the century. Babbage's proposals for scientific and technical education were entirely ignored[1]; so also were his plans for the systematic application of scientific method to British commerce and industry.

Recently, with the growing realization of the deep-seated nature of Britain's industrial problems and the attempt to restructure British industry on a scientific basis making extensive use of computing, there has been renewed interest in the earlier stages of Britain's industrial development. In the *Audit of War* (Macmillan, 1986) Correlli Barnet not only showed the disastrous state of British industry during the second world war, but traced Britain's industrial problems to the second half of the nineteenth century. However, it is clear that the roots of these problems reach much deeper into history: they are structural weaknesses which developed as the country built the world's first industrial economy. Britain's industry was created by practical men with scarcely any direct assistance from the higher reaches of science, but by the 1820s it was becoming clear that industry could not continue to develop satisfactorily without scientific method. Babbage and his friends fought the battle for the systematic development and application of science on a national level, and lost. The consequences of this defeat are the subject of the current debate.

During the 1830s a profound split opened between pure science, which became the province of the universities, and the mundane bread and butter technical and engineering problems of industry. The Royal Society was ineffectual for decades and the professional civil service was established almost entirely without scientific input. To this day the bulk of the British civil service, particularly in the senior grades, is scientifically illiterate, and the consequences, from defence procurement[2] to education planning and scientific research, have been disastrous. Britain's weakness in the new science-based industries which developed towards the end of the nineteenth century has been endemic since their inception. As the sons of the men who had created Britain's staple industries were educated at public schools, moving into the City and becoming country gentlemen, so the standard of management in industry declined.

The government had backed Babbage's first Difference Engine, which was designed to calculate tables according to the method of finite differences and then to print the tables. After 1833 government support ceased, though it was another decade before the project was formally abandoned, and the government's treatment of Babbage led Dickens to create the celebrated Circumlocution Office of *Little Dorrit*.

In 1834 Babbage turned his attention to designing calculating machinery for more general mathematical operations, and started on the work which led to his great series of Analytical Engines. Although Babbage's impact on industrial development was of great importance in the British context, for the world at large Babbage is thought of as the founder of computing. The Analytical Engines incorporated an extraordinary range of concepts which were to reappear in modern computing. Separate store and mill; binary-coded store-addressing, and fast carry mechanism; versatile input/output system with a range of output devices; program control; microprogram control of the small operations – all belong to the first half of the nineteenth century. Array processing followed in the 1860s.

During his main period of work on the Analytical Engines, from 1834 to 1847, Babbage was making feasibility studies and preparing plans. It is often asked whether an Analytical Engine would actually have worked if one had been built. Thus phrased the question is not obviously meaningful. Many things can go wrong during a project of such complexity, and Babbage would certainly have made many alterations and developments during construction. However the question can be put somewhat differently: would it have been feasible for Babbage and Whitworth to have constructed a working Analytical Engine during the 1850s? When I came to make a detailed study of the Analytical Engines about fifteen years ago I formed the opinion that there was no technical reason why an Analytical Engine should not have been constructed.[3] The problem was financial. For an Analytical Engine to have been made in the nineteenth century, Babbage would have required a private fortune like that of Henry Cavendish in the eighteenth century: public finance was out of the question. It might have been necessary to go round the loop twice, with a development model and a working engine, but technically the project in itself was perfectly feasible.

Two of the main works represented here, 'Babbage's Calculating Engines' (1834), and 'Sketch of the Analytical Engine' (1843), were written by others under Babbage's careful direction. They were the first extensive descriptions published of the uses of the Difference and Analytical Engines, respectively.

Notes

1 C.f. Prof. Margaret Gowing, Wilkins Lecture, *Notes and Records*, 32 (1), July 1977.
2 A recent example of the damage caused by having scientifically incompetent senior civil servants is waste disposal. Britain, with far laxer laws than its neighbours, is in danger of becoming the rubbish tip of North Europe. No doubt the Department of the Environment has its scientific advisers, but that is rarely sufficient when the top people are quite incapable of grasping the technical essence of the underlying issues.

A longer-term example is radar, where Britain appears to have been virtually without radar-controlled defences since the wartime systems became obsolete. This has been common gossip in the industry for decades, but the full horror of the situation was only brought out recently in Duncan Campbell's television series *The Secret State*. All the military espionage since the war must pale into triviality when compared with the damage caused by the technical incompetence of the defence procurement agencies, incompetence, it may be remarked, of which secrecy is itself a principal cause.

3 In a letter to *The Times* published on 20 December 1982 I proposed the construction of an Analytical Engine:

> The Manpower Services Commission has recently sponsored construction of a replica of Stephenson's Rocket. I should like to suggest that it would be appropriate, in Information Technology Year, to consider construction of working versions of Charles Babbage's calculating engines. It might be sensible to start with the second Difference Engine and then one of the more simple plans for Analytical Engines.
>
> This project would provide unique training in a combination of mechanical engineering and digital techniques. It would also make a fitting tribute to the Englishman who pioneered computing in the middle of the nineteenth century.

Manufacture of a complete first Difference Engine presents some problems as most of the drawings have been lost. After the government project to make the first Difference Engine had ended, the small completed portion was transferred to the museum at King's College in the Strand. This portion was shown at the Exhibition of 1862 and then removed to the Science Museum in South Kensington, as the authorities at King's College declined to receive it again. When the accompanying drawings were transferred to the Science Museum in recent years most were missing. Thus King's College has distinguished itself not only by rejecting the first Difference Engine but by losing the bulk of the historic drawings into the bargain.

One might add that as museums all around the world develop sections on the history of computing, and thus come to seek working models of Babbage's Engines, the construction of Difference and Analytical Engines might form the basis of a nice export trade, a consummation of which Babbage would have entirely approved.

Bibliography
Hodges, Andrew, *Alan Turing: The Enigma of Intelligence*, Burnett, 1983.
Hyman, Anthony, *Charles Babbage, Pioneer of the Computer*, OUP/Princeton, 1982.
Moore, Doris Langley, *Ada, Countess of Lovelace*, John Murray, 1977.

Family background and education

CHARLES BABBAGE came from old Devonshire stock and both his parents belonged to leading Totnes families. Bab, or Babb, is a very old Devon name, but Babbage and similar forms do not seem to appear before the end of the sixteenth century, while families were still changing their names from Bab to Babbage in the nineteenth. There was a family tradition that an ancestor of Babbage had been left behind in France after the fall of Calais, and that four brothers, his sons, had returned to Devon during the reign of Elizabeth. Babbage's forebears appear in Totnes rate books early in the seventeenth century under the name of Babbidge. Presumably the French could not wrap their tongues round the gutteral Bab, and the family had become *la famille Babbidge*.

In *Passages from the Life of a Philosopher*, p. 6, Babbage refers to a curious story involving a relative of his, one Richard Babbage who became involved with Bamfylde Moore Carew, the King of the Beggars. Richard's escapades led to complications about some property at Ashbrenton, a name which has caused some confusion. 'Ashbrenton' was actually an alternative old name for Ashprington near Totnes.

Babbage's paternal family were goldsmiths but his father, Benjamin, had become a banker and moved to London, where Charles was born on 26 December 1791. The family returned to Devonshire when Benjamin retired in about 1803. After a fragmentary schooling Charles Babbage entered Trinity College, Cambridge in 1810, migrating to Peterhouse in 1812.

Passages from the Life of a Philosopher, p. 34

Whilst I was an undergraduate, I lived probably in a greater variety of sets than any of my young companions. But my chief and choicest consisted of some ten or a dozen friends who usually breakfasted with me every Sunday after chapel; arriving at about nine, and remaining to between twelve and one o'clock. We discussed all knowable and many unknowable things.

6

Passages from the Life of a Philosopher, pp. 35–8

During the first part of my residence at Cambridge, I played at chess very frequently, often with D'Arblay and with several other good players. There was at that period a fellow-commoner at Trinity named Brande, who devoted almost his whole time to the study of chess. I was invited to meet him one evening at the rooms of a common friend for the purpose of trying our strength.

On arriving at my friend's rooms, I found a note informing me that he had gone to Newmarket, and had left coffee and the chessmen for us. I was myself tormented by great shyness, and my yet unseen adversary was, I understood, equally diffident. I was sitting before the chess-board when Brande entered. I rose, he advanced, sat down, and took a white and a black pawn from the board, which he held, one in either hand. I pointed with my finger to the left hand and won the move.

The game then commenced; it was rather a long one, and I won it: but not a word was exchanged until the end: when Brande uttered the first word. 'Another?' To this I nodded assent.

How that game was decided I do not now remember; but the first sentence pronounced by either of us, was a remark by Brande, that he had lost the first game by a certain move of his white bishop. To this I replied, that I thought he was mistaken, and that the real cause of his losing the game arose from the use I had made of my knight two moves previously to his white bishop's move.

We then immediately began to replace the men on the board in the positions they occupied at that particular point of the game when the white bishop's move was made. Each took up any piece indiscriminately, and placed it without hesitation on the exact square on which it had stood. It then became apparent that the effective move to which I had referred was that of my knight.

Brande, during his residence at Cambridge, studied chess regularly several hours each day, and read almost every treatise on the subject. After he left college he travelled abroad, took lessons from every celebrated teacher, and played with all the most eminent players on the Continent.

At intervals of three of four years I occasionally met him in London. After the usual greeting he always proposed that we should play a game of chess.

I found on these occasions, that if I played any of the ordinary openings, such as are found in the books, I was sure to be beaten. The only way in which I had a chance of winning, was by making early in the game a move so bad that it had not been mentioned in any treatise. Brande possessed, and had read, almost every book upon the subject.

Another set which I frequently joined were addicted to sixpenny whist. It consisted of Higman, afterwards Tutor of Trinity; Follet, afterwards Attorney-General; of a learned and accomplished Dean still living, and I have no doubt still playing an excellent rubber, and myself. We not unfrequently sat from chapel-time in the evening until the sound of the morning chapel bell again called us to our religious duties.

I mixed occasionally with a different set of whist players at Jesus College. They played high: guinea points, and five guineas on the rubber. I was always a most welcome visitor, not from my skill at the game; but because I never played more than shilling points and five shillings on the rubber. Consequently my partner had what they considered an advantage: namely, that of playing guinea points with one of our adversaries and pound points with the other.

Totally different in character was another set in which I mixed. I was very fond of boating, not of the manual labour of rowing, but the more intellectual art of sailing. I kept a beautiful light, London-built boat, and occasionally took long voyages down the river, beyond Ely into the fens. To accomplish these trips, it was necessary to have two or three strong fellows to row when the wind failed or was contrary. These were useful friends upon my aquatic expeditions, but not being of exactly the same calibre as my friends of the Ghost Club, were very cruelly and disrespectfully called by them 'my Tom fools.'

The plan of our voyage was thus:—I sent my servant to the apothecary for a thing called an aegrotat, which I understood, for I never saw one, meant a certificate that I was indisposed, and that it would be injurious to my health to attend chapel, or hall, or lectures. This was forwarded to the college authorities.

I also directed my servant to order the cook to send me a large well-seasoned meat pie, a couple of fowls, &c. These were packed in a hamper with three or four bottles of wine and one of noyeau. We sailed when the wind was fair, and rowed when there was none. Whittlesea Mere was a very favourite resort for sailing, fishing, and shooting. Sometimes we reached Lynn. After various adventures and five or six days of hard exercise in the open air, we returned with our health more renovated than if the best physician had prescribed for us.

During my residence at Cambridge, Smithson Tennant was the Professor of Chemistry, and I attended his lectures. Having a spare room, I turned it into a kind of laboratory, in which Herschel worked with me, until he set up a rival one of his own. We both occasionally assisted the Professor in preparing his experiments. The science of chemistry had not then assumed the vast development it has now attained. I gave up its practical pursuit soon after I resided in London, but I have never regretted the time I bestowed upon it at the commencement of my career.

When he went up to Cambridge Babbage was equally familiar with the three notations then commonly in use for the calculus: Newton's dots, Lagrange's dashes, and, by far the most powerful, the Leibniz 'd's. Passing through London on his journey from Devon, Babbage had purchased Lacroix's *Differential and Integral Calculus*, which used the Leibniz notation, for the huge price of seven guineas. No sooner was he installed in his rooms than Babbage buried himself in his new purchase. Hopefully Babbage asked his tutors to explain some difficulties, but discovered to his dismay that they were quite incapable of understanding the problems.

Passages from the Life of a Philosopher, pp. 27–9

I thus acquired a distaste for the routine of the studies of the place, and devoured the papers of Euler and other mathematicians, scattered through innumerable volumes of the academies of Petersburgh, Berlin, and Paris, which the libraries I had recourse to contained.

Under these circumstances it was not surprising that I should perceive and be penetrated with the superior power of the notation of Leibnitz.

At an early period, probably at the commencement of the second year of my residence at Cambridge, a friend of mine, Michael Slegg, of Trinity, was taking wine with me, discussing mathematical subjects, to which he also was enthusiastically attached. Hearing the chapel bell ring, he took leave of me, promising to return for a cup of coffee.

At this period Cambridge was agitated by a fierce controversy. Societies had been formed for printing and circulating the Bible. One party proposed to circulate it with notes, in order to make it intelligible; whilst the other scornfully rejected all explanations of the word of God as profane attempts to mend that which was perfect.

The walls of the town were placarded with broadsides, and posters were sent from house to house. One of the latter form of advertisement was lying upon my table when Slegg left me. Taking up the paper, and looking through it, I thought it, from its exaggerated tone, a good subject for a parody.

I then drew up the sketch of a society to be instituted for translating the small work of Lacroix on the Differential and Integral Calculus. It proposed that we should have periodical meetings for the propagation of d's; and consigned to perdition all who supported the heresy of dots. It maintained that the work of Lacroix was so perfect that any comment was unnecessary.

On Slegg's return from chapel I put the parody into his hands. My friend enjoyed the joke heartily, and at parting asked my permission to show the parody to a mathematical friend of his, Mr. Bromhead.

The next day Slegg called on me, and said that he had put the joke into the hand of his friend, who, after laughing heartily, remarked that it was too good a joke to be lost, and proposed seriously that we should form a society for the cultivation of mathematics.

The next day Bromhead called on me. We talked the subject over, and agreed to hold a meeting at his lodgings for the purpose of forming a society for the promotion of analysis.

At that meeting, besides the projectors, there were present Herschel, Peacock, D'Arblay, Ryan, Robinson, Frederick Maule, and several others. We constituted ourselves 'The Analytical Society;' hired a meeting-room, open daily; held meetings, read papers, and discussed them. Of course we were much ridiculed by the Dons; and, not being put down, it was darkly hinted that we were young infidels, and that no good would come of us.

In the meantime we quietly pursued our course, and at last resolved to publish a volume of our Transactions. Owing to the illness of one of the number, and to various other circumstances, the volume which was published was entirely contributed by Herschel and myself.

At last our work was printed, and it became necessary to decide upon a title. Recalling the slight imputation which had been made upon our faith, I suggested that the most appropriate title would be—

The Principles of pure D-ism in opposition to the Dot-age of the University.

Early mathematical work

Early mathematical work

DURING THE debates of the Analytical Society, Babbage and his friends developed a common pool of mathematical ideas on which they drew in their later work. For example the first paper read to the society, by Bromhead, was 'On Notation', one of Babbage's great interests and a theme which appears in the preface to the *Memoirs* of the Analytical Society. Another example is the important studies by Babbage and later by Peacock on algebra. The preface to the *Memoirs* was drafted by Babbage and in it we find, both in the observations about the importance of mathematical notation and the stress laid on the theory of functions, the earliest example of the authentic Babbage voice.

Memoirs of the Analytical Society, 1813, Preface

To examine the varied relations of necessary truth, and to trace through its successive developments, the simple principle to its ultimate result, is the peculiar province of Mathematical Analysis. Aided by that refined system, which the ingenuity of modern calculators has elicited, and to which the term Analytics is now almost exclusively appropriated, it pursues trains of reasoning, which, from their length and intricacy, would resist for ever the unassisted efforts of human sagacity. To what cause are we to attribute this surprising advantage? One, undoubtedly the most obvious, consists in the nature of the ideas themselves, whose relations form the object of investigation—and the accuracy with which they are defined. This is equally indeed the property of every branch of Mathematical enquiry. Three causes however chiefly appear to have given so vast a superiority to Analysis, as an instrument of reason. Of these, the accurate simplicity of its language claims the first place. An arbitrary symbol can neither convey, nor excite any idea foreign to its original definition. This immutability, no less than the symmetry of its notation, (which should ever be guarded with a jealousy commensurate to its vital importance,) facilitates the translation of an expression into common language at any stage of an operation,—disburdens the memory of all the load of the previous steps,—and at the same time, affords it a considerable assistance in retaining the results. Another, and perhaps not less considerable cause, is to be found in the conciseness of that notation. Every train of reasoning implies an exercise of the judgement, which, being an operation of the mind, deciding on the agreement or disagreement of ideas successively presented to it, it is reasonable to presume will be more correct, in proportion as the ideas compared follow each other more closely; provided the succession be not so rapid as to cause confusion. Were an Analytical operation of any complexity converted into common language, in all its detail, the mind, after acquiring a clear conception of one part of the related ideas, must suspend its decision until it could obtain an equally perspicuous one of the remainder of the proposition; and in so long an interval as this must occupy, the impression of the former ideas would necessarily have faded in some degree from the memory, unless fixed by an expense of time and attention, sufficient to deter any one from the employment of such means of discovery. It is the spirit of this symbolic language, by that mechanical tact, (so much in unison with all our faculties,) which carries the eye at one glance through the most intricate modifications of quantity, to condense pages into lines, and volumes into pages; shortening the road to discovery, and preserving the mind unfatigued by continued efforts of attention to the minor parts, that it may exert its whole vigor on those which are more important.

The last cause we have occasion to notice is, that Analysis, by separating the difficulties of a question, overcomes those which appear almost insuperable when combined, or at least, reducing each to its last terms, leaves them as the acknowledged landmarks of its progress,—open to approach on all sides, should ulterior discovery furnish any rational hope of their removal. Meanwhile that progress continues unimpeded. Simple relations are found to exist between the most refractory functions, and even when the difficulties themselves prove invincible, their nature at least becomes thoroughly understood, and means of evading them almost universally pointed out.

That the preceding observations are not founded on bare speculation, the whole history of Analytical Science will abundantly evince. It is our intention, in the following pages of this Preface, to give a general outline of that history up to the present time. From the space allotted to it, it is evident that little else than the most prominent points in so wide a field can be selected for observation. Faint, however, as it is, the subject cannot but communicate to it some portion of its interest; as well as the reflection, that, (with the exception of one branch of it*) the history of the more modern discoveries has hitherto unfortunately found little place in our language†.

Symbolic reasoning appears to have been ushered into the world under unfavourable auspices, and to have been regarded in its infancy with an eye of extreme jealousy. And, indeed, if we consider the rudeness of its first attempts, the poverty of its first resources, and the lavish want of œconomy in their employment, we shall find little reason to wonder, that for a long period the new methods were looked upon as inelegant, although serviceable auxiliaries of the ancient processes, to be regularly discarded after serving their turn. To employ as many symbols of operation and as few of quantity as possible, is a precept which is now found invariably to ensure elegance and brevity. The very reverse of this principle forms the character of symbolic analysis, up to within fifty years of the present date.

The first and most natural object of research in the Algebraic calculus, was the resolution of equations, involving simply the powers of the unknown quantity. As far as the fourth degree no difficulty occurred, but beyond this, not a step has yet been made. Almost every Analyst of eminence has applied his ingenuity to the accomplishment of the *general* problem, but without success; and after more than two centuries, during which every other branch of Analytics has been advancing with unrivalled rapidity, no progress whatever has been made in this. This, it must be allowed, presents little prospect of success to future researches on the subject; yet ought not the difficulty to be considered insurmountable, until opinion has been confirmed by demonstration. Delambre notices a Memoir presented to the National Institute by M. Ruffini, in which he proposes a proof of the impossibility of the resolution of equations above the fourth degree. If this demonstration be correct, it will render an important service to Algebraists, by diverting them from a pursuit which must necessarily be unsuccessful. The work, however, if yet published, has not arrived in this country. Recent French publications are not easily procured, nor is it surprising that to obtain those of the German Analysts is almost impossible, when Delambre regrets their scarcity even in France.

* The calculus of variations, the history of whose rise and progress has been ably combined with the exposition of its theory, in a late work, 'On Isoperimetrical Problems.'
† The admirable review of the *Mecanique Celeste* (Ed. Rev. No. 22.) will still be fresh in the minds of our readers. But it should be recollected, that the Author of that Essay confines his attention entirely to the subject of Analytical dynamics; referring to the discoveries in the integral calculus merely as connected with that subject, and that too very cursorily. *Our business is exclusively with the pure Analytics.*

Although to express in finite algebraic terms, the root of any proposed equation be impracticable, yet the inverse function of any expression, such as

$$a + bx + cx^2 + dx^3 + \&c.$$

may readily be exhibited in an infinite series. When the difficulty of solving equations above the fourth degree was perceived, it was natural to seek rapid and convenient approximations, and accordingly, three of our countrymen, Newton, Raphson, and Halley, produced nearly at the same time, modes of approximation which have since received various improvements. All such researches, when symbolically conducted, and without regard to the numerical value of the symbols, lead at the bottom to series of greater or less complexity. It seems to have been in following up this idea, that Lagrange was first conducted to that very general resolution of all equations in the series which bears his name; a series which has been productive of discovery wherever it has been applied, and whose fecundity appears yet far from being exhausted. It is thus that the most distant parts of Analysis hang together, nor is it possible to assign the point, however remote or unexpected, in which any proposed career of research may not ultimately terminate. This series made its first appearance in the *Mem. de l'Acad. Berlin*, 1767–8*. together with the most systematic method of approximating to the roots of numerical equations which has yet been given. The method has this considerable advantage over all others, that, in all cases where the root is an integer, the formula of approximation will give it exactly, and in many where it is a surd, the continued fraction employed will point out the rational number of which it is the root. Though the complete solutions of equations is nearly hopeless, it might perhaps be of some advantage, and certainly of little difficulty, supposing the roots of equations known, to investigate what change would take place if one or more of the coefficents were augmented or diminished by any small quantity.

To trace the history of the differential calculus through the cloud of dispute and national acrimony, which has been thrown over its origin, would answer little purpose. It is a lamentable consideration, that that discovery which has

---

* The demonstration there given is defective in rigour. A better was given in Note XI. of the 'Traitè de la resolution des equations numeriques.' Lagrange. But the most elegant is that of Laplace, to be found in the *Mem. de l'Acad. des Sciences*, Paris, 1777; in Lacroix's *Calc. Diff. et Int.* 2d edit. Art. 107; in the *Mecanique Celeste*, tom. I. page 172. Adopted (in principle at least) by Lagrange in the 'Theorie des Fonct.' *Analyt. Art.* 97, et suiv. And in our own language, in Mr. Woodhouse's *Trigonometry*, 2nd edition. Arbogast has also demonstrated this theorem. See his *Calcul des Derivations* Art. 282, et suiv.

most of any done honour to the genius of man, should nevertheless bring with it a train of reflections so little to the credit of his heart.

Discovered by Fermat, concinnated and rendered analytical by Newton, and enriched by Leibnitz with a powerful and comprehensive notation, it was presently seen that the new calculus might aspire to the loftiest ends. But, as if the soil of this country were unfavourable to its cultivation, it soon drooped and almost faded into neglect, and we have now to re-import the exotic, with nearly a century of foreign improvement, and to render it once more indigenous among us.

The most prominent feature of this calculus, is the theory of the development of functions. The theorem which has immortalized the name of Brook Taylor, forms its foundation. Elicited by its Author from a formula which at first sight seemed independent of it, by a method not remarkable for its rigour, it seems to have been long considered in the light of a very general formula of interpolation. Lagrange and Arbogast have, as it were, invented it anew, and established it as the true basis of the differential calculus. The theory of Lagrange is to be found in a Memoir among those of the Acad. de Berlin. 1772, which contains the independent demonstration of Taylor's theorem—in the 'Theorie des fonctions Analytiques,' wherein he exhibits its application to the various branches of the differential calculus, independently of any consideration of limits, infinitesimals or velocities—and lastly, in the *Journal de l'Ecole Polytechnique*. Cah. XII. (1802.) and in the 'Lecons sur le calcul des Fonctions.' The ideas of Arbogast are contained in a Manuscript presented to the Academy of Sciences in the year 1789, and of which the outline is given in the Preface to his celebrated work on Derivations. Such is the brief account of the greatest revolution which has yet taken place in Analytical Science.

The operations of the differential calculus once well understood, and rigorously demonstrated, may be employed in improving the theory which gives rise to them. The work of Arbogast just alluded to, has shewn to how vast an extent this application may be carried, and how great is the assistance thus rendered. The peculiar grace of Laplace's Analysis has no where been more beautifully exhibited, than in his improvement and extension of Lagrange's theorem already mentioned. Nor should the labours of Paoli in this field pass unnoticed. By the aid of a very remarkable series derived from reverting that of Taylor, he has been able to assign the development of any function of a quantity given by any equation whatever, in terms of a function any how composed of the remaining symbols which enter into that equation.

The developement of functions has lately been made, under the name of

'Calcul des fonctions generatrices,' the foundation of a most elegant theory of finite differences, of which more hereafter.

Soon after the discovery of the integral calculus, on the discussion of some problems, between Leibnitz and the Bernouillis, respecting the variation of the parameters of curves; there occurred certain equations, which, though they satisfied the conditions, were yet not contained in the complete integral of the equation whence they were derived. It is somewhat remarkable, considering the manner in which they first appeared, that their geometrical signification should have remained so long undiscovered. Brook Taylor, according to Lagrange, was the first who arrived at a particular solution by differentiating. Clairaut, in a Memoir presented to the Academy of Sciences at Paris, first remarked, that the equations so found, satisfy the geometrical conditions proposed. Euler styled them Analytical paradoxes, and shewed how in some cases they might be derived from the differential equations. But their theory remained unknown, till the year 1772*; when Laplace explained it in a *Memoir of the Academy of Sciences*, and pointed out the methods of discovering all the particular solutions of which an equation admits. This subject was still farther pursued in the *Berlin Memoirs*, by Lagrange, who there developes, with great perspicuity the whole theory both Analytical and Geometrical†. But the most complete exposition of the subject, which has yet appeared, is to be found in a paper read before the National Institute in the year 1806, by Poisson, in which the theory is extended to partial differential equations, and also to those of finite differences. He observes of certain partial differential equations, that they admit of particular solutions equally general with the complete integral. The Analytical theory, in its present state, is most elegant: still it requires some farther developements when applied to equations of partial differentials, and to those of differences, and might perhaps with advantage be applied to equations of mixed differences. Its Geometrical signification has been beautifully illustrated by Lagrange; but the meaning of particular solutions, when they occur in dynamical problems, which is a question of considerable importance, remains yet undecided. Poisson has shewn a case, in which the particular solution and the complete integral are both required, and has produced others, in which only one is necessary.

* J. Trembley, in the 5th vol. of the *Mem. de Turin*, has given a Paper on the derivation of the complete integral, having given the number of particular solutions. His method consists in multiplying the equation by these solutions, each raised to an indeterminate power.

† Fontaine first considered a differential equation as the result of the elimination of a constant between an equation and its differential; thus laying the foundation of the theory of equations, both differential, and of differences, and also of their particular solutions.

As the integration of expressions containing one variable is a matter of considerable importance, and the number of those which are capable of integration, is small, when compared with those which do not admit of it; some attention has been bestowed on the classification of those which are similar, and on the reduction of those which are absolutely different to the least number possible. When this is accomplished, all that remains for the perfection of this branch of Analysis, is to calculate tables which shall afford a value of the integral for any value of the variable. In general, all expressions which do not admit of complete integration, are denominated transcendants. Those which most frequently occur, are logarithmic and circular functions. Tables of these had been long calculated for trigonometrical purposes, and on the discovery of the integral calculus received a vast addition to their utility. It was next proposed to admit as known transcendants, all integrals which could be reduced to the rectification of the conic sections. But, besides the preposterous idea of limiting an Analytical expression by the properties of a curve, no tables had been constructed for them, and of course the determination of their arcs could only be performed by the actual calculation of the integral under consideration: nor, indeed, would it have been possible to form useful tables of any moderate length, without first discussing the properties of the transcendants themselves in the fullest manner.

The theory of the transcendant $\int \frac{Pdx}{R}$, where P is a rational and integral function of x, and R a quadratic radical of the form $\sqrt{(a + \beta x + \gamma x^2 + \delta x^3 + \epsilon x^4)}$ has at length by the successive labours of Fagnani, Euler, Landen, and Legendre, been brought to great perfection. All the transcendents comprised in this extensive formula, are reducible to three species. Those comprised in the first, are susceptible of multiplication or division, in the same way as the arcs of circles, by algebraic operations only. The transcendants of the second species, are susceptible of a similar multiplication or division, not simply, but when increased or diminished by an algebraic quantity. This algebraic quantity passes in the third species, into a transcendant of the logarithmic or circular kind. Landen has shewn, that an integral of the first species here enumerated may be reduced to two of the second: so that the number of distinct transcendants comprised in the formula

$\int \frac{Pdx}{R}$ is no more than two. A vast variety of integral formulae have, by dint of indefatigable research on all hands, been reduced to the evaluation of these functions; but to dwell longer on them, would lead us beyond our limits.

The only other species of transcendants of any considerable extent which have received much discussion, are those contained in the formulae

$$\int \epsilon^{-x^n}\cdot dx, \text{ and } \int^n \frac{\epsilon^x\cdot dx^n}{1+\epsilon^x}.$$ Kramp, at the end of his *Analyse des Refractions*, has given a table of the values of the first of these, (in the case of $n=2$), the integral being taken between the limits 0.00, . . . 3.00, ∞. The definite integrals dependent on the general form, we shall speak of hereafter. The second formula is (ultimately) that of the logarithmic transcendants, on the various orders of which Mr. Spence, in the year 1809*, published an Essay, which displays considerable ingenuity, and a depth of reading rarely to be met with among the Mathematical writers of this country. A general property there given of the transcendant $^nL(x)$, leads to the summation of some very extraordinary series, which are now in our possession, and which we cannot forbear mentioning.

Their general (or ith) terms are comprised in the formulae

$$\frac{f\{x\cdot\epsilon^{i\theta\sqrt{(-1)}}\}+f\{x\cdot\epsilon^{-i\theta\sqrt{(-1)}}\}}{i^{2n}}, \text{ and } \frac{f\{x\cdot\epsilon^{+i\theta\sqrt{(-1)}}\}-f\{x\cdot\epsilon^{-i\theta\sqrt{(-1)}}\}}{\sqrt{(-1)}\cdot i^{2n+1}}$$

where f is the characteristic of any function whatever†, developeable in integer powers, either positive or negative, or both.

Mr. Spence has also given tables (of some extent,) of the successive values of his functions, as we have before remarked of Kramp. Too much praise cannot be bestowed on such examples, which however there is little hope of seeing followed. The ingenious Analyst who has investigated the properties of some curious function, can feel little complaisance in calculating a table of its numerical values; nor is it for the interest of science, that he should *himself* be thus employed, though perfectly familiar with the method of operating on symbols; he may not perform extensive arithmetical operations with equal facility and accuracy; and even should this not be the case, his labours will at all events meet with little remuneration.

* Le Gendre published his *Exercises de Calcul Integral* in 1811. After having given Landen's and Euler's values of particular cases of the function $^nL(1+x)$ he adds, 'Jusqu'a present, on n'est pas allé plus loin dans la theorie de ces sortes des transendantes,' page 249. It is probable therefore, that, owing to our interrupted intercourse with the continent, he had not seen Mr. Spence's work.

† One singular result of these researches is, the evaluation in terms of the transcendants ϵ, and π, of the function

$$(\tan\theta)^{(\frac{1}{4})2n+1} \cdot (\cot 3\,\theta)^{(\frac{1}{3})2n+1} \quad (\tan 5\,\theta)^{(\frac{1}{5})2n+1} \quad \text{&c. ad infinitum.}$$

It sometimes happens, that the arbitrary constant does not continue the same throughout the whole extent of an integral. An instance of this, in the series

$$C - \frac{\theta}{2} = \frac{\sin \theta}{1} + \frac{\sin 2\theta}{2} + \&c.$$

was remarked by Daniel Bernouilli, in the Act. Petrop. Landen also notices, that this equation is false, when $\theta = 0$, but without explanation. Other instances occur in the 'Essay on logarithmic transcendants,' pp. 52–3; in Lacroix's *Traite de Calcul* &c. 4to. tom. III. p. 141, as well as some reflections on this difficult subject in page 483 of the same volume. The cause of these anomalies has not been satisfactorily explained. If we may hazard a conjecture, it must be looked for in the evanescent or infinite values of some of the differential coefficients of the function integrated, causing that function for an instant to change its form. Or, it may have some connection with exponentials, since all the instances which have hitherto been adduced depend on that species of function.

The discovery of partial differentials, has been generally attributed to D'Alembert. He certainly was the first who applied them to mechanical problems, and perceived their vast utility in all the more difficult applications of Analysis to physics. But, if he is to be considered as the inventor, who first solved an equation of the kind, and who, when their importance was acknowledged, contributed more than any other to the improvement and progress of this calculus; the glory of their discovery will undoubtedly belong to Euler. In the 7th vol. of the *Acta Acad. Petropolitanae*, is a Memoir of his, entitled, 'Methodus inveniendi aequationes pro infinitis curvis ejusdem generis.' (A.D. 1735.) In this paper, and more particularly in a supplement, are given the solutions of a number of partial differential equations, of which the most general is

$$q = Xz + pR$$

X being a function of x, and R a function of x and y.

In the latter part of the 'additamentum ad dissertationem,' he proceeds to integrate some equations of the second order. The 'Reflexions sur la cause generale des Vents,' which contains d'Alembert's first application of partial differential equations, was not published till the year 1747, and it gained the prize of the Academy of Berlin in 1746. It would lead us too far to trace, successively, the various improvements which the new theory underwent in the hands of Euler, Lagrange, Laplace, Monge, Parseval, and a multitude of great men, whom the vast importance of the subject incited to its prosecution. Notwithstanding every exertion, the theory however continues to present a multitude of difficulties. The analogy which was supposed to exist between the arbitrary functions, which enter into the integrals of partial differential equations, and the arbitrary constants in equations of total differentials, is found not to hold beyond the first degree; after which, even the number of these arbitrary functions is unknown. Instances have been adduced, where, besides the arbitrary functions, arbitrary constants also must be introduced to complete the integral*. The application of definite integrals to the integration of these equations, presents a wide field of research, as well as the promise of great discoveries. A very curious Memoir of Laplace is to be found in the *Mem. Acad. des Sciences* 1779, where this subject, among many others, is discussed with considerable success†.

In applying the test of integrability to differential equations, some were found, which could not be made to satisfy the equations of condition. These were for a long time deemed absurd, until about the year 1784, when Monge perceived their connection with the theory of curve surfaces, and demonstrated that these equations admit solutions, corresponding to the curves of double curvature, formed by the successive intersections of a curve surface, whose parameter varies; and discovered methods of transforming any equation of this kind into an equation of partial differentials, and also of solving the converse problem. From this, he proposed obtaining solutions of equations of partial differentials, which are not integrable by other methods. But the example he gives, shows, that to expect success in such an enquiry, we must be familiar with space, considered as of three dimensions, and also with a numerous collection of curves and curve surfaces situated in it; such considerations would but add intricacy to a subject already difficult: hints for the advancement of Analysis may be derived from foreign sources, but must always be improved and cultivated by its own powers. La Croix has given an excellent Analytical theory of this kind of equations.

To the practice which prevailed in the infancy of Analytics, of proposing to Geometers the solution of difficult problems, we owe many of its improvements. The method of variations, among others, is much indebted to this

* Monge; *Savans Etrangers*, vol. VII. p. 322.
† The great desideratum in the integration of an equation by definite integrals is, that whenever it is susceptible of actual resolution, these integrals should give it—a condition which does not always hold good. See Laplace. 'Mémoire sur divers points d'Analyse.' *Journ. de l'Ecole Polytechnique*, No. 15.

source. Obliged however to consult brevity, we must refer to the late treatise on 'Isoperimetrical Problems,' for a full account of its history: a work which being undoubtedly in the hands of the majority of our readers, must render superfluous all we could say on that subject. The chief difficulty now consists in distinguishing the maxima from the minima, which depends in general on the solution of difficult differential equations. The inverse of the method of variations does not appear to have received any attention; it would consist in solving such problems as the following: 'Given a curve, to find what properties of maxima and minima it possesses.'

The method of variations was applied by its inventor to differential equations, and also to those of finite differences. Cases may occur, in which it would be necessary to apply it to equations of mixed differences; these relate to a number of difficult problems, such, for instance, as this: 'What must be the nature of a curve, such, that drawing to any point, an ordinate and also a normal, and at the foot of this normal another ordinate; the curvilinear area intercepted between the first and last ordinate, may be a maximum or a minimum.'

There are but few instances in the history of Science, in which the path of the inventor has been the shortest and most direct. Thus it occurs, that the method of finite differences, which would most naturally have preceded that of the differential calculus, was not discovered until many years after. Its inventor, Brook Taylor, published it in a work, entitled, *Methodus Incrementorum directa et inversa*, a book noted for its obscurity. Montmort, Stirling, and Emerson, made several improvements, which may be found in the *Philosophical Transactions*, and also in their respective works. Moivre first investigated the nature of recurring series, on which, Laplace remarks, 'Sa theorie est une des choses les plus curieuses, et les plus utiles que l'on ait trouvees sur les suites.' It has been well observed, by the same author, that the first who summed a Geometrical or Arithmetical series, had really integrated an equation of finite differences. The same remark, as is well known, applies to any recurrent series. It was not, however, till Lagrange in the *Melanges de Turin*, vol. I. applied Alembert's method of indeterminate coefficients, to an equation of differences of the first degree, that this truth was perceived. In the fourth volume of the same work, Laplace published a Memoir, in which the two celebrated theorems of Lagrange, respecting equations of common differentials, are extended to those of differences, and those of partial differentials of a similar description, with constant coefficients. Returning to the subject, in a Memoir communicated to the Academy of Paris, he integrates a very extensive class of equations of partial differences, involving any number of variable indices,—and also a singular species of equations, frequent in the theory of chances, called by him, *equations rentrantes*.

Of equations beyond the first degree, very few have been solved, if we consider their amazing variety and importance. Monge has given a short paper on the subject, in the *Memoirs of the Academy of Sciences*, for 1783. Laplace also in the 15th number of the *Journal de l' Ecole Polytechnique*, has by a most happy combination of the equations of differences, with the discovery of Euler, respecting elliptic transcendants, integrated a few very difficult ones. The nature of the integrals obtained by Charles, requires a fuller investigation. They might perhaps receive considerable extension. When the variables are mixed in the indices, thus, u_{x+y}, u_{x-y+1}, u_{xy}, &c.; the subject seems to have passed altogether unnoticed. Many equations containing such expressions, are impossible or contradictory.

Euler first remarked, that the constant introduced by integrating an equation between u_x, u_{x+1}, &c. may be an arbitrary function of $\cos 2\pi x$, a remark which afterwards in the hands of Laplace, (*Savans Etrangers*, 1773,) became the foundation of a very general theory of determining functions from given conditions. To notice all the applications of the theory of finite differences, or all the profound researches which have enriched it, would occupy volumes. We cannot, however, pass over the theorems relating to the analogy of differences to powers, given first by Lagrange without demonstration, in *Mem. de Berlin.* 1772. A demonstration by Laplace, appeared in the *Mem. des Savans Etrangers* for the following year, and another in the *Mem. de l' Acad.* for 1777. In 1779, appeared that noted Memoir, in which the same author exhibited the principles of his powerful and elegant 'Calcul des fonctions generatrices.' In this*, he gives a far more systematic proof of the theorems, and extends them to any number of variables. Since that period, they have been demonstrated by Arbogast, in the 6th article of his, 'Calcul des derivations,' where, by a peculiarly elegant mode of separating the symbols of operation from those of quantity, and operating *upon them* as upon analytical symbols; he derives not only these, but many other much more general theorems with unparalleled conciseness. Brinkley has given a demonstration of the theorem

---

* This Memoir forms the greater part of the first Chapter of the *Theorie Analytique des Probabilites*. Its first principles, and the demonstration of the theorems, on the analogy of differences with powers, are given briefly in the 15th No. of the *Journal de l' Ecole Polytechnique*. A slight sketch of the method alluded to, is also to be found in the 9th book of the *Mécan. Cél.* tom. IV. p. 204.

$$\Delta^n u_x = \left\{ \epsilon^{\frac{\Delta x \cdot d}{dx}} - 1 \right\}^n \times u_x$$

of considerable elegance, and a simplicity truly elementary*.

We are now naturally led to say a few words on the 'Calcul des fonctions generatrices.' Its object may be best stated, in the words of its inventor†: 'C'est de ramener au simple developpement des fonctions, toutes les operations relatives aux différences et specialement l' integration des equations aux differences ordinaires et partielles,' and from the extreme facility with which all the known theorems flow from it, and its fecundity in affording new ones, it is, perhaps, in the present state of science, the best adapted of any, for explaining the general theory of differences, and the developement and transformation of series. At the same time, it must be confessed, that owing to its extreme generality, and the consequent complexity of many of its operations, particularly in what regards the transformation of series, it is an instrument to be placed only in the hands of an experienced Analyst. If we except the calculus of variations, it is the only method, perhaps, of any considerable importance, which has received its first and last touches from the same hand; and which first appeared in a state of perfection, very little short of what it at present possesses. The latest work which treats of this subject, is the *Theorie Analytique des Probabilites*; the first part of which is dedicated to a very full exposition of the method.

After the solution of differential equations, and those of finite differences, it was natural to consider those, in which the difference and differential of a quantity both occurred. These have been called equations of mixed differences. They were not attempted until about the year 1779, when Condorcet and Laplace obtained the integrals of some few particular cases. But problems which required their application had been proposed and resolved by several Geometers in the *Acta eruditorum*, and in the *Acta Acad. Petrop.* long before this time; their solutions were obtained by certain insulated artifices, dependent on the peculiar nature of the problems.

---

* *Philos. Transactions*, 1807. Part I. He has extended his researches to the actual expansion of the series themselves, to which these theorems lead; such, for instance, as

$$\Sigma u_x = \frac{1}{\Delta x} \cdot \int u_x dx - \frac{u_x}{2} + \Delta x \cdot^1 A_x + (\Delta x)^2 \cdot^2 A_x + \&c.$$

But in point of clearness and elegance, by no means with equal success: partly owing to an unfortunate notation, and partly to the perpetual employment of the theory of combinations. In his results, he has for the most part been anticipated by Laplace, and others.

† *Journal de l' Ecole Polyt.* Mém. sur divers points d' Analyse.

On this subject, a wide field is extended for investigation, and one which abounds with difficulties. The little that has yet been discovered, is chiefly contained in two papers, one by Biot, in the *Memoires de l' Institut*, and the other by Poisson, in the *Journal de l' Ecole Polytechnique*; the former treats chiefly of that kind of equations, called *equations successives*; the latter integrates a few particular equations, by employing a substitution used by Laplace, in integrating some equations of partial differentials.

There is no branch of mathematical science which has not received improvements, from the profound and original genius of Euler. Several are indebted to him for their existence; of this latter class is the knowledge of the nature, and use of definite integrals, a subject, to which the greatest Geometers of the present age look as the most probable source of future discoveries and improvements.

Legendre, in a work lately published, entitled *Exercises de Calcul Integral*, has collected all that has been discovered on this subject, and demonstrated the results with peculiar elegance; the greater part is extracted from the various works of Euler, and also a considerable portion from some Memoirs of Laplace. Its application to the solution of differential equations, from which so much is expected, does not enter into the plan of his work; this however has been well treated, by Lacroix, in the third volume of his *Traite de Calcul, &c.*

But the most elegant part of the theory of definite integrals is, their application to such problems in finite differences, as involve functions of very high numbers. In many cases (particularly in the theory of chances,) it has been well remarked, that the mere impracticability of the arithmetical operations requisite to obtain a result, (however simple its analytical expression,) must for ever preclude our advancement; were it not for some mode of approximation, which, grasping the prominent terms of an expression, in a formula easily reduced to numbers, should throw the minor ones into the back-ground, to be valued by a series converging the more rapidly the higher the numbers employed become. A more appropriate instance cannot be adduced, than the equation

$$1 \cdot 2 \cdot 3 \ldots s = s^{s+\frac{1}{2}} \cdot \epsilon^{-s} \cdot \sqrt{(2\pi)} \left\{ 1 + \frac{1}{12s} + \frac{1}{288s^2} + \&c. \right\}$$

in page 129 of the *Theorie Anal. des Probabilites*, or the method in which the author, in page 259 of the same work, computing the probability of a primitive cause influencing the inclinations of the cometary orbits, employs a definite integral to effect with conciseness, a calculation surpassing, without that assistance, the utmost limits of human patience and industry.

In this part of his career, Laplace stands unrivalled. Stirling indeed, and Moivre had seen, and in some cases obviated the difficulties, arising from the immensity of the numbers under consideration. The discoveries of Euler, gave a connection and unity to their results. But it was not till the labours of Lagrange, Condorcet, and Laplace had brought the theory of finite differences to considerable perfection, that the definite integrals were applied by the latter, to the solution of these equations, and a clear and strong light thrown over this most obscure part of the Mathematics. The whole of this interesting theory, has been digested into one work, (*Theor. de Prob.* above cited) which for comprehensive views, for depth of investigation, and the purity of its analysis, may justly be looked up to, as marking the highest point to which the science of abstract number has yet attained.

In analytical investigations, we frequently meet with a series of quantities connected together by multiplication, whose differences are constant. These have received various names from different Geometers; Vandermonde called them powers of the second order; Kramp adopted the appellation of Facultes numeriques; and Arbogast named them Factorials. Each of these writers treats them in a different manner, and employs a peculiar notation for them: that of Vandermonde is perhaps the best adapted to the subject, as it has a considerable resemblance to that of exponents, and possesses also an advantage, which is by no means inconsiderable, that of being capable of a ready extension to powers of all orders. Lacroix has adopted it, and by its help demonstrated many properties of powers of the second order. Those of the superior orders have not as yet been examined. Kramp has deduced from his 'Theorie des facultes numeriques,' some contradictions which require examination. One of his most useful theorems affords a method of transforming any power of the second order, into a series which converges *ad libitum*. It appears probable, that the theory of powers of different orders may afford a useful method of classing transcendents, as they can frequently be reduced to definite integrals, and by this means their value be obtained, when their index is fractional.

Interpolations were at first considered, as a branch of the method of finite differences, and as such they were usually treated of together. Wallis appears first to have applied this name, to the determination of the intermediate term of a series, whose law of formation is known. The extraction of roots is an interpolation of powers, and may be considered as an extension of the meaning of exponents, from whole numbers to fractions. Perhaps it might not be unworthy of consideration, whether the meaning of the indices of differentiation, could not be considerably extended. Euler seems to have had the first idea* of interpolating the series

$$dy, \ d^2y, \ d^3y, \ \&c.$$

Laplace has extended his researches on this subject to considerable length, and has given the value of such expressions as the following, by converging series and definite integrals

$$\frac{d^n \cdot x^m}{dx^n}, \qquad \Delta^n \cdot x^m$$

so as to allow of evaluation for fractional values of n. But the indices themselves might be supposed to vary *continuously*, and such expressions as these

$$\left\{ \frac{d^n \cdot \left(\frac{d^p y}{dx^p} \right)}{dp^n} \right\}, \qquad \left(\frac{d\Delta^n y_x}{dn} \right)$$

become the subject of Analytical investigation. Or the index of a function might vary, as in the following instance:

Let $f(x) = \sqrt{\left(\frac{2x}{1+x} \right)}$, $\quad f^2(x) = \sqrt{\left(\frac{2f(x)}{1+f(x)} \right)}$, $\quad$ and so on,

then we shall have,

$$\frac{df^n(\sec v)}{dn} = -\frac{v \cdot \log 2}{2^n} \cdot \tan \left(\frac{v}{2^n} \right) \cdot \sec \left(\frac{v}{2^n} \right)$$

and again, (making use of Arbogast's notation,) if $y = a^x$

$$\overset{p_n}{\underset{p_{n-1}}{D}} \cdot \overset{p_{n-1}}{\underset{p_{n-2}}{D}} \cdot \ \ldots \ \overset{p_2}{\underset{p_1}{D}} \cdot \overset{p_1}{D_x} \cdot y = a^x \cdot (\log a)^{p_1} \cdot (\log^2 a)^{p_2} \cdot \ \ldots \ (\log^n a)^{p_n}$$

where $\log^2 a = \log \log a$, $\log^3 a = \log \log \log a$, and so on.

It has been observed by Charles, that the equation

---

* Leibnitz, it is true, in a letter to J. Bernouilli, mentions fractional indices of differentiation. Euler, however, first determined the value of such an expression as $d^{\frac{1}{2}}y$, y being a certain function of x.

$$\Delta y_n = b \cdot \frac{dy_n}{dx}$$

may be transformed into an integral, in which the index of integration is variable. Its solution then is

$$y_{-n} = b^{-n} \epsilon^{-\frac{x}{b}} \cdot \int^n \epsilon^{\frac{x}{b}} \cdot y_0 \cdot dx^n$$

or, which comes to the same,

$$y_n = b^n \epsilon^{-\frac{x}{b}} \cdot D_x^n \{ y_0 \cdot \epsilon^{\frac{x}{b}} \}$$

y_0 being any function of x. Laplace, in the *Theorie Anal. des Prob.* gives other instances of the same kind*.

That such expressions are not merely analytical curiosities, but relate to the most difficult and important theories, is confirmed by the opinions of the most eminent Analysts; and though on mathematical subjects when proof can be produced, no weight must be allowed to authority; yet when the former is deficient, our judgement may surely be influenced by the latter.

The importance of adopting a clear and comprehensive notation did not, in the early period of analytical science, meet with sufficient attention; nor were the advantages resulting from it, duly appreciated. In proportion as science advanced, and calculations became more complex, the evil corrected itself, and each improvement in one, produced a corresponding change in the other. Perhaps no single instance of the improvement or extension of notation, better illustrates this opinion, than the happy idea of defining the result of every operation, that can be performed on quantity, by the general term of function, and expressing this generalization by a characteristic letter. It had the effect of

---

* The integral of the equation of mixed partial differences

$$u_x = A \cdot \left(\frac{d^\alpha u_{x-1}}{da^\alpha} \right) + B \cdot \left(\frac{d^\beta u_{x-2}}{db^\beta} \right) + C \cdot \left(\frac{d^\gamma u_{x-3}}{dc^\gamma} \right) + \&c.$$

may be easily shown to be

$$u_x = A^{\frac{x}{1}} \left(\frac{d^{\frac{\alpha x}{1}} \cdot \psi_1(a)}{da^{\frac{\alpha x}{1}}} \right) + B^{\frac{x}{2}} \cdot \left(\frac{d^{\frac{\beta x}{2}} \cdot \psi_2(b)}{db^{\frac{\beta x}{2}}} \right) + C^{\frac{x}{3}} \cdot \left(\frac{d^{\frac{\gamma x}{3}} \cdot \psi_3(c)}{dc^{\frac{\gamma x}{3}}} \right) + \&c.$$

$\psi_1, \psi_2, \&c.$ being the characteristics of arbitrary functions: an expression which except $\alpha, \beta, \gamma, \ldots$ are respectively multiples of $1, 2, 3, \ldots$ must necessarily involve fractional indices of differentiation for some values of x.

introducing into investigations, two qualities once deemed incompatible, generality and simplicity. It now points out a calculus* perhaps more general than any hitherto discovered, and which should be called the calculus of functions, a name that more naturally belongs to it, than to that which Lagrange has so classically treated in the work which bears this name, although this latter is a branch of it.

Its object would be in general, the determination of functions from given conditions of whatever nature, whether depending on the successive terms of their developements, or on a series of indices differing by unity; or lastly, on a species of equations depending on the successive *orders* of the same function, of which the first mention we believe is made in one of the papers which compose the present volume. The necessity of this calculus was perceived soon after the discovery of equations of partial differentials, when it became requisite to determine the arbitrary functions which enter into their integrals, so as to satisfy given conditions. Euler and Alembert determined a few particular cases. Lagrange, in a Memoir entitled, 'Solution de differens Problemes de calcul integral,' resolved the equation

$$T = a \cdot \phi\{t + a \cdot (h + kt)\} + \beta \cdot \phi\{t + b(h + kt)\} + \&c.$$

Monge also has given several papers upon the subject, in the Memoirs of the Societies of Turin and Paris. The method of Laplace for reducing an equation of the first order, where the difference of the independent variable is any function of the variable itself, to one wherein that difference is constant, is well known. It had not, however, hitherto been shewn to be possible to reduce *every* equation of the former kind, to one of the latter. This object is, however, accomplished in the following pages. Still, it appears by no means natural, to resolve these equations, by means of finite differences, and it were much to be wished, that some independent method could be discovered, by which they might be treated.

One of the most striking advantages of the theory of *functions*, is, that it seems equally adapted to the proof of the most elementary truths, and to that of the most complicated and abstruse theorems. The latter part of this assertion, no one will be inclined to deny. An example of the former may be found in Laplace's proof of the decomposition of forces in the *Mecanique Celeste*†.

---

* Quamobrem non solum in hoc negotio, sed in plurimis aliis casibus, maximé utile foret, si functionum doctrina magis perficeretur et excoleretur. Euler, *Act. Acad. Petropol.* tom. VII.

† This demonstration consists of two parts. In the first he proves, that the diagonal of a rectangle whose sides represent the separate forces, will on the same scale represent the

There are still many problems in the theory of functions, which analysis seems to afford no means of attacking directly. Among which, may be enumerated the greater part of those which lead to an equation, containing a definite integral, where the unknown function enters under the integral sign.

quantity of the resultant. The second is devoted to shew, that it represents also its direction. This part seems to be generally considered, as deficient in clearness and simplicity. What we have here to remark, is, that it is *redundant*. In fact, by the combination of Laplace's three equations,

$$x = z \cdot \phi(\theta); \qquad y = z \cdot \phi\left(\frac{\pi}{2} - \theta\right), \qquad x^2 + y^2 = z^2$$

we obtain

$$\left\{ \phi(\theta) \right\}^2 + \left\{ \phi\left(\frac{\pi}{2} - \theta\right) \right\}^2 = 1$$

an equation which suffices for determining the nature of the function ϕ, and from which, by known processes, combined with the conditions of the question; it is easy to obtain $\phi(\theta) = \cos \theta$, and $x = z \cdot \cos \theta$, which is the equation to be deduced.

Another very remarkable instance of the use of the theory of functions, in demonstrating elementary truths, may be found in the following demonstration of Euclid's 47th, which has generally been thought to admit of none, but a geometrical proof: Call a, b, c, the sides,

A, B, C, the opposite angles of a right-angled triangle, $C = \dfrac{\pi}{2}$. It is easy to see, that the following equations hold good:

$$b = c \cdot \phi(A); \qquad a = c \cdot \phi(B); \qquad\qquad (1)$$

drop a perpendicular p, dividing c into two parts x, y, and we shall have, in the same manner,

$$x = b \cdot \phi(A); \qquad y = a \cdot \phi(B)$$

and of course,

$$x + y = c = b \cdot \phi(A) + a \cdot \phi(B)$$

from which, eliminating $\phi(A)$, and $\phi(B)$, by equations (1)

$$c^2 = a^2 + b^2, \qquad \text{QED.}$$

Having proved from other principles, that $A + B + C = \pi$, we shall have the very same equation

$$1 = \left\{ \phi(A) \right\}^2 + \left\{ \phi\left(\frac{\pi}{2} - A\right) \right\}^2$$

and thus we obtain $b = c \cdot \cos A$, $a = c \cdot \cos B$. It is only by this way of proceeding, or some analogous one, that we can ever hope to see the elementary principles of Trigonometry, brought under the dominion of Analysis. But this is not the place to proceed farther with the subject. It may suffice to have thrown out a hint, which may be followed up at some future opportunity.

For instance, suppose it were required to find the form of a function $\phi(x)$, such that the integral

$$\int dx \cdot F\{a, x, \phi(f(x))\}$$

taken between the limits $x = 0$, $x = a$, should equal any assigned function of a, F and f being given characteristics.

The examination of the properties and relations of numbers, constitutes a distinct branch of mathematical enquiry, almost entirely of modern origin; so abstract, and apparently so far removed from the confines of utility, that it seems to have attracted little attention from the generality of those, who have dedicated themselves to the pursuits of science. To the few, however, who have thought it worth their while to explore its more profound recesses, it has proved a mine, fertile in the most brilliant produce. Euclid, in his 7th book, has given the elements of transcendental Arithmetic, (a name appropriated to it by Professor Gauss). In the work of Diophantus, notwithstanding the ingenuity of the author, we discover the infancy of science in the absence of that generalization so happily adopted by modern writers. It consists of a variety of insulated problems, relating to the solution of certain indeterminate equations, rather than to the properties of numbers. Indeed, the indeterminate analysis, and the theory of numbers, form two branches of enquiry, which, (however nearly connected), ought to be carefully distinguished, in any systematic arrangement of our knowledge. The former must be considered as a province of the pure Analytics. Its attainment is indeed necessary for the perfection of the latter, (which should rather be regarded in the light of an *application* of analysis,) but is by no means limited to this one object. Lacroix* has introduced it with a very elegant effect, in that part of the theory of curves, which relates to their construction by points.

The resolution of the indeterminate equation of the first degree, is said to be due to Bachet. Euler, in the Petersburg Commentaries, exhibited a method of obtaining any number of solutions of that of the second, provided one particular one be known; but it has been remarked, that his methods do not afford *all* the possible solutions. This, however, has been effected at length by Lagrange, in the *Mem. Acad. de Berlin* (1767 and 1768.) His method consists in reducing successively by a series of operations, the coefficients of the equation

$$ax^2 + by^2 = z^2$$

* *Traité de Calc. &c.* 2d edit. vol. I. Note to page 417.

(to which form, every equation of the second degree may be reduced,) till one of them becomes unity; in which case, the resolution is easy. Gauss also, in his *Disquisitiones Arithmeticae*, No 216. has, by a method entirely different, shewn how to obtain all the solutions, of which an equation of the second degree admits. It is extremely remarkable, that the Hindu Algebraist, Bhaskara Acharya, who flourished about the year 1188, had also succeeded so far in his attempts on this difficult problem, as to derive any number of solutions from one, previously known, in the case of $ax^2 + b = y^2$.

The theorems given by Fermat without demonstration, form, without doubt, the most remarkable era in the theory of numbers. Too general, and too extraordinary in their nature to escape notice, they seem to have been the principal cause of the advances, which have since been made in this theory, by drawing the attention of Mathematicans to their demonstration. Euler, than whom none ever entered with greater ardency into this career, has proved some of the principal. Lagrange has supplied the demonstrations to others: still, however, many remain, of which no proof has been offered. It has been suggested, that Fermat was indebted to the method of induction, for the discovery of many of his theorems; an opinion rendered probable by the observation of Euler, that one of them relating to prime numbers is not true. This method is perhaps more applicable to researches in the theory of numbers, than to any other branch of abstract investigation; but it is of dangerous use, and should be supported by a large number of instances.*

---

* The substitution of 0, 1, 2, . . . for x, in the expression $x^2 + x + 41$, gives a series of numbers, of which the 40 first terms are primes, as Euler has remarked: yet it is easy to shew, that no algebraic function of x can in all cases represent a prime. The reason of this singular circumstance, and of a variety of similar coincidences, has since been satisfactorily explained, and the property demonstrated *a priori*. Fermat, deceived by a similar induction, asserted that all the numbers contained in the formula $2^{2^n} + 1$, are primes which Euler has since (as above alluded to) shewn to fail, in the case of $n = 5$.

The following theorems are derived solely by induction:

1mo. An indefinite number of integer values of x may be found, which render $\dfrac{7^x - 1}{10^x}$ an integer, to which we may add, that the formula $\dfrac{7^{8.10^x} - 1}{10^{x+2}}$ is always an integer, as are also the formulae $\dfrac{3^{2^x} - 1}{2^x}$, $\dfrac{5^{2^x} - 1}{2^x}$ and, in general, $\dfrac{(2n-1)^{2^x} - 1}{2^x}$, which last may be easily shewn, *a priori*, as well as a variety of expressions of the same description.

2do. The expression $\dfrac{3^{10^x} - 1}{10^{x+1}}$ is an integer, and, it is somewhat remarkable, that this integer is always of the form $100n + 22$.

Among the later discoveries in the theory of numbers, we have to enumerate two of the most surprising, which perhaps are to be found in the whole circle of Analytics. The first is a formula of Lagrange, obtained by induction, for determining the number of primes contained between given limits. It has been demonstrated by Legendre, in his *Essai sur la Theorie des Nombres*, although not very rigorously, it must be confessed; such is the difficulty of the subject. Indeed, the author has given it only as an attempt. The other is that celebrated theorem of Gauss, given in his *Disquisitiones Arithmeticae*, on the resolution of the equation $x^n - 1 = 0$, n being any prime, viz. that this equation may be reduced to a equations of the degree a, β of the degree b, and so on, where $n - 1 = a^a \cdot b^\beta \ldots$ Of course, the resolution of the equation $x^n - 1 = 0$, where n is a prime of the form $2^m + 1$, requires only the application of quadratics. Thus the division of the circle into $17, 257, 65\,537$ parts, may be accomplished by the description only of circles and straight lines.

To enter into any account of the advances made in the mixed Analytics, would far exceed our limits. There is one point, however, which we cannot forbear cursorily touching upon, on account of the great difficulty of reducing its conditions into symbolic language. We allude to the geometry of situation. Like the theory of numbers, at the first glance it seems barren and useless, but on a nearer examination is found abounding with interesting relations. Like that theory too, its cultivators have hitherto been few, but eminent, distinguished for that restless spirit of enquiry, which is ever upon the wing in search of new truths, and that invention which knows how to extract them, from the most unpromising hints. Leibnitz, appears to have found its first application, in considering the game of solitaire. A similar case (the problem of the knight's move at chess,) occupied the attention of Euler, and afterwards of Vandermonde, who adapted to it a notation analogous to that, by which the position of a point in space is determined, by three rectangular co-ordinates. In a more advanced state, it might, perhaps, embrace problems of a much higher order of difficulty, such as the following: 'Given n points in space; to

30. If A be such, that 5^n is congruous to A, (modul. 10^k) then will also $5^{n+2^{k-2} \cdot i}$ be

congruous to the same A, to the same modulus, and consequently, $5^{n+2^{k-2} \cdot i} \equiv 5^n$. (modul. 10^k) i being any integer. In other words, the formula

$$\frac{5^{n+2^x \cdot i} - 5^n}{10^{x+2}}$$

is always an integer, x and i being integers.

find the course to be pursued, so that setting off from any one, and passing at least once through all the rest, on returning to the original position, the least possible space shall have been described.' Such is the brief account of a theory yet in its first infancy. On its basis some future LEIBNITZ may perhaps hereafter lay the foundation of a name great as that of its original inventor*.

The preceding pages have been devoted to a slight account of the history and present state of Analytical Science, that branch of human knowledge, of which Laplace has justly observed 'C'est le guide le plus sur qui peut nous conduire dans la recherche de la verite'. But some account will naturally be expected of the source itself, from which the present work emanates. Of this however, very little need be said, but, that it consists of a few individuals, perhaps too sanguine in their hopes of promoting their favourite science, and of adding at least some trifling aid to that spirit of enquiry, which seems lately to have been awakened in the minds of our countrymen, and which will no longer suffer them to receive discoveries in science at second hand, or to be thrown behind in that career, whose first impulse they so eminently partook. The time perhaps is not far distant, when such an attempt will be regarded in an honourable light, whatever may be its success.

Meanwhile the view we have taken of the subject, appears by no means to lead to the mortifying conclusion, deduced by a foreign Geometer of considerable eminence; 'que la puissance de notre analyse est a-peupres epuisee.' The golden age of mathematical literature is undoubtedly past. Another, 'less fine in carat,' may however yet succeed. The motive which could draw forth so severe a sentence on the success of future exertions we will forbear to enquire, but it must surely be looked for elsewhere, than in the real interest of science which can never be promoted by repressing the ardour of research, or extinguishing the hope of reward. The foundations of a vast edifice have been laid; some of its apartments have been finished; others yet remain incomplete: but the strength and solidity of the basis will justify the expectation of large additions to the superstructure.

Attentively to observe the operations of the mind in the discovery of new truths, and to retain at the same time those fleeting links, which furnish a momentary connection with distant ideas, the knowledge of whose existence

---

* In the '*Journal de l' Ecole Polytechnique*, An. 10.' is a Memoir on polygons and polyhedrons, by Poinsot, in which he shews, that there exist other regular polygons besides the equilateral triangle, the sum of whose angles is equal to two right angles, and also, that there are more than five regular polyhedrons. The whole Memoir relates to geometry of situation, and forms the introduction to some more considerable researches, which the author promises in a future paper.

we derive from reason rather than perception, are objects in whose pursuit nothing but the most patient assiduity can expect success. Powerful indeed, must be that mind, which can simultaneously carry on two processes, each of which requires the most concentrated attention. Yet these obstacles must be surmounted, before we can hope for the discovery of a philosophical theory of invention; a science which Lord Bacon reported to be wholly deficient two centuries ago, and which has made since that time but slight advances. Probably, the era which shall produce this discovery is yet far distant. The capital of science, however, from its very nature, must continue to increase by gradual yet permanent additions; at the same time that all such additions to the common stock yield an interest in the power they afford of multiplying our combinations, and examining old difficulties in new points of view. It is this connection with fresher sources, which can restore fertility to subjects apparently the most exhausted, and which cannot be too earnestly recommended to those who wish to enlarge the limits of analysis. The fire of improvement, however dormant, and seemingly extinct, may yet break forth at the contact of some external flame. The history of Mathematics affords too many instances of the most distant principles coming into play on the most unexpected occasions, to allow of our ever despairing of success in such enquiries.

One inconvenience however, results as a necessary consequence from the continued accumulation of indestructible knowledge. The beaten field of analysis, limited as it is when compared with the almost boundless extent which remains to be explored, is yet so considerable with respect to the powers of human reason, and (if we may be allowed to pursue the metaphor a little farther,) so intersected with the tracks of those who have traversed it in every direction, as to become bewildering and oppressive to the last degree. The labour of one life would be more than occupied in perusing those works on the subject which the labour of so many has been spent in composing. The multitude of different methods and artifices, which for the most part lead only to the same results, and whose power is limited by the same points of difficulty, is at length grown into a very serious evil. Our continental neighbours seem sensible of this, if we may judge from the number of works which have appeared within these few years, digesting various points into a systematic form. But there is still much to be done in this line. That man would render a most invaluable service to science, who would undertake the labour of reducing into a reasonable compass the whole essential part of analysis, with its applications, curtailing its superfluous luxuriance, rejecting its artificial difficulties, and giving connection and unity to its scattered members.

The first Difference Engine

BABBAGE CAME down in 1814 and married Georgiana Whitmore of Dudmaston in Shropshire. Dudmaston House now belongs to the National Trust, thanks to the munificence of Lady Labouchere, heiress of both the Whitmores of Dudmaston and the Darbys of Coalbrookdale. The Dudmaston estate includes William Shenstone's only extant grotto, once no doubt the scene of Babbage and Georgiana's courting. Through the Whitmores, Babbage met Lucien Bonaparte, then in exile at Thorngrove, and the Bonapartes were to have an important influence both on Babbage and on Wolryche Whitmore, Georgiana's elder brother, the MP who rose year after year in the House of Commons to propose repeal of the corn laws.

Georgiana and Babbage were married in Teignmouth and moved to London where they lived on Babbage's allowance and a small income of Georgiana's. Babbage gave a series of lectures on astronomy at the Royal Institution, drawing on the knowledge of Caroline and Sir William Herschel. The former Analyticals were still full of plans for reforming science, drawing on French ideas, and Babbage continued to work on functional equations and algebra. Georgiana was raising a family in Devonshire Street, St Marylebone, in North London, and they spent most of the summers in South Devon.

Babbage and his friends remained interested in almost anything with technical novelty or scientific interest.

Passages from the Life of a Philosopher, pp. 189–90

THE THAUMATROPE: ITS ORIGIN

One day Herschel, sitting with me after dinner, amusing himself by spinning a pear upon the table, suddenly asked whether I could show him the two sides of a shilling at the same moment.

I took out of my pocket a shilling, and holding it up before the looking-glass, pointed out *my* method. 'No,' said my friend, 'that won't do;' then spinning my shilling upon the table, he pointed out *his* method of seeing both sides at once. The next day I mentioned the anecdote to the late Dr. Fitton, who a few days after brought me a beautiful illustration of the principle. It consisted of a round disc of card suspended between the two pieces of sewing-silk. These threads being held between the finger and thumb of each hand, were then made to turn quickly, when the disc of card, of course, revolved also.

Upon one side of this disc of card was painted a bird; upon the other side, an empty bird-cage. On turning the thread rapidly, the bird appeared to have got inside the cage. We soon made numerous applications, as a rat on one side and a trap upon the other, &c. It was shown to Captain Kater, Dr. Wollaston, and many of our friends, and was, after the lapse of a short time, forgotten.

Some months after, during dinner at the Royal Society Club, Sir Joseph Banks being in the chair, I heard Mr. Barrow, then Secretary to the Admiralty, talking very loudly about a wonderful invention of Dr. Paris, the object of which I could not quite understand. It was called the thaumatrope, and was said to be sold at the Royal Institution, in Albermarle-street. Suspecting that it had some connection with our unnamed toy, I went the next morning and purchased, for seven shillings and sixpence, a thaumatrope, which I afterwards sent down to Slough to the late Lady Herschel. It was precisely the thing which her son and Dr. Fitton had contributed to invent, which amused all their friends for a time and had then been forgotten. There was however *one* additional thaumatrope made afterwards. It consisted of the usual disc of paper. On one side was represented a thaumatrope (the design upon it being a penny-piece) with the motto, 'How to turn a penny.'

On the other side was a gentleman in black, with his hands held out in the act of spinning a thaumatrope, the motto being, 'A new trick from Paris.'

Passages from the Life of a Philosopher, pp. 377–9

One of the most useful accomplishments for a philosophical traveller with which I am acquainted, I learned from a workman, who taught me how to punch a hole in a sheet of glass without making a crack in it.

The process is very simple. Two centre-punches, a hammer, an ordinary bench-vice, and an old file, are all the tools required. These may be found in any blacksmith's shop. Having decided upon the part of the glass in which you wish to make the hole, scratch a cross (X) upon the desired spot with the point of the old file; then turn the bit of glass over, and scratch on the other side a similar mark exactly opposite to the former.

Fix one of the small centre-punches with its point upwards in the vice. Let an assistant gently hold the bit of glass with its scratched point exactly resting upon the point of the centre-punch.

Take the other centre-punch in your own left hand and place its point in the centre of the upper scratch, which is of course nearly, if not exactly, above the fixed centre-punch. Now hit the upper centre-punch a *very* slight blow with the hammer: a mere touch is almost sufficient. This must be carefully repeated two or three times. The result of these blows will be to cause the centre of the cross to be, as it were, gently pounded.

Turn the glass over and let the slight cavity thus formed rest upon the fixed centre-punch. Repeat the light blows upon this side of the glass, and after turning it two or three times, a very small hole will be made through the glass. It not unfrequently happens that a small crack occurs in the glass; but with a little skill this can be cut out with the pane of the hammer.

The next process is to enlarge the hole and cut it into the required shape with the pane of the hammer. This is accomplished by supporting the glass upon the point of the fixed centre-punch, very close to the edge required to be cut. A light blow must then be struck with the pane of the hammer upon the edge to be broken. This must be repeated until the required shape is obtained.

The principles on which it depends are, that glass is a material breaking in every direction with a conchoidal fracture, and that the vibrations which would have caused cracking or fracture are checked by the support of the fixed centre-punch in close contiguity with the part to be broken off.

When by hastily performing this operation I have caused the glass to crack, I have frequently, by using more care, cut an opening all round the cracked part, and so let it drop out without spreading.

This process is rendered still more valuable by the use of the diamond. I usually carried in my travels a diamond mounted on a small circle of wood, so that I could easily cut out circles of glass with small holes in the centre. The description of this process is sufficient to explain it to an experienced workman; but if the reader should wish to employ it, his readiest plan would be to ask such a person to show him how to do it.

The above technical description will doubtless be rather dry and obscure to the general reader; so I hope to make him amends by one or two of the consequences which have resulted to me from having instructed others in the art.

In the year 1825, during a visit to Devonport, I had apartments in the house of a glazier, of whom one day I inquired whether he was acquainted with the art of punching a hole in glass, to which he answered in the negative, and expressed great curiosity to see it done. Finding that at a short distance there was a blacksmith whom he sometimes employed, we went together to pay him a visit, and having selected from his rough tools and centre-punches and the hammer, I proceeded to explain and execute the whole process, with which my landlord was highly delighted.

On the eve of my departure I asked for the landlord's account, which was duly sent up and quite correct, except the omission of the charge for the apartments which I had agreed for at two guineas a week. I added the four weeks for my lodgings, and the next morning, having placed the total amount upon the bill, I sent for my host in order to pay him, remarking that he had omitted the principal article of his account, which I had inserted.

He replied that he had intentionally omitted the lodgings, as he could not think of taking payment for them from a gentleman who had done him so great a service. Quite unconscious of having rendered him any service, I asked him to explain how I had done any good. He replied that he had the contract for the supply and repair of the whole of the lamps of Devonport, and that the art in which I had instructed him would save him more than twenty pounds a year. I found some difficulty in prevailing on my grateful landlord to accept what was justly his due.

———

During this period Babbage's principal scientific activity was the project to make a machine to calculate and print tables, his First Difference Engine.

Passages from the Life of a Philosopher, pp. 41–8

Calculating machines comprise various pieces of mechanism for assisting the human mind in executing the operations of arithmetic. Some few of these perform the whole operation without any mental attention when once the given numbers have been put into the machine.

Others require a moderate portion of mental attention: these latter are generally of much simpler construction than the former, and it may also be added, are less useful.

The simplest way of deciding to which of these two classes any calculating machine belongs is to ask its maker—Whether, when the numbers on which it is to operate are placed in the instrument, it is capable of arriving at its result by the mere motion of a spring, a descending weight, or any other constant force? If the answer be in the affirmative, the machine is really automatic; if otherwise, it is not self-acting.

Of the various machines I have had occasion to examine, many of those for Addition and Subtraction have been found to be automatic. Of machines for Multiplication and Division, which have fully come under my examination, I cannot at present recall one to my memory as absolutely fulfilling this condition.

The earliest idea that I can trace in my own mind of calculating arithmetical Tables by machinery arose in this manner:—

One evening I was sitting in the rooms of the Analytical Society, at Cambridge, my head leaning forward on the Table in a kind of dreamy mood, with a Table of logarithms lying open before me. Another member, coming into the room, and seeing me half asleep, called out, 'Well, Babbage, what are you dreaming about?' to which I replied, 'I am thinking that all these Tables (pointing to the logarithms) might be calculated by machinery.'

I am indebted to my friend, the Rev. Dr. Robinson, the Master of the Temple, for this anecdote. The event must have happened either in 1812 or 1813.

About 1819 I was occupied with devising means for accurately dividing astronomical instruments, and had arrived at a plan which I thought was likely to succeed perfectly. I had also at that time been speculating about making machinery to compute arithmetical Tables.

One morning I called upon the late Dr. Wollaston, to consult him about my plan for dividing instruments. On talking over the matter, it turned out that my system was exactly that which had been described by the Duke de Chaulnes, in the *Memoirs of the French Academy of Sciences*, about fifty or sixty years before. I then mentioned my other idea of computing Tables by machinery, which Dr. Wollaston thought a more promising subject.

I considered that a machine to execute the mere isolated operations of arithmetic, would be comparatively of little value, unless it were very easily set to do its work, and unless it executed not only accurately, but with great rapidity, whatever it was required to do.

On the other hand, the method of differences supplied a general principle by which *all* Tables might be computed through limited intervals, by one uniform process. Again, the method of differences required the use of mechanism for Addition only. In order, however, to insure accuracy in the printed Tables, it was necessary that the machine which computed Tables should also set them up in type, or else supply a mould in which stereotype plates of those Tables could be cast.

I now began to sketch out arrangements for accomplishing the several partial processes which were required. The arithmetical part must consist of two distinct processes—the power of adding one digit to another, and also of carrying the tens to the next digit, if it should be necessary.

The first idea was, naturally, to add each digit successively. This, however, would occupy much time if the numbers added together consisted of many places of figures.

The next step was to add all the digits of the two numbers each to each at the same instant, but reserving a certain mechanical memorandum, wherever a carriage became due. These carriages were then to be executed successively.

Having made various drawings, I now began to make models of some portions of the machine, to see how they would act. Each number was to be expressed upon wheels placed upon an axis; there being one wheel for each figure in the number operated upon.

Having arrived at a certain point in my progress, it became necessary to have teeth of a peculiar form cut upon these wheels. As my own lathe was not fit for this job, I took the wheels to a wheel-cutter at Lambeth, to whom I carefully conveyed my instructions, leaving with him a drawing as his guide.

These wheels arrived late one night, and the next morning I began putting them in action with my other mechanism, when, to my utter astonishment, I found they were quite unfit for their task. I examined the shape of their teeth, compared them with those in the drawings, and found they agreed perfectly; yet they could not perform their intended work. I had been so certain of the truth of my previous reasoning, that I now began to be somewhat uneasy. I reflected that, if the reasoning about which I had been so certain should prove to have been really fallacious, I could then no longer trust the power of my own reason. I therefore went over with my wheels to the artist who had formed the teeth, in order that I might arrive at some explanation of this extraordinary contradiction.

On conferring with him, it turned out that, when he had understood fully the peculiar form of the teeth of wheels, he discovered that his wheel-cutting engine had not got amongst its divisions that precise number which I had required. He therefore had asked me whether another number, which his machine possessed, would not equally answer my object. I had inadvertently replied in the affirmative. He then made arrangements for the precise number of teeth I required; and the new wheels performed their expected duty perfectly.

The next step was to devise means for printing the tables to be computed by this machine. My first plan was to make it put together moveable type. I proposed to make metal boxes, each containing 3000 types of one of the ten digits. These types were to be made to pass out one by one from the bottom of their boxes, when required by the computing part of the machine.

But here a new difficulty arose. The attendant who put the types into the boxes might, by mistake, put a wrong type in one or more of them. This cause of error I removed in the following manner:—There are usually certain notches in the side of the type. I caused these notches to be so placed that all the types of any given digit possessed the same characteristic notches, which no other type had. Thus, when the boxes were filled, by passing a small wire down these peculiar notches, it would be impeded in its passage, if there were included in the row a single wrong figure. Also, if any digit were accidentally turned upside down, it would be indicated by the stoppage of the testing wire.

One notch was reserved as common to every species of type. The object of this was that, before the types which the Difference Engine had used for its computation were removed from the iron platform on which they were placed, a steel wire should be passed through this common notch, and remain there. The tables, composed of moveable types, thus interlocked, could never have any of their figures drawn out by adhesion to the inking-roller, and then by possibility be restored in an inverted order. A small block of such figures tied together by a bit of string, remained unbroken for several years, although it was rather roughly used as a plaything by my children. One such box was finished, and delivered its type satisfactorily.

Another plan for printing the tables, was to place the ordinary printing type round the edges of wheels. Then, as each successive number was produced by the arithmetical part, the type-wheels would move down upon a plate of soft composition, upon which the tabular number would be impressed. This mould was formed of a mixture of plaster-of-Paris with other materials, so as to become hard in the course of a few hours.

The first difficulty arose from the impression of one tabular number on the mould being distorted by the succeeding one.

I was not then aware that a very slight depth of impression from the type would be quite sufficient. I surmounted the difficulty by previously passing a roller, having longitudinal wedge-shaped projections, over the plastic material. This formed a series of small depressions in the matrix between each line. Thus the expansion arising from the impression of one line partially filled up the small depression or ditch which occurred between each successive line.

The various minute difficulties of this kind were successively overcome; but subsequent experience has proved that the depth necessary for stereotype moulds is very small, and that even thick paper, prepared in a peculiar manner, is quite sufficient for the purpose.

Another series of experiments were, however, made for the purpose of punching the computed numbers upon copper plate. A special machine was contrived and constructed, which might be called a co-ordinate machine, because it moved the copper plate and steel punches in the direction of three rectangular co-ordinates. This machine was afterwards found very useful for many other purposes. It was, in fact, a general shaping machine, upon which many parts of the Difference Engine were formed.

Several specimens of surface and copper-plate printing, as well as of the copper plates, produced by these means, were exhibited at the Exhibition of 1862.

I have proposed and drawn various machines for the purpose of calculating a series of numbers forming Tables by means of certain system called 'The Method of Differences,' which it is the object of this sketch to explain.

The first Difference Engine with which I am acquainted comprised a few figures, and was made by myself, between 1820 and June 1822. It consisted of from six to eight figures. A much larger and more perfect engine was subsequently commenced in 1823 for the Government.

It was proposed that this latter Difference Engine should have six orders of differences, each consisting of about twenty places of figures, and also that it should print the Tables it computed.

The small portion of it which was placed in the International Exhibition of 1862 was put together nearly thirty years ago. It was accompanied by various parts intended to enable it to print the results it calculated, either as a single copy on paper—or by putting together moveable types—or by stereotype plates taken from moulds punched by the machine—or from copper plates impressed by it. The parts necessary for the execution of each of these processes were made, but these were not at that time attached to the calculating part of the machine.

A considerable number of the parts by which the printing was to be accomplished, as also several specimens of portions of tables punched on copper, and of stereotype moulds, were exhibited in a glass case adjacent to the Engine.

In 1834 Dr. Lardner published, in the *Edinburgh Review*,* a very elaborate description of this portion of the machine, in which he explained clearly the method of Differences.

It is very singular that two persons, one resident in London, the other in Sweden, should both have been struck, on reading this review, with the simplicity of the mathematical principle of differences as applied to the calculation of Tables, and should have been so fascinated with it as to have undertaken to construct a machine of the kind.

Mr. Deacon, of Beaufort House, Strand, whose mechanical skill is well

---

* *Edinburgh Review*, No. cxx., July, 1834.

known, made, for his own satisfaction, a small model of the calculating part of such a machine, which was shown only to a few friends, and of the existence of which I was not aware until after the Swedish machine was brought to London.

Mr Scheutz, an eminent printer at Stockholm, had far greater difficulties to encounter. The construction of mechanism, as well as the mathematical part of the question, was entirely new to him. He, however, undertook to make a machine having four differences, and fourteen places of figures, and capable of printing its own Tables.

After many years' indefatigable labour, and an almost ruinous expense, aided by grants from his Government, by the constant assistance of his son, and by the support of many enlightened members of the Swedish Academy, he completed his Difference Engine. It was brought to London, and some time afterwards exhibited at the great Exhibition at Paris. It was then purchased for the Dudley Observatory at Albany by an enlightened and public-spirited merchant of that city, John F. Rathbone, Esq.

An exact copy of this machine was made by Messrs. Donkin and Co., for the English Government, and is now in use in the Registrar-General's Department at Somerset House. It is very much to be regretted that this specimen of English workmanship was not exhibited in the International Exhibition.

––––––––

Babbage's plans for making a calculating engine were launched publicly.

Letter to Sir Humphry Davy, p. 3–12

A LETTER TO SIR HUMPHRY DAVY, BART.

My Dear Sir,

The great interest you have expressed in the success of that system of contrivances which has lately occupied a considerable portion of my attention, induces me to adopt this channel for stating more generally the principles on which they proceed, and for pointing out the probable extent and important consequences to which they appear to lead. Acquainted as you were with this inquiry almost from its commencement, much of what I have now to say cannot fail to have occurred to your own mind: you will however permit me to re-state it for the consideration of those with whom the principles and the machinery are less familiar.

The intolerable labour and fatiguing monotony of a continued repetition of similar arithmetical calculations, first excited the desire, and afterwards suggested the idea, of a machine, which, by the aid of gravity of any other moving power, should become a substitute for one of the lowest operations of human intellect. It is not my intention in the present Letter to trace the progress of this idea, or the means which I have adopted for its execution; but I propose stating some of their general applications, and shall commence with describing the powers of several engines which I have contrived: of that part which is already executed I shall speak more in the sequel.

The first engine of which drawings were made was one which is capable of computing any table by the aid of differences, whether they are positive or negative, or of both kinds. With respect to the number of the order of differences, the nature of the machinery did not in my own opinion, nor in that of a skilful mechanic whom I consulted, appear to be restricted to any very limited number; and I should venture to construct one with ten or a dozen orders with perfect confidence. One remarkable property of this machine is, that the greater the number of differences the more the engine will outstrip the most rapid calculator.

By the application of certain parts of no great degree of complexity, this may be converted into a machine for extracting the roots of equations, and consequently the roots of numbers: and the extent of the approximation depends on the magnitude of the machine.

Of a machine for multiplying any number of figures (m) by any other number (n) I have several sketches; but it is not yet brought to that degree of perfection which I should wish to give it before it is to be executed.

I have also certain principles by which, if it should be desirable, a table of prime numbers might be made, extending from o to ten millions.

Another machine, whose plans are much more advanced than several of those just named, is one for constructing tables which have no order of differences constant.

A vast variety of equations of finite differences may by its means be solved, and a variety of tables, which could be produced in successive parts by the first machine I have mentioned, could be calculated by the latter one with a still less exertion of human thought. Another and very remarkable point in the structure of this machine is, that it will calculate tables governed by laws which have not been hitherto shown to be explicitly determinable, or that it will solve equations for which analytical methods of solution have not yet been contrived.

Supposing these engines executed, there would yet be wanting other means to ensure the accuracy of the printed tables to be produced by them.

The errors of the persons employed to copy the figures presented by the engines would first interfere with their correctness. To remedy this evil, I have contrived means by which the machines themselves shall take from several boxes containing type, the numbers which they calculate, and place them side by side; thus becoming at the same time a substitute for the compositor and the computer: by which means all error in copying as well as in printing is removed.

There are, however, two sources of error which have not yet been guarded against. The ten boxes with which the engine is provided contain each about three thousand types; any box having of course only those of one number in it. It may happen that the person employed in filling these boxes shall accidentally place a wrong type in some of them; as for instance, the number 2 in the boxes which ought only to contain 7s. When these boxes are delivered to the superintendant of the engine, I have provided a simple and effectual means by which he shall in less than half an hour ascertain whether, amongst these 30 000 types, there be any individual misplaced or even inverted. The other cause of error to which I have alluded, arises from the type falling out when the page has been set up: this I have rendered impossible by means of a similar kind.

The quantity of errors from carelessness in correcting the press, even in tables of the greatest credit, will scarcely be believed, except by those who have had constant occasion for their use. A friend of mine, whose skill in practical as well as theoretical astronomy is well known, produced to me a copy of the tables published by order of the French Board of Longitude, containing those of the Sun by Delambre and of the Moon by Burg, in which he had corrected above *five hundred errors:* most of these appear to be errors of the press; and it is somewhat remarkable, that in turning over the leaves in the fourth page I opened we observed a new error before unnoticed. These errors are so much the more dangerous, because independent computers using the same tables will agree in the same errors.

To bring to perfection the various machinery which I have contrived, would require an expense both of time and money which can be known only to those who have themselves attempted to execute mechanical inventions. Of the greater part of that which has been mentioned, I have at present contented myself with sketches on paper, accompanied by short memorandums, by which I might at any time more fully develop the contrivances; and where any new principles are introduced I have had models executed in order to examine their actions. For the purpose of demonstrating the practicability of these views, I have chosen the engine for differences, and have constructed one of them which will produce any tables whose second differences are constant. Its size is the same as that which I should propose for any more extensive one of the same kind: the chief difference would be, that in one intended for use there would be a greater repetition of the same parts in order to adapt it to the calculation of a larger number of figures. Of the action of this engine, you have yourself had opportunities of judging, and I will only at present mention a few trials which have since been made by some scientific gentlemen to whom it has been shown, in order to determine the rapidity with which it calculates. The computed table is presented to the eye at two opposite sides of the machine; and a friend having undertaken to write down the numbers as they appeared, it proceeded to make a table from the formula $x^2 + x + 41$. In the earlier numbers my friend, in writing quickly, rather more than kept pace with the engine; but as soon as four figures were required, the machine was at least equal in speed to the writer.

In another trial it was found that thirty numbers of the same table were calculated in two minutes and thirty seconds: as these contained eighty-two figures, the engine produced thirty-three every minute.

In another trial it produced figures at the rate of forty-four in a minute. As the machine may be made to move uniformly by a weight, this rate might be maintained for any length of time, and I believe few writers would be found to copy with equal speed for many hours together. Imperfect as a first machine generally is, and suffering as this particular one does from great defect in the workmanship, I have every reason to be satisfied with the accuracy of its computations; and by the few, skilful mechanics to whom I have in confidence shown it, I am assured that its principles are such that it may be carried to any extent. In fact, the parts of which it consists are few but frequently repeated, resembling in this respect the arithmetic to which it is applied, which, by the aid of a few digits often repeated, produces all the wide variety of number. The wheels of which it consists are numerous, but few move at the same time; and I have employed a principle by which any small error that may arise from accident or bad workmanship is corrected as soon as it is produced, in such a manner as effectually to prevent any accumulation of small errors from producing a wrong figure in the calculation.

Of those contrivances by which the composition is to be effected, I have made many experiments and several models; the results of these leave me no reason to doubt of success, which is still further confirmed by a working model that is just finished.

As the engine for calculating tables by the method of differences is the only one yet completed, I shall in my remarks on the utility of such instruments confine myself to a statement of the powers which that method possesses.

I would however premise, that if any one shall be of opinion, notwithstanding all the precaution I have taken and means I have employed to guard against the occurrence of error, that it may still be possible for it to arise, the method of differences enables me to determine its existence. Thus, if proper numbers are placed at the outset in the engine, and if it has composed a page of any kind of table, then by comparing the last number it has set up with that number previously calculated, if they are found to agree, the whole page must be correct: should any disagreement occur, it would scarcely be worth the trouble of looking for its origin, as the shortest plan would be to make the engine recalculate the whole page, and nothing would be lost but a few hours' labour of the moving power.

Of the variety of tables which such an engine could calculate, I shall mention but a few. The tables of powers and products published at the expense of the Board of Longitude, and calculated by Dr. Hutton, were solely executed by the method of differences; and other tables of the roots of numbers have been calculated by the same gentleman on similar principles.

As it is not my intention in the present instance to enter into the theory of differences, a field far too wide for the limits of this letter, and which will probably be yet further extended in consequence of the machinery I have contrived, I shall content myself with describing the course pursued in one of the most stupendous monuments of arithmetical calculation which the world has yet produced, and shall point out the mode in which it was conducted and what share of mental labour would have been saved by the employment of such an engine as I have contrived.

The tables to which I allude are those calculated under the direction of M. Prony by order of the French Government,—a work which will ever reflect the highest credit on the nation which patronized and on the scientific men who executed it. The tables computed were the following.

1. The natural sines of each 10 000 of the quadrant calculated to twenty-five figures with seven or eight orders of differences.
2. The logarithmic sines of each 100 000 of the quadrant calculated to fourteen decimals with five orders of differences.
3. The logarithm of the ratios of the sines to their arcs of the first 5000 of the 100,000ths of the quadrant calculated to fourteen decimals with three orders of differences.
4. The logarithmic tangents corresponding to the logarithmic sines calculated to the same extent.

5. The logarithms of the ratios of the tangents to their arcs calculated in the same manner as the logarithms of the ratios of the sines to their arcs.
6. The logarithms of numbers from 1 to 10 000 calculated to nineteen decimals.
7. The logarithms of all numbers from 10 000 to 200 000 calculated to fourteen figures with five orders of differences.

Such are the tables which have been calculated, occupying in their present state seventeen large folio volumes. It will be observed that the trigonometrical tables are adapted to the decimal system, which has not been generally adopted even by the French, and which has not been at all employed in this country. But, notwithstanding this objection, such was the opinion entertained of their value, that a distinguished member of the English Board of Longitude was not long since commissioned by our Government to make a proposal to the Board of Longitude of France to print an abridgement of these tables at the joint expense of the two countries; and five thousand pounds were named as the sum our Government was willing to advance for this purpose. It is gratifying to record this disinterested offer, so far above those little jealousies which frequently interfere between nations long rivals, and manifesting so sincere a desire to render useful to mankind the best materials of science in whatever country they might be produced. Of the reasons why this proposal was declined by our neighbours, I am at present uninformed: but, from a personal acquaintance with many of the distinguished foreigners to whom it was referred, I am convinced that it was received with the same good feelings as those which dictated it.

I will now endeavour shortly to state the manner in which this enormous mass of computation was executed; one table of which (that of the logarithms of numbers) must contain about eight millions of figures.

The calculators were divided into three sections. The first section comprised five or six mathematicians of the highest merit, amongst whom were M. Prony and M. Legendre. These were occupied entirely with the analytical part of the work; they investigated and determined on the formulae to be employed.

The second section consisted of seven or eight skilful calculators habituated both to analytical and arithmetical computations. These received the formulae from the first section, converted them into numbers, and furnished to the third section the proper differences at the stated intervals.

They also received from that section the calculated results, and compared the two sets, which were computed independently for the purpose of verification.

The third section, on whom the most labourious part of the operations devolved, consisted of from sixty to eighty persons, few of them possessing a knowledge of more than the first rules of arithmetic: these received from the second class certain numbers and differences, with which, by additions and subtractions in a prescribed order, they completed the whole of the tables above mentioned.

I will now examine what portion of this labour might be dispensed with, in case it should be deemed advisable to compute these or any similar tables of equal extent by the aid of the engine I have referred to.

In the first place, the labour of the first section would be considerably reduced, because the formulae used in the great work I have been describing have already been investigated and published. One person, or at the utmost two, might therefore conduct it.

If the persons composing the second section, instead of delivering the numbers they calculate to the computers of the third section, were to deliver them to the engine, the whole of the remaining operations would be executed by machinery, and it would only be necessary to employ people to copy down as fast as they were able the figures presented to them by the engine. If, however, the contrivances for printing were brought to perfection and employed, even this labour would be unnecessary, and a few superintendents would manage the machine and receive the calculated pages set up in type. Thus the number of calculators employed, instead of amounting to ninety-six, would be reduced to twelve. This number might however be considerably diminished, because when an engine is used the intervals between the differences calculated by the second section may be greatly enlarged. In the tables of logarithms M. Prony caused the differences to be calculated at intervals of two hundred, in order to save the labour of the third section: but as that would now devolve on machinery, which would scarcely move the slower for its additional burthen, the intervals might properly be enlarged to three or four times that quantity. This would cause a considerable diminution in the labour of the second section. If to this diminution of mental labour we add that which arises from the whole work of the compositor being executed by the machine, and the total suppression of that most annoying of all literary labour, the correction of the errors of the press*, I think I am justified in presuming

---

* I have been informed that the publishers of a valuable collection of mathematical tables, now re-printing, pay to the gentleman employed in correcting the press at the rate of three guineas a sheet, a sum by no means too large for the faithful execution of such a laborious duty.

that if engines were made purposely for this object, and were afterwards useless, the tables could be produced at a much cheaper rate; and of their superior accuracy there could be no doubt. Such engines would however be far from useless: containing within themselves the power of generating to an almost unlimited extent tables whose accuracy would be unrivalled, at an expense comparatively moderate, they would become active agents in reducing the abstract inquiries of geometry to a form and an arrangement adapted to the ordinary purposes of human society.

I should be unwilling to terminate this Letter without noticing another class of tables of the greatest importance, almost the whole of which are capable of being calculated by the method of differences. I refer to all astronomical tables for determining the positions of the sun or planets: it is scarcely necessary to observe that the constituent parts of these are of the form $a \sin \theta$, where a is a constant quantity, and θ is what is usually called the argument. Viewed in this light they differ but little from a table of sines, and like it may be computed by the method of differences.

I am aware that the statements contained in this Letter may perhaps be viewed as something more than Utopian, and that the philosophers of Laputa may be called up to dispute my claim to originality. Should such be the case, I hope the resemblance will be found to adhere to the nature of the subject rather than to the manner in which it has been treated. Conscious, from my own experience, of the difficulty of convincing those who are but little skilled in mathematical knowledge, of the possibility of making a machine which shall perform calculations, I was naturally anxious, in introducing it to the public, to appeal to the testimony of one so distinguished in the records of British science. Of the extent to which the machinery whose nature I have described may be carried, opinions will necessarily fluctuate, until experiment shall have finally decided their relative value: but of that engine which already exists I think I shall be supported, both by yourself and by several scientific friends who have examined it, in stating that it performs with rapidity and precision all those calculations for which it was designed.

Whether I shall construct a larger engine of this kind, and bring to perfection the others I have described, will in a great measure depend on the nature of the encouragement I may receive.

Induced, by a conviction of the great utility of such engines, to withdraw for some time my attention from a subject on which it has been engaged during several years, and which possesses charms of a higher order, I have now arrived at a point where success is no longer doubtful. It must however be attained at a very considerable expense, which would not probably be replaced, by the works it might produce, for a long period of time, and which is an undertaking I should feel unwilling to commence, as altogether foreign to my habits and pursuits.

I remain, my dear Sir,

With the greatest respect,

Faithfully yours,

C. BABBAGE.

Devonshire Street, Portland Place,
    July 3rd, 1822.

––––––––

Babbage continued to work on his first Difference Engine until 1833. Under his detailed direction Dionysius Lardner prepared a description of the Engine.

'Babbage's Calculating Engine,' *Edinburgh Review*, cxx, pp. 263–327

There is no position in society more enviable than that of the few who unite a moderate independence with high intellectual qualities. Liberated from the necessity of seeking their support by a profession, they are unfettered by its restraints, and are enabled to direct the powers of their minds, and to concentrate their intellectual energies on those objects exclusively to which they feel that their powers may be applied with the greatest advantage to the community, and with the most lasting reputation to themselves. On the other hand, their middle station and limited income rescue them from those allurements to frivolity and dissipation, to which rank and wealth ever expose their possessors. Placed in such favourable circumstances, Mr Babbage selected science as the field of his ambition; and his mathematical researches have conferred on him a high reputation, wherever the exact sciences are studied and appreciated. The suffrages of the mathematical world have been ratified in his own country, where he has been elected to the Lucasian Professorship in his own University—a chair, which, though of inconsiderable emolument, is one on which Newton has conferred everlasting celebrity. But it has been the fortune of this mathematican to surround himself with fame of another and more popular kind, and which rarely falls to the lot of those who devote their lives to the cultivation of the abstract sciences. This distinction he owes to the announcement, some years since, of his celebrated project of a Calculating Engine. A proposition to reduce arithmetic to the dominion of mechanism,—to substitute an automaton for a compositor,—to throw the powers of thought into wheel-work could not fail to awaken the attention of the world. To bring the practicability of such a project within the compass of popular belief was not easy: to do so by bringing it within the compass of popular comprehension was not possible. It transcended the imagination of the public in general to conceive its possibility; and the sentiments of wonder with which it was received, were only prevented from merging into those of incredulity, by the faith reposed in the high attainments of its projector. This extraordinary undertaking was, however, viewed in a very different light by the small section of the community, who, being sufficiently versed in mathematics, were acquainted with the principle upon which it was founded. By reference to that principle, they perceived at a glance the practicability of the project; and being enabled by the nature of their attainments and pursuits to appreciate the immeasurable importance of its results, they regarded the invention with a proportionately profound interest. The production of numerical tables, unlimited in quantity and variety, restricted to no particular species, and limited by no particular law;— extending not merely to the boundaries of existing knowledge, but spreading their powers over the undefined regions of future discovery—were results, the magnitude and the value of which the community in general could neither comprehend nor appreciate. In such a case, the judgment of the world could only rest upon the authority of the philosophical part of it; and the fiat of the scientific community swayed for once political councils. The British Government, advised by the Royal Society, and a committee formed of the most eminent mechanicians and practical engineers, determined on constructing the projected mechanism at the expense of the nation, to be held as national property.

Notwithstanding the interest with which this invention has been regarded in every part of the world, it has never yet been embodied in a written, much less in a published form. We trust, therefore, that some credit will be conceded to us for having been the first to make the public acquainted with the object, principle, and structure of a piece of machinery, which, though at present unknown (except as to a few of its probable results), must, when completed, produce important effects, not only on the progress of science, but on that of civilization.

The calculating machinery thus undertaken for the public gratuitously (so far as Mr Babbage is concerned), has now attained a very advanced stage towards completion; and a portion of it has been put together, and performs various calculations;—affording a practical demonstration that the anticipations of those, under whose advice Government has acted, have been well founded.

There are nevertheless many persons who, admitting the great ingenuity of the contrivance, have, notwithstanding, been accustomed to regard it more in the light of a philosophical curiosity, than an instrument for purposes practically useful. This mistake (than which it is not possible to imagine a greater) has arisen mainly from the ignorance which prevails of the extensive utility of those numerical tables which it is the purpose of the engine in question to produce. There are also some persons who, not considering the time requisite to bring any invention of this magnitude to perfection in all its details, incline to consider the delays which have taken place in its progress as presumptions against its practicability. These persons should, however, before they arrive at such a conclusion, reflect upon the time which was necessary to bring to perfection engines infinitely inferior in complexity and mechanical difficulty. Let them remember that —not to mention the *invention* of that machine—the *improvements* alone introduced into the steam-engine by the celebrated Watt, occupied a period of not less than twenty years of the life of that distinguished person, and involved an expenditure of capital amounting to L.50,000.* The calculating machinery is a contrivance new even in its details. Its inventor did not take it up already imperfectly formed, after having received the contributions of human ingenuity exercised upon it for a century or more. It has not, like almost all other great machanical inventions, been gradually advanced to its present state through a series of failures, through difficulties encountered and overcome by a succession of projectors. It is not an object on which the light of various minds has thus been shed. It is, on the contrary, the production of solitary and individual thought, – begun, advanced through each successive stage of improvement, and brought to perfection by one mind. Yet this creation of genius, from its first rude conception to its present state, has cost little more than half the time, and not one-third of the expense, consumed in bringing the steam-engine (previously far advanced in the course of improvement) to that state of comparative perfection in which it was left by Watt. Short as the period of time has been which the inventor has devoted to this enterprise, it has, nevertheless, been demonstrated, to the satisfaction of

---

* Watt commenced his investigations respecting the steam-engine in 1763, between which time, and the year 1782 inclusive, he took out several patents for improvements in details. Bolton and Watt had expended the above sum on their improvements before they began to receive any return.

many scientific men of the first eminence, that the design in all its details, reduced, as it is, to a system of mechanical drawings, is complete; and requires only to be constructed in conformity with those plans, to realize all that its inventor has promised.

With a view to remove and correct erroneous impressions, and at the same time to convert the vague sense of wonder at what seems incomprehensible, with which this project is contemplated by the public in general, into a more rational and edifying sentiment, it is our purpose in the present article,

First, To show the immense importance of any method by which numerical tables, absolutely accurate in every individual copy, may be produced with facility and cheapness. This we shall establish by conveying to the reader some notion of the number and variety of tables published in every country of the world to which civilisation has extended, a large portion of which have been produced at the public expense; by showing also, that they are nevertheless rendered inefficient, to a greater or less[er] extent, by the prevalence of errors in them; that these errors pervade not merely tables produced by individual labour and enterprise, but that they vitiate even those on which national resources have been prodigally expended, and to which the highest mathematical ability, which the most enlightened nations of the world could command, has been unsparingly and systematically directed.

Secondly, To attempt to convey to the reader a general notion of the mathematical principle on which the calculating machinery is founded, and of the manner in which this principle is brought into practical operation, both in the process of calculating and printing. It would be incompatible with the nature of this review, and indeed impossible without the aid of numerous plans, sections, and elevations, to convey clear and precise notions of the details of the means by which the process of reasoning is performed by inanimate matter, and the abitrary and capricious evolutions of the fingers of typographical compositors are reduced to a system of wheel-work. We are, nevertheless, not without hopes of conveying, even to readers unskilled in mathematics, some satisfactory notions of a general nature on this subject.

Thirdly, To explain the actual state of the machinery at the present time; what progress has been made towards its completion; and what are the probable causes of those delays in its progress, which must be a subject of regret to all friends of science. We shall indicate what appears to us the best and most practicable course to prevent the unnecessary recurrence of such obstructions for the future, and to bring this noble project to a speedy and successful issue.

Viewing the infinite extent and variety of the tables which have been calculated and printed, from the earliest periods of human civilisation to the present time, we feel embarrassed with the difficulties of the task which we have imposed on ourselves;—that of attempting to convey to readers unaccustomed to such speculations, any thing approaching to an adequate idea of them. These tables are connected with the various sciences, with almost every department of the useful arts, with commerce in all its relations; but above all, with Astronomy and Navigation. So important have they been considered, that in many instances large sums have been appropriated by the most enlightened nations in the production of them; and yet so numerous and insurmountable have been the difficulties attending the attainment of this end, that after all, even navigators, putting aside every other department of art and science, have, until very recently, been scantily and imperfectly supplied with the tables indispensably necessary to determine their position at sea.

The first class of tables which naturally present themselves, are those of Multiplication. A great variety of extensive multiplication tables have been published from an early period in different countries; and especially tables of *Powers*, in which a number is multiplied by itself successively. In Dodson's *Calculator* we find a table of multiplication extending as far as 10 times 1000.* In 1775, a still more extensive table was published to 10 times 10,000. The Board of Longitude subsequently employed the late Dr Hutton to calculate and print various numerical tables, and among others, a multiplication table extending as far as 100 times 1000; tables of the squares of numbers, as far as 25,400; tables of cubes, and of the first ten powers of numbers, as far as 100.† In 1814, Professor Barlow, of Woolwich, published, in an octavo volume, the squares, cubes, square roots, cube roots, and reciprocals of all numbers from 1 to 10,000; a table of the first ten powers of all numbers from 1 to 100, and of the fourth and fifth powers of all numbers from 100 to 1000.

Tables of Multiplication to a still greater extent have been published in France. In 1785, was published an octavo volume of tables of the squares, cubes, square roots, and cube roots of all numbers from 1 to 10 000; and similar tables were again published in 1801. In 1817, multiplication tables were published in Paris by Voisin; and similar tables, in two quarto volumes, in 1824, by the French Board of Longitude, extending as far as a thousand

* Dodson's *Calculator*. 4to. London: 1747.
† Hutton's *Tables of Products and Powers*, Folio, London: 1781.

times a thousand. A table of squares was published in 1810, in Hanover; in 1812, at Leipzig; in 1825, at Berlin; and in 1827, at Ghent. A table of cubes was published in 1827, at Eisenach; in the same year a similar table at Ghent; and one of the squares of all numbers as far as 10,000, was published in that year, in quarto, at Bonn. The Prussian Government has caused a multiplication table to be calculated and printed, extending as far as 1000 times 1000. Such are a few of the tables of this class which have been published in different countries.

This class of tables may be considered as purely arithmetical, since the results which they express involve no other relations than the arithmetical dependence of abstract numbers upon each other. When numbers, however, are taken in a concrete sense, and are applied to express peculiar modes of quantity,—such as angular, linear, superficial, and solid magnitudes,—a new set of numerical relations arise, and a large number of computations are required.

To express angular magnitude, and the various relations of linear magnitude with which it is connected, involves the consideration of a vast variety of Geometrical and Trigonometrical tables; such as tables of the natural sines, co-sines, tangents, secants, co-tangents, &c. &c.; tables of arcs and angles in terms of the radius; tables for the immediate solution of various cases of triangles, &c. Volumes without number of such tables have been from time to time computed and published. It is not sufficient, however, for the purposes of computation to tabulate these immediate trigonometrical functions. Their squares* and higher powers, their square roots, and other roots, occur so frequently, that it has been found expedient to compute tables for them, as well as for the same functions of abstract numbers.

The measurement of linear, superficial, and solid magnitudes, in the various forms and modifications in which they are required in the arts, demands another extensive catalogue of numerical tables. The surveyor, the architect, the builder, the carpenter, the miner, the gauger, the naval architect, the engineer, civil and military, all require the aid of peculiar numerical tables, and such have been published in all countries.

The increased expedition and accuracy which was introduced into the art of computation by the invention of Logarithms, greatly enlarged the number of tables previously necessary. To apply the logarithmic method, it was not

---

* The squares of the sines of angles are extensively used in the calculations connected with the theory of the tides. Not aware that tables of these squares existed, Bouvard, who calculated the tides for Laplace, underwent the labour of calculating the square of each individual sine in every case in which it occurred.

merely necessary to place in the hands of the computist extensive tables of the logarithms of the natural numbers, but likewise to supply him with tables in which he might find already calculated the logarithms of those arithmetical, trigonometrical, and geometrical functions of numbers, which he has most frequent occasion to use. It would be a circuitous process, when the logarithm of a sine or co-sine of an angle is required, to refer, first to the table of sines, or co-sines, and thence to the table of the logarithms of natural numbers. It was therefore found expedient to compute distinct tables of the logarithms of the sines, co-sines, tangents, &c., as well as of various other functions frequently required, such as sums, differences, &c.

Great as is the extent of the tables we have just enumerated, they bear a very insignificant proportion to those which remain to be mentioned. The above are, for the most part, general in their nature, not belonging particularly to any science or art. There is a much greater variety of tables, whose importance is no way inferior, which are, however, of a more special nature: Such are, for example, tables of interest, discount, and exchange, tables of annuities, and other tables necessary in life insurances; tables of rates of various kinds necessary in general commerce. But the science in which, above all others, the most extensive and accurate tables are indispensable, is Astronomy; with the improvement and perfection of which is inseparably connected that of the kindred art of Navigation. We scarcely dare hope to convey to the general reader any thing approaching to an adequate notion of the multiplicity and complexity of the tables necessary for the purposes of the astronomer and navigator. We feel, nevertheless, that the truly national importance which must attach to any perfect and easy means of producing those tables cannot be at all estimated, unless we state some of the previous calculations necessary in order to enable the mariner to determine, with the requisite certainty and precision, the place of his ship.

In a word, then, all the purely arithmetical, trigonometrical, and logarith-mic tables already mentioned, are necessary, either immediately or remotely, for this purpose. But in addition to these, a great number of tables, exclusively astronomical, are likewise indispensable. The predictions of the astronomer, with respect to the positions and motions of the bodies of the firmament, are the means, and the only means, which enable the mariner to prosecute his art. By these he is enabled to discover the distance of his ship from the Line, and the extent of his departure from the meridian of Greenwich, or from any other meridian to which the astronomical predictions refer. The more numerous, minute, and accurate these predictions can be made, the greater will be the facilities which can be furnished to the mariner. But the computation of those tables, in which the future position of celestial objects are registered, depend themselves upon an infinite variety of other tables which never reach the hands of the mariner. It cannot be said that there is any table whatever, necessary for the astronomer, which is unnecessary for the navigator.

The purposes of the marine of a country whose interests are so inseparably connected as ours are with the improvement of the art of navigation, would be very inadequately fulfilled, if our navigators were merely supplied with the means of determining by *Nautical Astronomy* the position of a ship at sea. It has been well observed by the Committee of the Astronomical Society, to whom the recent improvement of the *Nautical Almanac* was confided, that it is not by those means merely by which the seaman is enabled to determine the position of his vessel at sea, that the full intent and purpose of what is usually called *Nautical Astronomy* are answered. This object is merely a part of that comprehensive and important subject; and might be attained by a very cheap publication, and without the aid of expensive instruments. A not less important and much more difficult part of nautical science has for its object to determine the precise position of various interesting and important points on the surface of the earth,—such as remarkable headlands, ports, and islands; together with the general trending of the coast between well-known harbours. It is not necessary to point out here how important such knowledge is to the mariner. This knowledge, which may be called *Nautical Geography*, cannot be obtained by the methods of observation used on board ship, but requires much more delicate and accurate instruments, firmly placed upon the solid ground, besides all the astronomical aid which can be afforded by the best tables, arranged in the most convenient form for immediate use. This was Dr Maskelyne's view of the subject, and his opinion has been confirmed by the repeated wants and demands of those distinguished navigators who have been employed in several recent scientific expeditions.*

Among the tables *directly* necessary for navigation, are those which predict the position of the centre of the sun from hour to hour. These tables include the sun's right ascension and declination, daily, at noon, with the hourly change in these quantities. They also include the equation of time, together with its hourly variation.

Tables of the moon's place for every hour, are likewise necessary, together with the change of declination for every ten minutes. The lunar method of determining the longitude depends upon tables containing the predicted

* Report of the Committee of the Astronomical Society prefixed to the Nautical Almanac for 1834.

distances of the moon from the sun, the principal planets, and from certain conspicuous fixed stars; which distances being observed by the mariner, he is enabled thence to discover the *time* at the meridian from which the longitude is measured; and, by comparing that time with the time known or discoverable in his actual situation, he infers his longitude. But not only does the prediction of the position of the moon, with respect to these celestial objects, require a vast number of numerical tables, but likewise the observations necessary to be made by the mariner, in order to determine the lunar distances, also require several tables. To predict the exact position of any fixed star, requires not less than ten numerical tables peculiar to that star; and if the mariner be furnished (as is actually the case) with tables of the predicted distances of the moon from one hundred such stars, such predictions must require not less than a thousand numerical tables. Regarding the range of the moon through the firmament, however, it will readily be conceived that a hundred stars form but a scanty supply; especially when it is considered that an accurate method of determining the longitude, consists in observing the extinction of a star by the dark edge of the moon. Within the limits of the lunar orbit there are not less than one thousand stars, which are so situated as to be in the moon's path, and therefore to exhibit, at some period or other, those desirable occultations. These stars are also of such magnitudes, that their occultations may be distinctly observed from the deck, even when subject to all the unsteadiness produced by an agitated sea. To predict the occultations of such stars, would require not less than ten thousand tables. The stars from which lunar distances might be taken are still more numerous; and we may safely pronounce, that, great as has been the improvement effected recently in our *Nautical Almanac*, it does not yet furnish more than a small fraction of that aid to navigation (in the large sense of that term), which, with greater facility, expedition, and economy in the calculation and printing of tables, it might be made to supply.

Tables necessary to determine the places of the planets are not less necessary than those for the sun, moon, and stars. Some notion of the number and complexity of these tables may be formed, when we state that the positions of the two principal planets, (and these the most necessary for the navigator,) Jupiter and Saturn, require each not less than one hundred and sixteen tables. Yet it is not only necessary to predict the position of these bodies, but it is likewise expedient to tabulate the motions of the four satellites of Jupiter, to predict the exact times at which they enter his shadow, and at which their shadows cross his disc, as well as the times at which they are interposed between him and the Earth, and he between them and the Earth.

Among the extensive classes of tables here enumerated, there are several which are in their nature permanent and unalterable, and would never require to be recomputed, if they could once be computed with perfect accuracy on accurate data; but the data on which such computations are conducted, can only be regarded as approximations to truth, within limits the extent of which must necessarily vary with our knowledge of astronomical science. It has accordingly happened, that one set of tables after another has been superseded with each advance of astronomical science. Some striking examples of this may not be uninstructive. In 1765, the Board of Longitude paid to the celebrated Euler the sum of L.300, for furnishing general formulae for the computation of lunar tables. Professor Mayer was employed to calculate the tables upon these formulae, and the sum of L.3000 was voted for them by the British Parliament, to his widow, after his decease. These tables had been used for ten years, from 1766 to 1776, in computing the *Nautical Almanac*, when they were superseded by new and improved tables, composed by Mr Charles Mason, under the direction of Dr Maskelyne, from calculations made by order of the Board of Longitude, on the observations of Dr Bradley. A farther improvement was made by Mason in 1780; but a much more extensive improvement took place in the lunar calculations by the publication of the tables of the Moon, by M. Bürg, deduced from Laplace's theory, in 1806. Perfect, however, as Bürg's tables were considered, at the time of their publication, they were, within the short period of six years, superseded by a more accurate set of tables published by Burckhardt in 1812; and these also have since been followed by the tables of Damoiseau. Professor Shumacher has calculated by the latter tables his ephemeris of the Planetary Lunar Distances, and astronomers will hence be enabled to put to the strict test of observation the merits of the tables of Burckhardt and Damoiseau.*

The solar tables have undergone, from time to time, similar changes. The solar tables of Mayer were used in the computation of the *Nautical Almanac*, from its commencement in 1767, to 1804 inclusive. Within the six years immediately succeeding 1804, not less than three successive sets of solar tables appeared, each improving on the other; the first by Baron de Zach, the second by Delambre, under the direction of the French Board of Longitude, and the third by Carlini. The last, however, differ only in arrangement from those of Delambre.

Similar observations will be applicable to the tables of the principal planets. Bouvard published, in 1808, tables of Jupiter and Saturn; but from the

---

* A comparison of the results for 1834, will be found in the *Nautical Almanac* for 1835.

improved state of astronomy, he found it necessary to recompute these tables in 1821.

Although it is now about thirty years since the discovery of the four new planets, Ceres, Pallas, Juno, and Vesta, it was not till recently that tables of their motions were published. They have lately appeared in Encke's *Ephemeris*.

We have thus attempted to convey some notion (though necessarily a very inadequate one) of the immense extent of numerical tables which it has been found necessary to calculate and print for the purposes of the arts and sciences. We have before us a catalogue of the tables contained in the library of one private individual, consisting of not less than one hundred and forty volumes. Among these there are no duplicate copies; and we observe that many of the most celebrated voluminous tabular works are not contained among them. They are confined exclusively to arithmetical and trigonometrical tables; and, consequently, the myriad of astronomical and nautical tables are totally excluded from them. Nevertheless, they contain an extent of printed surface covered with figures amounting to above sixteen thousand square feet. We have taken at random forty of these tables, and have found that the number of errors *acknowledged* in the respective errata, amounts to above *three thousand seven hundred*.

To be convinced of the necessity which has existed for accurate numerical tables, it will only be necessary to consider at what an immense expenditure of labour and of money even the imperfect ones which we possess have been produced.

To enable the reader to estimate the difficulties which attend the attainment even of a limited degree of accuracy, we shall now explain some of the expedients which have been from time to time resorted to for the attainment of numerical correctness in calculating and printing them.

Among the scientific enterprises which the ambition of the French nation aspired to during the Republic, was the construction of a magnificent system of numerical tables. Their most distinguished mathematicians were called upon to contribute to the attainment of this important object; and the superintendence of the undertaking was confided to the celebrated Prony, who co-operated with the government in the adoption of such means as might be expected to ensure the production of a system of logarithmic and trigonometric tables, constructed with such accuracy that they should form a monument of calculation the most vast and imposing that had ever been executed, or even conceived. To accomplish this gigantic task, the principle of the division of labour, found to be so powerful in manufactures, was resorted to with singular success. The persons employed in the work were divided into three sections: the first consisted of half a dozen of the most eminent analysts. Their duty was to investigate the most convenient mathematical formulae, which should enable the computers to proceed with the greatest expedition and accuracy by the method of Differences, of which we shall speak more fully hereafter. These formulae, when decided upon by this first section, were handed over to the second section, which consisted of eight or ten properly qualified mathematicians. It was the duty of this second section to convert into numbers certain general or algebraical expressions which occurred in the formulae, so as to prepare them for the hands of the computers. Thus prepared, these formulae were handed over to the third section, who formed a body of nearly one hundred computers. The duty of this numerous section was to compute the numbers finally intended for the tables. Every possible precaution was of course taken to ensure the numerical accuracy of the results. Each number was calculated by two or more distinct and independent computers, and its truth and accuracy determined by the coincidence of the results thus obtained.

The body of tables thus calculated occupied in manuscript *seventeen* folio volumes.*

As an example of the precautions which have been considered necessary to guard against errors in the calculation of numerical tables, we shall further state those which were adopted by Mr Babbage, previously to the publication of his tables of logarithms. In order to render the terminal figure of tables in which one or more decimal places are omitted as accurate as it can be, it has been the practice to compute one or more of the succeeding figures; and if the first omitted figure be greater than 4, then the terminal figure is always increased by 1, since the value of the tabulated number is by such means brought nearer to the truth.† The tables of Callet, which were among the most

---

* These tables were never published. The printing of them was commenced by Didot, and a small portion was actually stereotyped, but never published. Soon after the commencement of the undertaking, the sudden fall of the assignats rendered it impossible for Didot to fulfil his contract with the government. The work was accordingly abandoned, and has never since been resumed. We have before us a copy of 100 pages folio of the portion which was printed at the time the work was stopped, given to a friend on a late occasion by Didot himself. It was remarked in this, as in other similar cases, that the computers who committed fewest errors were those who understood nothing beyond the process of addition.

† Thus suppose the number expressed at full length were 3.1415927. If the table extend to no more than four places of decimals, we should tabulate the number 3.1416 and not 3.1415. The former would be evidently nearer to the true number 3.1415927.

accurate published logarithms, and which extended to seven places of decimals, were first carefully compared with the tables of Vega, which extended to ten places, in order to discover whether Callet had made the above correction of the final figure in every case where it was necessary. This previous precaution being taken, and the corrections which appeared to be necessary being made in a copy of Callet's tables, the proofs of Mr Babbage's tables were submitted to the following test: They were first compared, number by number, with the corrected copy of Callet's logarithms; secondly, with Hutton's logarithms; and thirdly, with Vega's logarithms. The corrections thus suggested being marked in the proofs, corrected revises were received back. These revises were then again compared, number by number, first with Vega's logarithms; secondly, with the logarithms of Callet; and thirdly, as far as the first 20 000 numbers, with the corresponding ones in Briggs's logarithms. They were now returned to the printer, and were stereotyped; proofs were taken from the stereotyped plates, which were put through the following ordeal: They were first compared once more with the logarithms of Vega as far as 47 500; they were then compared with the whole of the logarithms of Gardner; and next with the whole of Taylor's logarithms; and as a last test, they were transferred to the hands of a different set of readers, and were once more compared with Taylor. That these precautions were by no means superfluous may be collected from the following circumstances mentioned by Mr Babbage: In the sheets read immediately previous to stereotyping, thirty-two errors were detected; after stereotyping, eight more were found, and corrected in the plates.

By such elaborate and expensive precautions many of the errors of computation and printing may certainly be removed; but it is too much to expect that in general such measures can be adopted; and we accordingly find by far the greater number of tables disfigured by errors, the extent of which is rather to be conjectured than determined. When the nature of a numerical table is considered,—page after page densely covered with figures, and with nothing else,—the chances against the detection of any single error will be easily comprehended; and it may therefore be fairly presumed, that for one error which may happen to be detected, there must be a great number which escape detection. Notwithstanding this difficulty, it is truly surprising how great a number of numerical errors have been detected by individuals no otherwise concerned in the tables than in their use. Mr Baily states that he has himself detected in the solar and lunar tables, from which our *Nautical Almanac* was for a long period computed, more than five hundred errors. In the multiplication table already mentioned, computed by Dr Hutton for the Board of

Longitude, a single page was examined and recomputed: it was found to contain about forty errors.

In order to make the calculations upon the numbers found in the Ephemeral Tables published in the *Nautical Almanac*, it is necessary that the mariner should be supplied with certain permanent tables. A volume of these, to the number of about thirty, was accordingly computed, and published at national expense, by order of the Board of Longitude, entitled 'Tables requisite to be used with the Nautical Ephemeris for finding the latitude and longitude at sea.' In the first edition of these requisite tables, there were detected, by one individual, above a thousand errors.

The tables published by the Board of Longitude for the correction of the observed distances of the moon from certain fixed stars, are followed by a table of acknowledge errata, extending to seven folio pages, and containing more than eleven hundred errors. Even this table of errata itself is not correct: a considerable number of errors have been detected in it, so that errata upon errata have become necessary.

One of the tests most frequently resorted to for the detection of errors in numerical tables, has been the comparison of tables of the same kind, published by different authors. It has been generally considered that those numbers in which they are found to agree must be correct; inasmuch as the chances are supposed to be very considerable against two or more independent computers falling into precisely the same errors. How far this coincidence may be safely assumed as a test of accuracy we shall presently see.

A few years ago, it was found desirable to compute some very accurate logarithmic tables for the use of the great national survey of Ireland, which was then, and still is in progress; and on that occasion a careful comparison of various logarithmic tables was made. Six remarkable errors were detected, which were found to be common to several apparently independent sets of tables. This singular coincidence led to an unusually extensive examination of the logarithmic tables published both in England and in other countries; by which it appeared that thirteen sets of tables, published in London between the years 1633 and 1822, all agreed in these six errors. Upon extending the enquiry to foreign tables, it appeared that two sets of tables published at Paris, one at Gouda, one at Avignon, one at Berlin, and one at Florence, were infected by exactly the same six errors. The only tables which were found free from them were those of Vega, and the more recent impressions of Callet. It happened that the Royal Society possessed a set of tables of logarithms printed in the Chinese character, and on Chinese paper, consisting of two volumes:

these volumes contained no indication or acknowledgment of being copied from any other work. They were examined; and the result was the detection in them of the same six errors.*

It is quite apparent that this remarkable coincidence of error must have arisen from the various tables being copied successively one from another. The earliest work in which they appeared was Vlacq's *Logarithms*, (folio, Gouda, 1628); and from it, doubtless, those which immediately succeeded it in point of time were copied; from which the same errors were subsequently transcribed into all the other, including the Chinese logarithms.

The most certain and effectual check upon errors which arise in the process of computation, is to cause the same computations to be made by separate and independent computers; and this check is rendered still more decisive if they make their computations by different methods. It is, nevertheless, a remarkable fact, that several computers, working separately and independently, do frequently commit precisely the same error; so that falsehood in this case assumes that character of consistency, which is regarded as the exclusive attribute of truth. Instances of this are familiar to most persons who have had the management of the computation of tables. We have reason to know, that M. Prony experienced it on many occasions in the management of the great French tables, when he found three, and even a greater number of computers, working separately and independently, to return him the same numerical result, and *that result wrong*. Mr Stratford, the conductor of the *Nautical Almanac*, to whose talents and zeal that work owes the execution of its recent improvements, has more than once observed a similar occurrence. But one of the most signal examples of this kind, of which we are aware, is related by Mr Baily. The catalogue of stars published by the Astronomical Society was computed by two separate and independent persons, and was afterwards compared and examined with great care and attention by Mr Stratford. On examining this catalogue, and recalculating a portion of it, Mr Baily discovered an error in the case of the star, κ. Cephei. Its right ascension was calculated *wrongly*, and yet *consistently*, by two computers working separately. Their numerical results agreed precisely in every figure; and Mr Stratford, on examining the catalogue, failed to detect the error. Mr Baily having reason, from some discordancy which he observed, to suspect an error, recomputed the place of the star with a view to discover it; and he himself, in the first instance, obtained precisely *the same erroneous numerical result*. It was only on

* *Memoirs Ast. Soc.* vol. iii., p. 65.

going over the operation a second time that he *accidentally* discovered that all had inadvertently committed the same error.*

It appears, therefore, that the coincidence of different tables, even when it is certain that they could not have been copied one from another, but must have been computed independently, is not a decisive test of their correctness, neither is it possible to ensure accuracy by the device of separate and independent computation.

Besides the errors incidental to the process of computation, there are further liabilities in the process of *transcribing* the final results of each calculation into the fair copy of the table designed for the printer. The next source of error lies with the compositor, in transferring this copy into type. But the liabilities to error do not stop even here; for it frequently happens, that after the press has been fully corrected, errors will be produced in the process of printing. A remarkable instance of this occurs in one of the six errors detected in so many different tables already mentioned. In one of these cases, the last five figures of two successive numbers of a logarithmic table were the following:—

$$35875$$
$$10436.$$

Now, both of these are erroneous; the figure 8 in the first line should be 4, and the figure 4 in the second should be 8. It is evident that the types, as first composed, were correct; but in the course of printing, the two types 4 and 8 being loose, adhered to the inking-balls, and were drawn out: the pressmen in replacing them transposed them, putting the 8 *above* and the 4 *below*, instead of *vice versâ*. It would be a curious enquiry, were it possible to obtain all the copies of the original edition of Vlacq's Logarithms, published at Gouda in 1628, from which this error appears to have been copied in all the subsequent tables, to ascertain whether it extends through the entire edition. It would probably, nay almost certainly, be discovered that some of the copies of that edition are correct in this number, while others are incorrect; the former having been worked off before the transposition of the types.

It is a circumstance worthy of notice, that this error in Vlacq's tables has produced a corresponding error in a variety of other tables deduced from them, *in which nevertheless the erroneous figures in Vlacq are omitted*. In no less than sixteen sets of tables published at various times since the publication of Vlacq, in which the logarithms extend only to seven places of figures, the error just mentioned in the *eighth place* in Vlacq causes a corresponding error in the

* *Memoirs Ast. Soc.* vol. iv., p. 290.

seventh place. When the last three figures are omitted in the first of the above numbers, the seventh figure should be 5, inasmuch as the first of the omitted figures is under 5: the erroneous insertion, however, of the figure 8 in Vlacq has caused the figure 6 to be substituted for 5 in the various tables just alluded to. For the same reason, the erroneous occurrence of 4 in the second number has caused the adoption of a 0 instead of a 1 in the seventh place in the other tables. The only tables in which this error does not occur are those of Vega, the more recent editions of Callet, and the still later Logarithms of Mr Babbage.

The *Opus Palatinum*, a work published in 1596, containing an extensive collection of trigonometrical tables, affords a remarkable instance of a tabular error; which, as it is not generally known, it may not be uninteresting to mention here. After that work had been for several years in circulation in every part of Europe, it was discovered that the commencement of the table of co-tangents and co-secants was vitiated by an error of considerable magnitude. In the first co-tangent the last nine places of figures were incorrect; but from the manner in which the numbers of the table were computed, the error was gradually, though slowly, diminished, until at length it became extinguished in the eighty-sixth page. After the detection of this extensive error, Pitiscus undertook the recomputation of the eighty-six erroneous pages. His corrected calculation was printed, and the erroneous part of the remaining copies of the *Opus Palatinum* was cancelled. But as the corrected table of Pitiscus was not published until 1607,—thirteen years after the original work,—the erroneous part of the volume was cancelled in comparatively few copies, and conse-quently correct copies of the work are now exceedingly rare. Thus, in the collection of tables published by M. Schulze,* the whole of the erroneous part of the *Opus Palatinum* has been adopted; he having used the copy of that work which exists in the library of the Academy of Berlin, and which is one of those copies in which the incorrect part was not cancelled. The corrected copies of this work may be very easily distinguished at present from the erroneous ones: it happened that the former were printed with a very bad and worn-out type, and upon paper of a quality inferior to that of the original work. On comparing the first eighty-six pages of the volume with the succeeding ones, they are, therefore, immediately distinguishable in the corrected copies. Besides this test, there is another, which it may not be uninteresting to point out:—At the bottom of page 7 in the corrected copies, there is an error in the position of the words *basis* and *hypothenusa*, their places being interchanged. In the original uncorrected work this error does not exist.

* *Recueil des Tables Logarithmiques et Trigonometriques.* Par J. C. Shulze. 2 vols. Berlin: 1778.

At the time when the calculation and publication of Taylor's *Logarithms* were undertaken, it so happened that a similar work was in progress in France; and it was not until the calculation of the French work was completed, that its author was informed of the publication of the English work. This circumstance caused the French calculator to relinquish the publication of his tables. The manuscript subsequently passed into the library of Delambre, and, after his death, was purchased at the sale of his books, by Mr Babbage, in whose possession it now is. Some years ago it was thought advisable to compare these manuscript tables with Taylor's *Logarithms*, with a view to ascertain the errors in each, but especially in Taylor. The two works were peculiarly well suited for the attainment of this end; as the circumstances under which they were produced, rendered it quite certain that they were computed independently of each other. The comparison was conducted under the direction of the late Dr Young, and the result was the detection of the following nineteen errors in Taylor's *Logarithms*. To enable those who used Taylor's *Logarithms* to make the necessary corrections in them, the corrections of the detected errors appeared as follows in the *Nautical Almanac* for 1832.

Errata, detected in Taylor's Logarithms. *London: 4to, 1792.*

			° ′ ″				
1	*E*	Co–tangent of	1.35.55	*for*	43671	*read*	42671
2	*M*	Co–tangent of	4. 4.49	—	66976	—	66979
3		Sine of	4.23.38	—	43107	—	43007
4		Sine of	4.23.39	—	43381	—	43281
5	*S*	Sine of	6.45.52	—	10001	—	11001
6	*Kk*	Co–sine of	14.18. 3	—	3398	—	3298
7	*Ss*	Tangent of	18. 1.56	—	5064	—	6064
8	*Aaa*	Co–tangent of	21.11.14	—	6062	—	5962
9	*Ggg*	Tangent of	23.48.19	—	6087	—	5987
10		Co–tangent of	23.48.19	—	3913	—	4013
11	*Iii*	Sine of	25. 5. 4	—	3173	—	3183
12		Sine of	25. 5. 5	—	3218	—	3228
13		Sine of	25. 5. 6	—	3263	—	3273
14		Sine of	25. 5. 7	—	3308	—	3318
15		Sine of	25. 5. 8	—	3353	—	3363
16		Sine of	25. 5. 9	—	3398	—	3408
17	*Qqq*	Tangent of	28.19.39	—	6302	—	6402
18	*4H*	Tangent of	35.55.51	—	1681	—	1581
19	*4K*	Co–sine of	37.29. 2	—	5503	—	5603

An error being detected in this list of ERRATA, we find, in the *Nautical Almanac* for the year 1833, the following ERRATUM of the ERRATA of Taylor's *Logarithms*:—

In the list of ERRATA detected in Taylor's *Logarithms*, for cos. 4° 18′ 3″, *read* cos. 14° 18′ 2″.

Here, however, confusion is worse confounded; for a new error, not before existing, and of much greater magnitude, is introduced! It will be necessary, in the *Nautical Almanac* for 1836, (that for 1835 is already published,) to introduce the following

ERRATUM of the ERRATUM of the ERRATA of TAYLOR's *Logarithms*. For cos. 4° 18′ 3″, *read* cos. 14° 18′ 3″.

If proof were wanted to establish incontrovertibly the utter impracticability of precluding numerical errors in works of this nature, we should find it in this succession of error upon error, produced, in spite of the universally acknowledged accuracy and assiduity of the persons at present employed in the construction and management of the *Nautical Almanac*. It is only by the *mechanical fabrication of tables* that such errors can be rendered impossible.

On examining this list with attention, we have been particularly struck with the circumstances in which these errors appear to have originated. It is a remarkable fact, that of the above nineteen errors, eighteen have arisen from mistakes in *carrying*. Errors 5, 7, 10, 11, 12, 13, 14, 15, 16, 17, 19, have arisen from a carriage being neglected; and errors 1, 3, 4, 6, 8, 9, and 18, from a carriage being made where none should take place. In four cases, namely, errors 8, 9, 10, and 16, this has caused *two* figures to be wrong. The only error of the nineteen which appears to have been a press error is the second; which has evidently arisen from the type 9 being accidentally inverted, and thus becoming a 6. This may have originated with the compositor, but more probably it took place in the press-work; the type 9 being accidentally drawn out of the form by the inking-ball, as mentioned in a former case, and on being restored to its place, inverted by the pressman.

There are two cases among the above errata, in which an error, committed in the calculation of one number, has evidently been the cause of other errors. In the third erratum, a wrong carriage was made, in computing the sine of 4° 23′ 38″. The next number of the table was vitiated by this error; for we find the next erratum to be in the sine of 4° 23′ 39″, in which the figure similarly placed is 1 in excess. A still more extensive effect of this kind appears in errata 11, 12, 13, 14, 15, 16. A carriage was neglected in computing the sine of

25°5′4″, and this produced a corresponding error in the five following numbers of the table, which are those corrected in the five following errata.

This frequency of errors arising in the process of carrying, would afford a curious subject of metaphysical speculation respecting the operation of the faculty of memory. In the arithmetical process, the memory is employed in a twofold way;—in ascertaining each successive figure of the calculated result by the recollection of a table committed to memory at an early period of life; and by another act of memory, in which the number *carried* from column to column is retained. It is a curious fact, that this latter circumstance, occurring only the moment before, and being in its nature little complex, is so much more liable to be forgotten or mistaken than the results of rather complicated tables. It appears, that among the above errata, the errors 5, 7, 10, 11, 17, 19, have been produced by the computer forgetting a carriage; while the errors 1, 3, 6, 8, 9, 18, have been produced by his making a carriage improperly. Thus, so far as the above list of errata affords grounds for judging, it would seem, (contrary to what might be expected,) that the error by which improper carriages are made is as frequent as that by which necessary carriages are overlooked.

We trust that we have succeeded in proving, first, the great national and universal utility of numerical tables, by showing the vast number of them, which have been calculated and published; secondly, that more effectual means are necessary to obtain such tables suitable to the present state of the arts, sciences and commerce, by showing that the existing supply of tables, vast as it certainly is, is still scanty, and utterly inadequate to the demands of the community;—that it is rendered inefficient, not only in quantity, but in quality, by its want of numerical correctness; and that such numerical correctness is altogether unattainable until some more perfect method be discovered, not only of calculating the numerical results, but of tabulating these,—of reducing such tables to type, and of printing that type so as to intercept the possibility of error during the press-work. Such are the ends which are proposed to be attained by the calculating machinery invented by Mr Babbage.

The benefits to be derived from this invention cannot be more strongly expressed than they have been by Mr Colebrooke, President of the Astronomical Society, on the occasion of presenting the gold medal voted by that body to Mr Babbage:—

In no department of science, or of the arts, does this discovery promise to be so eminently useful as in that of astronomy, and its kindred sciences, with the various arts dependent on them. In none are computations more operose than those which astronomy in particular requires;—in none are preparatory facilities more needful;—in none is error more detrimental. The practical astronomer is interrupted in his pursuit, and diverted from his task of observation by the irksome labours of computation, or his diligence in observing becomes ineffectual for want of yet greater industry of calculation. Let the aid which tables previously computed afford, be furnished to the utmost extent which mechanism has made attainable through Mr Babbage's invention, and the most irksome portion of the astronomer's task is alleviated, and a fresh impulse is given to astronomical research.

The first step in the progress of this singular invention was the discovery of some common principle which pervaded numerical tables of every description; so that by the adoption of such a principle as the basis of the machinery, a corresponding degree of generality would be conferred upon its calculations. Among the properties of numerical functions, several of a general nature exist; and it was a matter of no ordinary difficulty, and requiring no common skill, to select one which might, in all respects, be preferable to the others. Whether or not that which was selected by Mr Babbage affords the greatest practical advantages, would be extremely difficult to decide—perhaps impossible, unless some other projector could be found possessed of sufficient genius, and sustained by sufficient energy of mind and character, to attempt the invention of calculating machinery on other principles. The principle selected by Mr Babbage as the basis of that part of the machinery which calculates, is the Method of Differences; and he has in fact literally thrown this mathematical principle into wheel-work. In order to form a notion of the nature of the machinery, it will be necessary, first to convey to the reader some idea of the mathematical principle just alluded to.

A numerical table, of whatever kind, is a series of numbers which possess some common character, and which proceed increasing or decreasing according to some general law. Supposing such a series continually to increase, let us imagine each number in it to be subtracted from that which follows it, and the remainders thus successively obtained to be ranged beside the first, so as to form another table: these numbers are called the *first differences*. If we suppose these likewise to increase continually, we may obtain a third table from them by a like process, subtracting each number from the succeeding one: this series is called the *second differences*. By adopting a like method of proceeding, another series may be obtained, called the *third differences*; and so on. By continuing this process, we shall at length obtain a series of differences, of some order, more or less high, according to the nature of the original table, in which we shall find the same number constantly repeated, to whatever extent the original table may have been continued; so that if the next series of differences had been obtained in the same manner as the preceding ones, every term of it would be 0. In some cases this would continue to whatever extent the original table might be carried; but in all cases a series of differences would be obtained, which would continue constant for a very long succession of terms.

As the successive serieses of differences are derived from the original table, and from each other, by *subtraction*, the same succession of series may be reproduced in the other direction by *addition*. But let us suppose that the first number of the original table, and of each of the series of differences, including the last, be given: all the numbers of each of the series may thence be obtained by the mere process of addition. The second term of the original table will be obtained by adding to the first the first term of the first difference series; in like manner, the second term of the first difference series will be obtained by adding to the first term, the first term of the third difference series, and so on. The second terms of all the serieses being thus obtained, the third terms may be obtained by a like process of addition; and so the series may be continued. These observations will perhaps be rendered more clearly intelligible when illustrated by a numerical example. The following is the commencement of a series of the fourth powers of the natural numbers:—

No	Table
1	1
2	16
3	81
4	256
5	625
6	1296
7	2401
8	4096
9	6561
10	10 000
11	14 641
12	20 736
13	28 561

By subtracting each number from the succeeding one in this series, we obtain the following series of first differences:—

15
65
175
369

671
1105
1695
2465
3439
4641
6095
7825

In like manner, subtracting each term of this series from the succeeding one, we obtain the following series of second differences:—

50
110
194
302
434
590
770
974
1202
1454
1730

Proceeding with this series in the same way, we obtain the following series of third differences:—

60
84
108
132
156
180
204
228
252
276

Proceeding in the same way with these, we obtain the following for the series of fourth differences:—

24
24
24

$$24$$
$$24$$
$$24$$
$$24$$
$$24$$
$$24$$

It appears, therefore, that in this case the series of fourth differences consists of a constant repetition of the number 24. Now, a slight consideration of the succession of arithmetical operations by which we have obtained this result, will show, that by reversing the process, we could obtain the table of fourth powers by the mere process of addition. Beginning with the first numbers in each successive series of differences, and designating the table and the successive differences by the letters T, D^1 D^2 D^3 D^4, we have then the following to begin with:—

T	D^1	D^2	D^3	D^4
1	15	50	60	24

Adding each number to the number on its left, and repeating 24, we get the following as the second terms of the several series:—

T	D^1	D^2	D^3	D^4
16	65	110	84	24

And, in the same manner, the third and succeeding terms as follows:—

No.	T	D^1	D^2	D^3	D^4
1	1	15	50	60	24
2	16	65	110	84	24
3	81	175	194	108	24
4	256	369	302	132	24
5	625	671	434	156	24
6	1296	1105	590	180	24
7	2401	1695	770	204	24
8	4096	2465	974	228	24
9	6561	3439	1202	252	24
10	10000	4641	1454	276	
11	14641	6095	1730		
12	20736	7825			
13	28561				

There are numerous tables in which, as already stated, to whatever order of differences we may proceed, we should not obtain a series of rigorously constant differences; but we should always obtain a certain number of differences which to a given number of decimal places would remain constant for a long succession of terms. It is plain that such a table might be calculated by addition in the same manner as those which have a difference rigorously and continuously constant; and if at every point where the last difference requires an increase, that increase be given to it, the same principle of addition may again be applied for a like succession of terms, and so on.

By this principle it appears, that all tables in which each series of differences continually increases, may be produced by the operation of addition alone; provided the first terms of the table, and of each series of differences, be given in the first instance. But it sometimes happens, that while the table continually increases, one or more serieses of differences may continually diminish. In this case, the series of differences are found by subtracting each term of the series, not from that which follows, but from that which precedes it; and consequently, in the re-production of the several serieses, when their first terms are given, it will be necessary in some cases to obtain them by *addition*, and in others by *subtraction*. It is possible, however, still to perform all the operations by addition alone: this is effected in performing the operation of subtraction, by substituting for the subtrahend its *arithmetical complement*, and adding that, omitting the unit of the highest order in the result. This process, and its principle, will be readily comprehended by an example. Let it be required to subtract 357 from 768.

The common process would be as follows:—

From	768
Subtract	357
Remainder	411

The *arithmetical complement* of 357, or the number by which it falls short of 1000, is 643. Now, if this number be added to 768, and the first figure on the left be struck out of the sum, the process will be as follows:—

To	768
Add	643
Sum	1411
Remainder sought	411

The principle on which this process is founded is easily explained. In the latter process we have first added 643, and then subtracted 1000. On the whole, therefore, we have subtracted 357, since the number actually subtracted exceeds the number previously added by that amount.

Since, therefore, subtraction may be effected in this manner by addition, it follows that the calculation of all serieses, so far as an order of differences can be found in them which continues constant, may be conducted by the process of addition alone.

It also appears from what has been stated, that each addition consists only of two operations. However numerous the figures may be of which the several pairs of numbers to be thus added may consist, it is obvious that the operation of adding them can only consist of repetitions of the process of adding one digit to another; and of carrying one from the column of inferior units to the column of units next superior when necessary. If we would therefore reduce such a process to machinery, it would only be necessary to discover such a combination of moving parts as are capable of performing these two processes of *adding* and *carrying* on two single figures; for, this being once accomplished, the process of adding two numbers, consisting of any number of digits, will be effected by repeating the same mechanism as often as there are pairs of digits to be added. Such was the simple form to which Mr Babbage reduced the problem of discovering the calculating machinery; and we shall now proceed to convey some notion of the manner in which he solved it.

For the sake of illustration, we shall suppose that the table to be calculated shall consist of numbers not exceeding six places of figures; and we shall also suppose that the difference of the fifth order is the constant difference. Imagine, then, six rows of wheels, each wheel carrying upon it a dial-plate like that of a common clock, but consisting of *ten* instead of *twelve* divisions; the several divisions being marked 1, 2, 3, 4, 5, 6, 7, 8, 9, 0. Let these dials be supposed to revolve whenever the wheels to which they are attached are put in motion, and to turn in such a direction that the series of increasing numbers shall pass under the index which appears over each dial:—thus, after 0 passes the index, 1 follows, then 2, 3, and so on, as the dial revolves. In Fig. 1 are represented six horizontal rows of such dials.

The method of differences, as already explained, requires, that in proceeding with the calculation, this apparatus should perform continually the addition of the number expressed upon each row of dials, to the number expressed upon the row immediately above it. Now, we shall first explain how this process of addition may be conceived to be performed by the motion of the dials; and in doing so, we shall consider separately the processes of addition and carriage, considering the addition first, and then the carriage.

Let us first suppose the line D^I to be added to the line T. To accomplish this, let us imagine that while the dials on the line D^I are quiescent, the dials on the line T are put in motion, in such a manner, that as many divisions on each dial shall pass under its index, as there are units in the number at the index immediately below it. It is evident that this condition supposes that if o be at any index on the line D^I, the dial immediately above it in the line T shall not move. Now the motion here supposed, would bring under the indices on the line T such a number as would be produced by adding the number D^I to T, neglecting all the carriages; for a carriage should have taken place in every case in which the figure 9 of any dial in the line T had passed under the index during the adding motion. To accomplish this carriage, it would be necessary that the dial immediately on the left of any dial in which 9 passes under the index, should be advanced one division, independently of those divisions which it may have been advanced by the addition of the number immediately below it. This effect may be conceived to take place in either of two ways. It may be either produced at the moment when the division between 9 and o of

Fig. 1

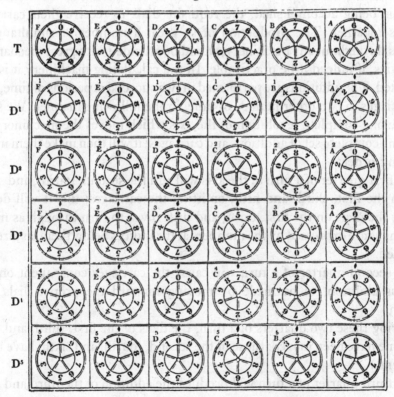

any dial passes under the index; in which case the process of carrying would go on simultaneously with the process of adding; or the process of carrying may be postponed in every instance until the process of addition, without carrying, has been completed; and then by another distinct and independent motion of the machinery, a carriage may be made by advancing one division all those dials on the right of which a dial had, during the previous addition, passed from 9 to 0 under the index. The latter is the method adopted in the calculating machinery, in order to enable its inventor to construct the carrying machinery independent of the adding mechanism.

Having explained the motion of the dials by which the addition, excluding the carriages of the number on the row D^1, may be made to the number on the row T, the same explanation may be applied to the number on the row D^2 to the number on the row D^1; also, of the number D^3 to the number on the row D^2, and so on. Now it is possible to suppose the additions of all the rows, except the first, to be made to all the rows except the last, simultaneously; and after these additions have been made, to conceive all the requisite carriages to be also made by advancing the proper dials one division forward. This would suppose all the dials in the scheme to receive their adding motion together; and, this being accomplished, the requisite dials to receive their carrying motions together. The production of so great a number of simultaneous motions throughout any machinery, would be attended with great mechanical difficulties, if indeed it be practicable. In the calculating machinery it is not attempted. The additions are performed in two successive periods of time, and the carriages in two other periods of time, in the following manner. We shall suppose one complete revolution of the axis which moves the machinery, to make one complete set of additions and carriages; it will then make them in the following order:—

The first quarter of a turn of the axis will add the second, fourth, and sixth rows to the first, third, and fifth, omitting the carriages; this it will do by causing the dials on the first, third, and fifth rows, to turn through as many divisions as are expressed by the numbers at the indices below them, as already explained.

The second quarter of a turn will cause the carriages consequent on the previous addition, to be made by moving forward the proper dials one division.

(During these two quarters of a turn, the dials of the first, third, and fifth row alone have been moved; those of the second, fourth, and sixth, have been quiescent.)

The third quarter of a turn will produce the addition of the third and fifth rows to the second and fourth, omitting the carriages; which it will do by causing the dials of the second and fourth rows to turn through as many divisions as are expressed by the numbers at the indices immediately below them.

The fourth and last quarter of a turn will cause the carriages consequent on the previous addition, to be made by moving the proper dials forward one division.

This evidently completes one calculation, since all the rows except the first have been respectively added to all the rows except the last.

To illustrate this: let us suppose the table to be computed to be that of the fifth powers of the natural numbers, and the computation to have already proceeded so far as the fifth power of 6, which is 7776. This number appears, accordingly, in the highest row, being the place appropriated to the number of the table to be calculated. The several differences as far as the fifth, which is in this case constant, are exhibited on the successive rows of dials in such a manner, as to be adapted to the process of addition by alternate rows, in the manner already explained. The process of addition will commence by the motion of the dials in the first, third, and fifth rows, in the following manner: The dial A, Fig. 1, must turn through one division, which will bring the number 7 to the index; the dial B must turn through three divisions, which will bring 0 to the index; this will render a carriage necessary, but that carriage will not take place during the present motion of the dial. The dial C will remain unmoved, since 0 is at the index below it; the dial D must turn through nine divisions; and as, in doing so, the division between 9 and 0 must pass under the index, a carriage must subsequently take place upon the dial to the left; the remaining dials of the row T, Fig. 1, will remain unmoved. In the row D^2 the dial A^2 will remain unmoved, since 0 is at the index below it; the dial B^2 will be moved through five divisions, and will render a subsequent carriage on the dial to the left necessary; the dial C^2 will be moved through five divisions; the dial D^2 will be moved through three divisions, and the remaining dials of this row will remain unmoved. The dials of the row D^4 will be moved according to the same rules; and the whole scheme will undergo a change exhibited in Fig. 2; a mark (*) being introduced on those dials to which a carriage is rendered necessary by the addition which has just taken place.

The second quarter of a turn of the moving axis, will move forward through one division all the dials which in Fig. 2 are marked (*), and the scheme will be converted into the scheme expressed in Fig. 3.

In the third quarter of a turn, the dial A^1, Fig. 3, will remain unmoved, since 0 is at the index below it; the dial B^1 will be moved forward through three divisions; C¹ through nine divisions, and so on; and in like manner the dials of the row D³ will be moved forward through the number of divisions expressed at the indices in the row D⁴. This change will convert the arrangement into that expressed in Fig. 4, the dials to which a carriage is due, being distinguished as before by (*).

Fig. 2

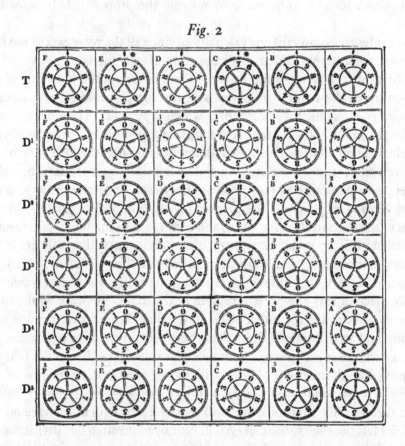

The fourth quarter of a turn of the axis will move forward one division all the dials marked (*); and the arrangement will finally assume the form exhibited in Fig. 5, in which the calculation is completed. The first row T in this expresses the fifth power of 7; and the second expresses the number which must be added to the first row, in order to produce the fifth power of 8; the numbers in each row being prepared for the change which they must undergo, in order to enable them to continue the computation according to the method of alternate addition here adopted.

Having thus explained what it is that the mechanism is required to do, we shall now attempt to convey at least a general notion of some of the mechanical contrivances by which the desired ends are attained. To simplify the explanation, let us first take one particular instance—the dials B and B^I, Fig. 1, for example. Behind the dial B^I is a bolt, which, at the commencement of the process, is shot between the teeth of a wheel which drives the dial B: during the first quarter of a turn this bolt is made to revolve, and if it continued to be engaged in the teeth of the said wheel, it would cause the dial B to make a complete revolution; but it is necessary that the dial B should only move through three divisions, and, therefore, when three divisions of this dial have passed under its index, the aforesaid bolt must be withdrawn: this is accomplished by a small wedge, which is placed in a fixed position on the wheel behind the dial B^I, and that position is such that this wedge will press upon the bolt in such a manner, that at the moment when three divisions of the dial B have passed under the index, it shall withdraw the bolt from the teeth of the wheel which it drives. The bolt will continue to revolve during the remainder

Fig. 3

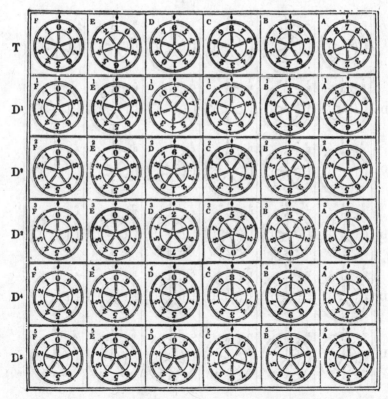

of the first quarter of a turn of the axis, but it will no longer drive the dial B, which will remain quiescent. Had the figure at the index of the dial B^I been any other, the wedge which withdraws the bolt would have assumed a different position, and would have withdrawn the bolt at a different time, but at a time always corresponding with the number under the index of the dial B^I: thus, if 5 had been under the index of the dial B^I, then the bolt would have been withdrawn from between the teeth of the wheel which it drives, when five divisions of the dial B had passed under the index, and so on. Behind each dial in the row D^I there is a similar bolt and a similar withdrawing wedge, and the action upon the dial above is transmitted and suspended in precisely the same manner. Like observations will be applicable to all the dials in the scheme here referred to, in reference to their adding actions upon those above them.

There is, however, a particular case which here merits notice: it is the case in which 0 is under the index of the dial from which the addition is to be

Fig. 4

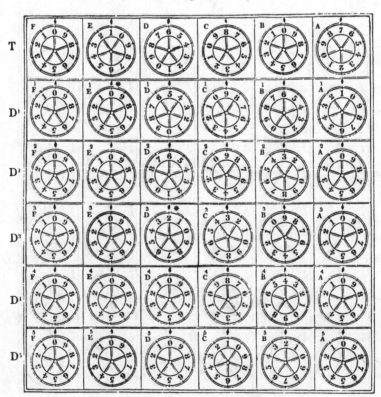

Fig. 5

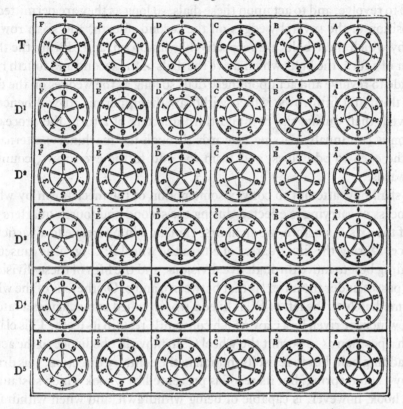

transmitted upwards. As in that case nothing is to be added, a mechanical provision should be made to prevent the bolt from engaging in the teeth of the wheel which acts upon the dial above: the wedge which causes the bolt to be withdrawn, is thrown into such a position as to render it impossible that the bolt should be shot, or that it should enter between the teeth of the wheel, which in other cases it drives. But inasmuch as the usual means of shooting the bolt would still act, a strain would necessarily take place in the parts of the mechanism, owing to the bolt not yielding to the usual impulse. A small shoulder is therefore provided, which puts aside, in this case, the piece by which the bolt is usually struck, and allows the striking implement to pass without encountering the head of the bolt or any other obstruction. This mechanism is brought into play in the scheme, Fig. 1, in the cases of all those dials in which o is under the index.

Such is a general description of the nature of the mechanism by which the adding process, apart from the carriages, is effected. During the first quarter of a turn, the bolts which drive the dials in the first, third, and fifth rows, are caused to revolve, and to act upon these dials, so long as they are permitted by the position of the several wedges on the second, fourth, and sixth rows of dials, by which these bolts are respectively withdrawn; and, during the third quarter of a turn, the bolts which drive the dials of the second and fourth rows are made to revolve and act upon these dials so long as the wedges on the dials of the third and fifth rows, which withdraw them, permit. It will hence be perceived, that, during the first and third quarters of a turn, the process of addition is continually passing upwards through the machinery; alternately from the even to odd rows, and from the odd to the even rows, counting downwards.

We shall now attempt to convey some notion of the mechanism by which the process of carrying is effected during the second and fourth quarters of a turn of the axis. As before, we shall first explain it in reference to a particular instance. During the first quarter of a turn the wheel B^2, Fig. 1, is caused by the adding bolt to move through five divisions; and the fifth of these divisions, which passes under the index, is that between 9 and 0. On the axis of the wheel C^2, immediately to the left of B^2, is fixed a wheel, called in mechanics a ratchet wheel, which is driven by a claw which constantly rests in its teeth. This claw is in such a position as to permit the wheel C^2 to move in obedience to the action of the adding bolt, but to resist its motion in the contrary direction. It is drawn back by a spiral spring, but its recoil is prevented by a hook which sustains it; which hook, however, is capable of being withdrawn, and when withdrawn, the aforesaid spiral spring would draw back the claw, and make it fall through one tooth of the ratchet wheel. Now, at the moment that the division between 9 and 0 on the dial B^2 passes under the index, a thumb placed on the axis of this dial touches a trigger which raises out of the notch the hook which sustains the claw just mentioned, and allows it to fall back by the recoil of the spring, and to drop into the next tooth of the ratchet wheel. This process, however, produces no immediate effect upon the position of the wheel C^2, and is merely preparatory to an action intended to take place during the second quarter of a turn of the moving axis. It is in effect a memorandum taken by the machine of a carriage to be made in the next quarter of a turn.

During the second quarter of a turn, a finger placed on the axis of the dial B^2 is made to revolve, and it encounters the heel of the above-mentioned claw. As it moves forward it drives the claw before it; and this claw, resting in the teeth of the ratchet wheel fixed upon the axis of the dial C^2, drives forward that wheel, and with it the dial. But the length and position of the finger which drives the claw limits its action, so as to move the claw forward through such a space only as will cause the dial C^2 to advance through a single division; at which point it is again caught and retained by the hook. This will be added to the number under its index, and the requisite carriage from B^2 to C^2 will be accomplished.

In connexion with every dial is placed a similar ratchet wheel with a similar claw, drawn by a similar spring, sustained by a similar hook, and acted upon by a similar thumb and trigger; and therefore the necessary carriages, throughout the whole machinery, take place in the same manner and by similar means.

During the second quarter of a turn, such of the carrying claws as have been allowed to recoil in the first, third, and fifth rows, are drawn up by the fingers on the axes of the adjacent dials; and, during the fourth quarter of a turn, such of the carrying claws on the second and fourth rows have been allowed to recoil during the third quarter of a turn, are in like manner drawn up by the carrying fingers on the axes of the adjacent dials. It appears that the carriages proceed alternately from right to left along the horizontal rows during the second and fourth quarters of a turn; in the one, they pass along the first, third, and fifth rows, and in the other, along the second and fourth.

There are two systems of waves of mechanical action continually flowing from the bottom to the top; and two streams of similar action constantly passing from the right to the left. The crests of the first system of adding waves fall upon the last difference, and upon every alternate one proceeding upwards; while the crests of the other system touch upon the intermediate differences. The first stream of carrying action passes from right to left along the highest row and every alternate row, while the second stream passes along the intermediate rows.

Such is a very rapid and general outline of this machinery. Its wonders, however, are still greater in its details than even in its broader features. Although we despair of doing it justice by any description which can be attempted here, yet we should not fulfil the duty we owe to our readers, if we did not call their attention at least to a few of the instances of consummate skill which are scattered, with a prodigality characteristic of the highest order of inventive genius, throughout this astonishing mechanism.

In the general description which we have given of the mechanism for *carrying*, it will be observed, that the preparation for every carriage is stated to be made during the previous addition, by the disengagement of the carrying claw before mentioned, and by its consequent recoil, urged by the spiral spring with which it is connected; but it may, and does, frequently happen, that though the process of addition may not have rendered a carriage necessary, one carriage may itself produce the necessity for another. This is a contingency not provided against in the mechanism as we have described it: the case would occur in the scheme represented in Fig. 1, if the figure under the index of C^2 were 4 instead of 3. The addition of the number 5 at the index of C^3 would, in this case, in the first quarter of a turn, bring 9 to the index of C^2: this would obviously render no carriage necessary, and of course no preparation would be made for one by the mechanism—that is to say, the carrying claw of the wheel D^2 would not be detached. Meanwhile a carriage upon C^2 has been rendered necessary by the addition made in the first quarter of a turn to B^2. This carriage takes place in the ordinary way, and would cause the dial C^2, in the second quarter of a turn, to advance from 9 to 0: this would make the necessary preparation for a carriage from C^2 to D^2. But unless some special arrangement was made for the purpose, that carriage would not take place during the second quarter of a turn. This peculiar contingency is provided against by an arrangement of singular mechanical beauty, and which, at the same time, answers another purpose—that of equalizing the resistance opposed to the moving power by the carrying mechanism. The fingers placed on the axes of the several dials in the row D^2, do not act at the same instant on the carrying claws adjacent to them; but they are so placed, that their action may be distributed throughout the second quarter of a turn in regular succession. Thus the finger on the axis of the dial A^2 first encounters the claw upon B^2, and drives it through one tooth immediately forwards; the finger on the axis of B^2 encounters the claw upon C^2, and drives it through one tooth; the action of the finger on C^2 on the claw on D^2 next succeeds, and so on. Thus, while the finger on B^2 acts on C^2, and causes the division from 9 to 0 to pass under the index, the thumb on C^2 at the same instant acts on the trigger, and detaches the carrying claw on D^2, which is forthwith encountered by the carrying finger on C^2, and driven forward one tooth. The dial D^2 accordingly moves forward one division, and 5 is brought under the index. This arrangement is beautifully effected by placing the several fingers, which act upon the carrying claws, *spirally* on their axes, so that they come into action in regular succession.

We have stated that, at the commencement of each revolution of the moving axis, the bolts which drive the dials of the first, third, and fifth rows, are shot. The process of shooting these bolts must therefore have taken place during the last quarter of the preceding revolution; but it is during that quarter of a turn that the carriages are effected in the second and fourth rows. Since the bolts which drive the dials of the first, third, and fifth rows, have no mechanical connexion with the dials in the second and fourth rows, there is nothing in the process of shooting those bolts incompatible with that of moving the dials of the second and fourth rows: hence these two processes may both take place during the same quarter of a turn. But in order to equalize the resistance to the moving power, the same expedient is here adopted as that already described in the process of carrying. The arms which shoot the bolts of each row of dials are arranged *spirally*, so as to act successively throughout the quarter of a turn. There is, however, a contingency which, under certain circumstances, would here produce a difficulty which must be provided against. It is possible, and in fact does sometimes happen, that the process of carrying causes a dial to move under the index from 0 to 1. In that case, the bolt, preparatory to the next addition, ought not to be shot until after the carriage takes place; for if the arm which shoots it passes its point of action before the carriage takes place, the bolt will be moved out of its sphere of action, and will not be shot, which, as we have already explained, must always happen when 0 is at the index: therefore no addition would in this case take place during the next quarter of a turn of the axis; whereas, since 1 is brought to the index by the carriage, which immediately succeeds the passage of the arm which ought to bolt, 1 should be added during the next quarter of a turn. It is plain, accordingly, that the mechanism should be so arranged, that the action of the arms, which shoot the bolts successively, should immediately follow the action of those fingers which raise the carrying claws successively; and therefore either a separate quarter of a turn should be appropriated to each of those movements, or if they be executed in the same quarter of a turn, the mechanism must be so constructed, that the arms which shoot the bolts successively, shall severally follow immediately after those which raise the carrying claws successively. The latter object is attained by a mechanical arrangement of singular felicity, and partaking of that elegance which characterizes all the details of this mechanism. Both sets of arms are spirally arranged on their respective axes, so as to be carried through their period in the same quarter of a turn; but the one spiral is shifted a few degrees, in angular position, behind the other, so that each pair of corresponding arms succeed each other in the most regular order,—equalizing the resistance, economizing time, harmonizing the mechanism, and giving to the whole mechanical action the utmost practical perfection.

The system of mechanical contrivances by which the results, here attempted to be described, are attained, form only one order of expedients adopted in this machinery;—although such is the perfection of their action, that in any ordinary case they would be regarded as having attained the ends in view with an almost superfluous degree of precision. Considering, however, the immense importance of the purposes which the mechanism was destined to fulfil, its inventor determined that a higher order of expedients should be superinduced upon those already described; the purpose of which should be to obliterate all small errors or inequalities which might, even by remote possibility, arise, either from defects in the original formation of the mechanism, from inequality of wear, from casual strain or derangement,—or, in short, from any other cause whatever. Thus the movements of the first and principal parts of the mechanism were regarded by him merely as a first, though extremely nice approximation, upon which a system of small corrections was to be subsequently made by suitable and independent mechanism. This supplementary system of mechanism is so contrived, that if one or more of the moving parts of the mechanism of the first order be slightly out of their places, they will be forced to their exact position by the action of the mechanical expedients of the second order to which we now allude. If a more considerable derangement were produced by any accidental disturbance, the consequence would be that the supplementary mechanism would cause the whole system to become locked, so that not a wheel would be capable of moving; the impelling power would necessarily lose all its energy, and the machine would stop. The consequence of this exquisite arrangement is, that the machine will either calculate rightly, or not at all.

The supernumerary contrivances which we now allude to, being in a great degree unconnected with each other, and scattered through the machinery to a certain extent, independent of the mechanical arrangement of the principal parts, we find it difficult to convey any distinct notion of their nature or form.

In some instances they consist of a roller resting between certain curved surfaces, which has but one position of stable equilibrium, and that position the same, however the roller or the curved surfaces may wear. A slight error in the motion of the principal parts would make this roller for the moment rest on one of the curves; but, being constantly urged by a spring, it would press on the curved surface in such a manner as to force the moving piece on which that curved surface is formed, into such a position that the roller may rest between the two surfaces; that position being the one which the mechanism should have. A greater derangement would bring the roller to the crest of the curve, on which it would rest in instable equilibrium; and the machine would either become locked, or the roller would throw it as before into its true position.

In other instances a similar object is attained by a solid cone being pressed into a conical seat; the position of the axis of the cone and that of its seat being necessarily invariable, however the cone may wear; and the action of the cone upon the seat being such, that it cannot rest in any position except that in which the axis of the cone coincides with the axis of its seat.

Having thus attempted to convey a notion, however inadequate, of the calculating section of the machinery, we shall proceed to offer some explanation of the means whereby it is enabled to print its calculations in such a manner as to preclude the possibility of error in any individual printed copy.

On the axle of each of the wheels which express the calculated number of the table T, there is fixed a solid piece of metal, formed into a curve, not unlike the wheel in a common clock, which is called the *snail*. This curved surface acts against the arm of a lever, so as to raise that arm to a higher or lower point according to the position of the dial with which the snail is connected. Without entering into a more minute description, it will be easily understood that the snail may be so formed that the arm of the lever shall be raised to ten different elevations, corresponding to the ten figures of the dial which may be brought under the index. The opposite arm of the lever here described puts in motion a solid arch, or sector, which carries ten punches; each punch bearing on its face a raised character of a figure, and the ten punches bearing the ten characters, 1, 2, 3, 4, 5, 6, 7, 8, 9, 0. It will be apparent from what has been just stated, that this *type sector* (as it is called) will receive ten different attitudes, corresponding to the ten figures which may successively be brought under the index of the dial-plate. At a point over which the type sector is thus moved, and immediately under a point through which it plays, is placed a frame, in which is fixed a plate of copper. Immediately over a certain point through which the type sector moves, is likewise placed a *bent lever*, which, being straightened, is forcibly pressed upon the punch which has been brought under it. If the type sector be moved, so as to bring under the bent lever one of the steel punches above mentioned, and be held in that position for a certain time, the bent lever, being straightened, acts upon the steel punch, and drives it against the face of the copper beneath, and thus causes a sunken impression of the character upon the punch to be left upon the copper. If the copper be now shifted slightly in its position, and the type sector be also shifted so as to bring another punch under the bent lever, another character may be engraved on the copper by straightening the bent lever, and pressing it on the punch as before. It will be evident, that if the copper was shifted from right to left through a space equal to two figures of a number, and, at the same time, the type sector so shifted as to bring the punches corresponding to the figures of the number successively under the bent lever, an engraved impression of the number might thus be obtained upon the copper by the continued action of the bent lever. If, when one line of figures is thus obtained, a provision be made to shift the copper in a direction at right angles to its former motion, through a space equal to the distance between two lines of figures, and at the same time to shift it through a space in the other direction equal to the length of an entire line, it will be evident that another line of figures might be printed below the first in the same manner.

The motion of the type sector, here described, is accomplished by the action of the snail upon the lever already mentioned. In the case where the number calculated is that expressed in Fig. 1, the process would be as follows:—The snail of the wheel F^1, acting upon the lever, would throw the type sector into such an attitude, that the punch bearing the character o would come under the bent lever. The next turn of the moving axis would cause the bent lever to press on the tail of the punch, and the character o would be impressed upon the copper. The bent lever being again drawn up, the punch would recoil from the copper by the action of a spring; the next turn of the moving axis would shift the copper through the interval between two figures, so as to bring the point destined to be impressed with the next figure under the bent lever. At the same time, the snail of the wheel E would cause the type sector to be thrown into the same attitude as before, and the punch o would be brought under the bent lever; the next turn would impress the figure o beside the former one, as before described. The snail upon the wheel D would now come into action, and throw the type sector into that position in which the punch bearing the character 7 would come under the bent lever, and at the same time the copper would be shifted through the interval between two figures; the straightening of the lever would next follow, and the character 7 would be engraved. In the same manner, the wheels, C, B, and A would successively act by means of their snails; and the copper being shifted, and the lever allowed to act, the number 007776 would be finally engraved upon the copper: this being accomplished, the calculating machinery would next be called into action, and another calculation would be made, producing the next number of the Table exhibited in Fig. 5. During this process the machinery would be engaged in shifting the copper both in the direction of its length and its breadth, with a view to commence the printing of another line; and this change of position would be accomplished at the moment when the next calculation would be completed: the printing of the next number would go on like the former, and the operation of the machine would proceed in the same manner, calculating and printing alternately. It is not, however, at all necessary—though we have here supposed it, for the sake of simplifying the explanation—that the calculating part of the mechanism should have its action suspended while the printing part is in operation, or *vice versa*; it is not intended, in fact, to be so suspended in the actual machinery. The same turn of the axis by which one number is printed, executes a part of the movements necessary for the succeeding calculation; so that the whole mechanism will be simultaneously and continuously in action.

Of the mechanism by which the position of the copper is shifted from figure to figure, from line to line, we shall not attempt any description. We feel that it would be quite vain. Complicated and difficult to describe as every other part of this machinery is, the mechanism for moving the copper is such as it would be quite impossible to render at all intelligible, without numerous illustrative drawings.

The engraved plate of copper obtained in the manner above described, is designed to be used as a mould from which a stereotyped plate may be cast; or, if deemed advisable, it may be used as the immediate means of printing. In the one case we should produce a table, printed from type, in the same manner as common letter-press printing; in the other an engraved table. If it be thought most advisable to print from the stereotyped plates, then as many stereotyped plates as may be required may be taken from the copper mould; so that when once a table has been calculated and engraved by the machinery, the whole world may be supplied with stereotyped plates to print it, and may continue to be so supplied for an unlimited period of time. There is no practical limit to the number of stereotyped plates which may be taken from the engraved copper; and there is scarcely any limit to the number of printed copies which may be taken from any single stereotyped plate. Not only, therefore, is the numerical table by these means engraved and stereotyped with infallible accuracy, but such stereotyped plates are producible in unbounded quantity. Each plate, when produced, becomes itself the means of producing printed copies of the table, in accuracy perfect, and in number without limit.

Unlike all other machinery, the calculating mechanism produces, not the object of consumption, but the machinery by which that object may be made. To say that it computes and prints with infallible accuracy, is to understate its merits:—it computes and fabricates *the means* of printing with absolute correctness and in unlimited abundance.

For the sake of clearness, and to render ourselves more easily intelligible to the general reader, we have in the preceding explanation thrown the mechanism into an arrangement somewhat different from that which is really adopted. The dials expressing the numbers of the tables of the successive differences are not placed, as we have supposed them, in horizontal rows, and read from right to left, in the ordinary way; they are, on the contrary, placed vertically, one below the other, and read from top to bottom. The number of the table occupies the first vertical column on the right, the units being expressed on the lowest dial, and the tens on the next above that, and so on. The first difference occupies the next vertical column on the left; and the numbers of the succeeding differences occupy vertical columns, proceeding regularly to the left; the constant difference being on the last vertical column. It is intended in the machine now in progress to introduce six orders of differences, so that there will be seven columns of dials; it is also intended that the calculations shall extend to eighteen places of figures: thus each column will have eighteen dials. We have referred to the dials as if they were inscribed upon the faces of wheels, whose axes are horizontal and planes vertical. In the actual machinery the axes are vertical and the planes horizontal, so that the edges of the *figure wheels*, as they are called, are presented to the eye. The figures are inscribed, not upon the dial-plate, but around the surface of a small cylinder or barrel, placed upon the axis of the figure wheel, which revolves with it; so that as the figure wheel revolves, the figures on the barrel are successively brought to the front, and pass under an index engraved upon a plate of metal immediately above the barrel. This arrangement has the obvious practical advantage, that, instead of each figure wheel having a separate axis, all the figure wheels of the same vertical column revolve on the same axis; and the same observation will apply to all the wheels with which the figure wheels are in mechanical connexion. This arrangement has the further mechanical advantage over that which has been assumed for the purposes of explanation, that the friction of the wheel-work on the axes is less in amount, and more uniformly distributed, than it could be if the axes were placed in the horizontal position.

A notion may therefore be formed of the front elevation of the calculating part of the mechanism, by conceiving seven steel axes erected, one beside another, on each of which shall be placed eighteen wheels,* five inches in diameter, having cylinders or barrels upon them an inch and a half in height, and inscribed, as already stated, with the ten arithmetical characters. The entire elevation of the machinery would occupy a space measuring ten feet broad, ten feet high, and five feet deep. The process of calculation would be observed by the alternate motion of the figure wheels on the several axes. During the first quarter of a turn, the wheels on the first, third, and fifth axes would turn, receiving their addition from the second, fourth, and sixth; during the second quarter of a turn, such of the wheels on the first, third, and fifth axes, to which carriages are due, would be moved forward one additional figure; the second, fourth, and sixth columns of wheels being all this time quiescent. During the third quarter of a turn, the second, fourth, and sixth columns would be observed to move, receiving their additions from the third,

---

* The wheels, and every other part of the mechanism except the axes, springs, and such parts as are necessarily of steel, are formed of an alloy of copper with a small portion of tin.

fifth, and seventh axes; and during the fourth quarter of a turn, such of these wheels to which carriages are due, would be observed to move forward one additional figure; the wheels of the first, third, and fifth columns being quiescent during this time.

It will be observed that the wheels of the seventh column are always quiescent in this process; and it may be asked, of what use they are, and whether some mechanism of a fixed nature would not serve the same purpose? It must, however, be remembered, that for different tables there will be different constant differences; and that when the calculation of a table is about to commence, the wheels on the seventh axis must be moved by the hand, so as to express the constant difference, whatever it may be. In tables, also, which have not a difference rigorously constant, it will be necessary, after a certain number of calculations, to change the constant difference by the hand; and in this case the wheels of the seventh axis must be moved when occasion requires. Such adjustment, however, will only be necessary at very distant intervals, and after a considerable extent of printing and calculation has taken place; and when it is necessary, a provision is made in the machinery by which notice will be given by the sounding of a bell, so that the machine may not run beyond the extent of its powers of calculation.

Immediately behind the seven axes on which the figure wheels revolve, are seven other axes; on which are placed, first, the wheels already described as driven by the figure wheels, and which bear upon them the wedge which withdraws the bolt immediately over these latter wheels, and on the same axis is placed the adding bolt. From the bottom of this bolt there projects downwards the pin, which acts upon the unbolting wedge by which the bolt is withdrawn: from the upper surface of the bolt proceeds a tooth, which, when the bolt is shot, enters between the teeth of the adding wheels, which turns on the same axis, and is placed immediately above the bolt: its teeth, on which the bolt acts, are like the teeth of a crown wheel, and are presented downwards. The bolt is fixed upon this axis, and turns with it; but the adding wheel above the bolt, and the unbolting wheel below it, both turn upon the axis, and independently of it. When the axis is made to revolve by the moving power, the bolt revolves with it; and so long as the tooth of the bolt remains inserted between those of the adding wheel, the latter is likewise moved; but when the lower pin of the bolt encounters the unbolting wedge on the lower wheel, the tooth of the bolt is withdrawn, and the motion of the adding wheel is stopped. This adding wheel is furnished with spur teeth, besides the crown teeth just mentioned; and these spur teeth are engaged with those of that unbolting wheel which is in connexion with the adjacent figure wheel to which the addition is to be made. By such an arrangement it is evident that the revolution of the bolt will necessarily add to the adjacent figure wheel the requisite number.

It will be perceived, that upon the same axis are placed an unbolting wheel, a bolt, and an adding wheel, one above the other, for every figure wheel; and as there are eighteen figure wheels there will be eighteen tiers; each tier formed of an unbolting wheel, a bolt, and an adding wheel, placed one above the other; the wheels on this axis all revolving independent of the axis, but the bolts being all fixed upon it. The same observations, of course, will apply to each of the seven axes.

At the commencement of every revolution of the adding axes, it is evident that the several bolts placed upon them must be shot in order to perform the various additions. This is accomplished by a third set of seven axes, placed at some distance behind the range of the wheels, which turn upon the adding axes: these are called *bolting axes*. On these bolting axes are fixed, so as to revolve with them, a bolting finger opposite to each bolt: as the bolting axis is made to revolve by the moving power, the bolting finger is turned, and as it passes near the bolt, it encounters the shoulder of a hammer or lever, which strikes the heel of the bolt, and presses it forward so as to shoot its tooth between the crown teeth of the adding wheel. The only exception to this action is the case in which o happens to be at the index of the figure wheel; in that case, the lever or hammer, which the bolting finger would encounter, is, as before stated, lifted out of the way of the bolting finger, so that it revolves without encountering it. It is on the bolting axes that the fingers are spirally arranged so as to equalize their action, as already explained.

The same axes in the front of the machinery on which the figure wheels turn, are made to serve the purpose of *carrying*. Each of these bear a series of fingers which turn with them, and which encounter a carrying claw, already described, so as to make the carriage: these carrying fingers are also spirally arranged on their axes, as already described.

Although the absolute accuracy which appears to be ensured by the mechanical arrangements here described is such as to render further precautions nearly superfluous, still it may be right to state, that, supposing it were possible for an error to be produced in calculation, this error could be easily and speedily detected in the printed tables: it would only be necessary to calculate a number of the table taken at intervals, through which the mechanical action of the machine has not been suspended, and during which it has received no adjustment by the hand: if the computed number be found to agree with those printed, it may be taken for granted that all the intermediate numbers are correct; because, from the nature of the mechanism, and the principle of computation, an error occurring in any single number of the table would be unavoidably entailed, in an increasing ratio, upon all the succeeding numbers.

We have hitherto spoken merely of the practicability of executing by the machinery, when completed, that which its inventor originally contemplated—namely, the calculating and printing of all numerical tables, derived by the method of differences from a constant difference. It has, however, happened that the actual powers of the machinery greatly transcend those contemplated in its original design:—they not only have exceeded the most sanguine anticipations of its inventor, but they appear to have an extent to which it is utterly impossible, even for the most acute mathematical thinker, to fix a probable limit. Certain subsidiary mechanical inventions have, in the progress of the enterprise, been, by the very nature of the machinery, suggested to the mind of the inventor, which confer upon it capabilities which he had never foreseen. It would be impossible even to enumerate, within the limits of this article, much less to describe in detail, those extraordinary mechanical arrangements, the effects of which have not failed to strike with astonishment every one who has been favoured with an opportunity of witnessing them, and who has been enabled, by sufficient mathematical attainments, in any degree to estimate their probable consequences.

As we have described the mechanism, the axes containing the several differences are successively and regularly added one to another; but there are certain mechanical adjustments, and these of a very simple nature, which being thrown into action, will cause a difference of any order to be added any number of times to a difference of any other order; and that either proceeding backwards or forwards, from a difference of an inferior to one of a superior order, and *vice versa.**

Among other peculiar mechanical provisions in the machinery is one by which, when the table for any order of difference amounts to a certain number, a certain arithmetical change would be made in the constant difference. In this way a series may be tabulated by the machine, in which the constant difference is subject to periodical change; or the very nature of the table itself may be subject to periodical change, and yet to one which has a regular law.

Some of these subsidiary powers are peculiarly applicable to calculations

---

* The machine was constructed with the intention of tabulating the equation $\Delta^7 u_z = 0$, but, by the means above alluded to, it is capable of tabulating such equations as the following: $\Delta^7 u = a\Delta u$, $\Delta^7 u = a\Delta^3 u$, $\Delta^7 u =$ units figure of Δu.

required in astronomy, and are therefore of eminent and immediate practical utility: others there are by which tables are produced, following the most extraordinary, and apparently capricious, but still regular laws. Thus a table will be computed, which, to any required extent, shall coincide with a given table, and which shall deviate from that table for a single term, or for any required number of terms, and then resume its course, or which shall permanently alter the law of its construction. Thus the engine has calculated a table which agreed precisely with a table of square numbers, until it attained the hundred and first term, which was not the square of 101, nor were any of the subsequent numbers squares. Again, it has computed a table which coincided with a series of natural numbers, as far as 100 000 001, but which subsequently followed another law. This result was obtained, not by working the engine through the whole of the first table, for that would have required an enormous length of time; but by showing, from the arrangement of the mechanism, that it must continue to exhibit the succession of natural numbers, until it would reach 100 000 000. To save time, the engine was set by the hand to the number 99 999 995, and was then put in regular operation. It produced successively the following numbers.*

$$99\ 999\ 996$$
$$99\ 999\ 997$$
$$99\ 999\ 998$$
$$99\ 999\ 999$$
$$100\ 000\ 000$$
$$100\ 010\ 002$$
$$100\ 030\ 003$$
$$100\ 060\ 004$$
$$100\ 100\ 005$$
$$100\ 150\ 006$$

&c. &c.

Equations have been already tabulated by the portion of the machinery which has been put together, which are so far beyond the reach of the present power of mathematics, that no distant term of the table can be predicted, nor any function discovered capable of expressing its general law. Yet the very fact of the table being produced by mechanism of an invariable form, and including a distinct principle of mechanical action, renders it quite manifest

* Such results as this suggest a train of reflection on the nature and operation of general laws, which would lead to very curious and interesting speculations. The natural philosopher and astronomer will be hardly less struck with them than the metaphysician and theologian.

that *some* general law must exist in every table which it produces. But we must dismiss these speculations: we feel it impossible to stretch the powers of our own mind, so as to grasp the probable capabilities of this splendid production of combined mechanical and mathematical genius; much less can we hope to enable others to appreciate them, without being furnished with such means of comprehending them as those with which we have been favoured. Years must in fact elapse, and many enquirers direct their energies to the cultivation of the vast field of research thus opened, before we can fully estimate the extent of this triumph of matter over mind. 'Nor is it,' says Mr Colebrooke,

> among the least curious results of this ingenious device, that it affords a new opening for discovery, since it is applicable, as has been shown by its inventor, to surmount novel difficulties of analysis. Not confined to constant differences, it is available in every case of differences that follow a definite law, reducible therefore to an equation. An engine adjusted to the purpose being set to work, will produce any distant term, or succession of terms, required—thus presenting the numerical solution of a problem, even though the analytical solution be yet undetermined.

That the future path of some important branches of mathematical enquiry must now in some measure be directed by the dictates of mechanism, is sufficiently evident; for who would toil on in any course of analytical enquiry, in which he must ultimately depend on the expensive and fallible aid of human arithmetic, with an instrument in his hands, in which all the dull monotony of numerical computation is turned over to the untiring action and unerring certainty of mechanical agency?

It is worth notice, that each of the axes in front of the machinery on which the figure wheels revolve, is connected with a bell, the tongue of which is governed by a system of levers, moved by the several figure wheels; an adjustment is provided by which the levers shall be dismissed, so as to allow the hammer to strike against the bell, whenever any proposed number shall be exhibited on the axis. This contrivance enables the machine to give notice to its attendants at any time that an adjustment may be required.

Among a great variety of curious accidental properties (so to speak) which the machine is found to possess, is one by which it is capable of solving numerical equations which have rational roots. Such an equation being reduced (as it always may be) by suitable transformations to that state in which the roots shall be whole numbers, the values 0, 1, 2, 3, &c., are substituted for the unknown quantity, and the corresponding values of the equation ascertained. From these a sufficient number of differences being derived, they are set upon the machine. The machine being then put in motion, the table axis will exhibit the successive values of the formula, corresponding to the substitutions of the successive whole numbers for the unknown quantity: at length the number exhibited on the table axis will be 0, which will evidently correspond to a root of the equation. By previous adjustment, the bell of the table axis will in this case ring and give notice of the exhibition of the value of the root in another part of the machinery.

If the equation have imaginary roots, the formula being necessarily a maximum or minimum on the occurrence of such roots, the first difference will become nothing; and the dials of that axis will under such circumstances present 0 to the respective indices. By previous adjustment, the bell of this axis would here give notice of a pair of imaginary roots.

Mr Colebrooke speculates on the probable extension of these powers of the machine:

> It may not therefore be deemed too sanguine an anticipation when I express the hope that an instrument which, in its simpler form, attains to the extraction of roots of numbers, and approximates to the roots of equations, may, in a more advanced state of improvement, rise to the approximate solution of algebraic equations of elevated degrees. I refer to solutions of such equations proposed by La Grange, and more recently by other annalists, which involve operations too tedious and intricate for use, and which must remain without efficacy, unless some mode be devised of abridging the labour, or facilitating the means of its perform-ance. In any case this engine tends to lighten the excessive and accumulating burden of arithmetical application of mathematical formulae, and to relieve the progress of science from what is justly termed by the author of this invention, the overwhelming encumbrance of numerical detail.

Although there are not more than eighteen figure wheels on each axis, and therefore it might be supposed that the machinery was capable of calculating only to the extent of eighteen decimal places; yet there are contrivances connected with it, by which, in two successive calculations, it will be possible to calculate even to the extent of thirty decimal places. Its powers, therefore, in this respect, greatly exceed any which can be required in practical science. It is also remarkable, that the machinery is capable of producing the calculated results *true to the last figure*. We have already explained, that when the figure which would follow the last is greater than 4, then it would be necessary to increase the last figure by 1; since the excess of the calculated number above the true value would in such case be less than its defect from it would be, had the regularly computed final figure been adopted: this is a precaution neces-sary in all numerical tables, and it is one which would hardly have been expected to be provided for in the calculating machinery.

As might be expected in a mechanical undertaking of such complexity and novelty, many practical difficulties have since its commencement been encountered and surmounted. It might have been foreseen, that many expedients would be adopted and carried into effect, which farther experiments would render it necessary to reject; and thus a large source of additional expense could scarcely fail to be produced. To a certain extent this has taken place; but owing to the admirable system of mechanical drawings, which in every instance Mr Babbage has caused to be made, and owing to his own profound acquaintance with the practical working of the most complicated mechanism, he has been able to predict in every case what the result of any contrivance would be, as perfectly from the drawing, as if it had been reduced to the form of a working model. The drawings, consequently, form a most extensive and essential part of the enterprise. They are executed with extraordinary ability and precision, and may be considered as perhaps the best specimens of mechanical drawings which have ever been executed. It has been on these, and on these only, that the work of invention has been bestowed. In these, all those progressive modifications suggested by consideration and study have been made; and it was not until the inventor was fully satisfied with the result of any contrivance, that he had it reduced to a working form. The whole of the loss which has been incurred by the necessarily progressive course of invention, has been the expense of rejected drawings. Nothing can perhaps more forcibly illustrate the extent of labour and thought which has been incurred in the production of this machinery, than the contemplation of the working drawings which have been executed previously to its construction: these drawings cover above a thousand square feet of surface, and many of them are of the most elaborate and complicated description.

One of the practical difficulties which presented themselves at a very early stage in the progress of this undertaking, was the impossibility of bearing in mind all the variety of motions propagated similtaneously through so many complicated trains of mechanism. Nothing but the utmost imaginable harmony and order among such a number of movements, could prevent obstructions arising from incompatible motions encountering each other. It was very soon found impossible, by a mere act of memory, to guard against such an occurrence; and Mr Babbage found, that, without some effective expedient by which he could at a glance see what every moving piece in the machinery was doing at each instant of time, such inconsistencies and obstructions as are here alluded to must continually have occurred. This difficulty was removed by another invention of even a more general nature than the calculating machinery itself, and pregnant with results probably of higher importance. This invention consisted in the contrivance of a scheme of *mechanical notation* which is generally applicable to all machinery whatsoever; and which is exhibited on a table or plan consisting of two distinct sections. In the first is traced, by a peculiar system of signs, the origin of every motion which takes place throughout the machinery; so that the mechanist or inventor is able, by moving his finger along a certain line, to follow out the motion of every piece from effect to cause, until he arrives at the prime mover. The same sign which thus indicates the *source* of motion indicates likewise the *species* of motion, whether it be continuous or reciprocating, circular or progressive, &c. The same system of signs further indicates the nature of the mechanical connexion between the mover and the thing moved, whether it be permanent and invariable (as between the two arms of a lever), or whether the mover and the moved are separate and independent pieces, as is the case when a pinion drives a wheel; also whether the motion of one piece necessarily implies the motion of another; or when such motion in the one is interrupted, and in the other continuous, &c.

The second section of the table divides the time of a complete period of the machinery into any required number of parts; and it exhibits in a map, as it were, that which every part of the machine is doing at each moment of time. In this way, incompatibility in the motions of different parts is rendered perceptible at a glance. By such means the contriver of machinery is not merely prevented from introducing into one part of the mechanism any movement inconsistent with the simultaneous action of the other parts; but when he finds that the introduction of any particular movement is necessary for his purpose, he can easily and rapidly examine the whole range of the machinery during one of its periods, and can find by inspection whether there is any [period], and what portion of time [that is], at which no motion exists incompatible with the desired one, and thus discover a *niche*, as it were, in which to place the required movement. A further and collateral advantage consists in placing it in the power of the contriver to exercise the utmost possible economy of *time* in the application of his moving power. For example without some instrument of mechanical enquiry equally powerful with that now described, it would be scarcely possible, at least in the first instance, so to arrange the various movements that they should be all executed in the least possible number of revolutions of the moving axis. Additional revolutions would almost inevitably be made for the purpose of producing movements and changes which it would be possible to introduce in some of the phases of previous revolutions; and there is no one acquainted with the history of mechanical invention who must not be aware, that in the progressive contrivance of almost every machine the earliest arrangements are invariably defective in this respect; and that it is only by a succession of improvements, suggested by long experience, that that arrangement is at length arrived at, which accomplishes all the necessary motions in the shortest possible time. By the application of the mechanical notation, however, absolute perfection may be arrived at in this respect; even before a single part of the machinery is constructed, and before it has any other existence than that which it obtains upon paper.

Examples of this class of advantages derivable from the notation will occur to the mind of every one acquainted with the history of mechanical invention. In the common suction-pump, for example, the effective agency of the power is suspended during the descent of the piston. A very simple contrivance, however, will transfer to the descent the work to be accomplished in the next ascent; so that the duty of four strokes of the piston may thus be executed in the time of two. In the earlier applications of the steam-engine, that machine was applied almost exclusively to the process of pumping; and the power acted only during the descent of the piston, being suspended during its ascent. When, however, the notion of applying the engine to the general purposes of manufacture occurred to the mind of Watt, he saw that it would be necessary to cause it to produce a continued rotatory motion; and, therefore, that the intervals of intermission must be filled up by the action of the power. He first proposed to accomplish this by a second cylinder working alternately with the first; but it soon became apparent that the blank which existed during the upstroke in the action of the power, might be filled up by introducing the steam at both ends of the cylinder alternately. Had Watt placed before him a scheme of mechanical notation such as we allude to, this expedient would have been so obtruded upon him that he must have adopted it from the first.

One of the circumstances from which the mechanical notation derives a great portion of its power as an instrument of investigation and discovery, is that it enables the inventor to dismiss from his thoughts, and to disencumber his imagination of the arrangement and connexion of the mechanism; which, when it is very complex (and it is in that case that the notation is most useful), can only be kept before the mind by an embarrassing and painful effort. In this respect the powers of the notation may not inaptly be illustrated by the facilities derived in complex and difficult arithmetical questions from the use of the language and notation of algebra. When once the peculiar conditions of the question are translated into algebraical signs, and 'reduced to an equation,' the computist dismisses from his thoughts all the circumstances of the question, and is relieved from the consideration of the complicated relations of the quantities of various kinds which may have entered it. He deals with the algebraical symbols, which are representatives of those quantities and relations, according to certain technical rules of a general nature, the truth of which he has previously established; and, by a process almost mechanical, he arrives at the required result. What algebra is to arithmetic, the notation we now allude to is to mechanism. The various parts of the machinery under consideration being once expressed upon paper by proper symbols, the enquirer dismisses altogether from his thoughts the mechanism itself, and attends only to the symbols; the management of which is so extremely simple and obvious, that the most unpractised person, having once acquired an acquaintance with the signs, cannot fail to comprehend their use.

A remarkable instance of the power and utility of this notation occurred in a certain stage of the invention of the calculating machinery. A question arose as to the best method of producing and arranging a certain series of motions necessary to print and calculate a number. The inventor, assisted by a practical engineer of considerable experience and skill, had so arranged these motions, that the whole might be performed by twelve revolutions of the principal moving axis. It seemed, however, desirable, if possible, to execute these motions by a less number of revolutions. To accomplish this, the engineer sat down to study the complicated details of a part of the machinery which had been put together; the inventor at the same time applied himself to the consideration of the arrangement and connexion of the symbols in his scheme of notation. After a short time, by some transposition of symbols, he caused the received motions to be completed by eight turns of the axis. This he accomplished by transferring the symbols which occupied the last four divisions of his scheme, into such blank spaces as he could discover in the first eight divisions; due care being taken that no symbols should express actions at once simultaneous and incompatible. Pushing his enquiry, however, still further, he proceeded to ascertain whether his scheme of symbols did not admit of a still more compact arrangement, and whether eight revolutions were not more than enough to accomplish what was required. Here the powers of the practical engineer completely broke down. By no effort could he bring before his mind such a view of the complicated mechanism as would enable him to decide upon any improved arrangement. The inventor, however, without any extraordinary mental exertion, and merely by sliding a bit of ruled pasteboard up and down his plan, in search of a vacancy where the different motions might be placed, at length contrived to *pack* all the motions, which had previously occupied eight turns of the handle, into five turns. The symbolic instrument with which he conducted the investigation, now informed him of the impossibility of reducing the action of the machine to a more condensed form. This appeared by the fulness of every space along the lines of compatible action. It was, however, still possible, by going back to the actual machinery, to ascertain whether movements, which, under existing arrangements, were incompatible, might not be brought into harmony. This he accordingly did, and succeeded in diminishing the number of incompatible conditions, and thereby rendered it possible to make actions simultaneous which were before necessarily successive. The notation was now again called into requisition, and a new disposition of the parts was made. At this point of the investigation, this extraordinary instrument of mechanical analysis put forth one of its most singular exertions of power. It presented to the eye of the engineer two currents of mechanical action, which, from their nature, could not be simultaneous; and each of which occupied a complete revolution of the axis, except about a twentieth; the one occupying the last nineteen-twentieths of a complete revolution of the axis, and the other occupying the first nineteen-twentieths of a complete revolution. One of these streams of action was, the successive picking up by the carrying fingers of the successive carrying claws; and the other was, the successive shooting of nineteen bolts by the nineteen bolting fingers. The notation rendered it obvious, that as the bolting action commenced a small space below the commencement of the carrying, and ended an equal space below the termination of the carrying, the two streams of action could be made to flow after one another in one and the same revolution of the axis. He thus succeeded in reducing the period of completing the action to four turns of the axis; when the notation again informed him that he had again attained a limit of condensed action, which could not be exceeded without a further change in the mechanism. To the mechanism he again recurred, and soon found that it was possible to introduce a change which would cause the action to be completed in three revolutions of the axis. An odd number of revolutions, however, being attended with certain practical inconveniences, it was considered more advantageous to execute the motions in four turns; and here again the notation put forth its powers, by informing the inventor, *through the eye*, almost independent of his mind, what would be the most elegant, symmetrical, and harmonious disposition of the required motions in four turns. This application of an almost metaphysical system of abstract signs, by which the motion of the hand performs the office of the mind, and of profound practical skill in mechanics alternately, to the construction of a most complicated engine, forcibly reminds us of a parallel in another science, where the chemist with difficulty succeeds in dissolving a refractory mineral, by the alternate action of the most powerful acids, and the most caustic alkalies, repeated in long-continued succession.

This important discovery was explained by Mr Babbage, in a short paper read before the Royal Society, and published in the *Philosophical Transactions* in 1826.* It is to us more a matter of regret than surprise, that the subject did not receive from scientific men in this country that attention to which its importance in every practical point of view so fully entitled it. To appreciate it would indeed have been scarcely possible, from the very brief memoir which its inventor presented, unaccompanied by any observations or arguments of a nature to force it upon the attention of minds unprepared for it by the nature of their studies or occupations. In this country, science has been generally separated from practical mechanics by a wide chasm. It will be easily admitted, that an assembly of eminent naturalists and physicians, with a sprinkling of astronomers, and one or two abstract mathematicians, were not precisely the persons best qualified to appreciate such an instrument of mechanical investigation as we have here described. We shall not therefore be understood as intending the slightest disrespect for these distinguished persons, when we express our regret, that a discovery of such paramount practical value, in a country preeminently conspicuous for the results of its machinery, should fall still-born and inconsequential through their hands, and be buried unhonoured and undiscriminated in their miscellaneous transactions. We trust that a more auspicious period is at hand; that the chasm which has separated practical from scientific men will speedily close; and that that combination of knowledge will be effected, which can only be obtained when we see the men of science more frequently extending their observant eye over the wonders of our factories, and our great practical manufacturers, with a reciprocal ambition, presenting themselves as active and useful members of our scientific associations. When this has taken place, an order of scientific men will spring up, which will render impossible an oversight so little creditable to the country as that which has been committed respecting the mechanical notation.† This notation has recently undergone very considerable extension and improvement. An additional section has been introduced into it; designed to express the process of circulation in machines, through which fluids, whether liquid or gaseous, are moved. Mr Babbage, with the assistance of a friend who happened to be conversant with the structure and

---

* *Phil. Trans.* 1826, Part III. p. 250, on a method of expressing by signs the action of machinery.
† This discovery has been more justly appreciated by scientific men abroad. It was, almost immediately after its publication, adopted as the topic of lectures, in an institution on the Continent for the instruction of Civil Engineers.

operation of the steam-engine, has illustrated it with singular felicity and success in its application to that machine. An eminent French surgeon, on seeing the scheme of notation thus applied, immediately suggested the advantages which must attend it as an instrument for expressing the structure, operation, and circulation of the animal system; and we entertain no doubt of its adequacy for that purpose. Not only the mechanical connexion of the solid members of the bodies of men and animals, but likewise the structure and operation of the softer parts, including the muscles, integuments, membranes, &c.; the nature, motion, and circulation of the various fluids, their reciprocal effects, the changes through which they pass, the deposits which they leave in various parts of the system; the functions of respiration, digestion, and assimilation,—all would find appropriate symbols and representatives in the notation, even as it now stands, without those additions of which, however, it is easily susceptible. Indeed, when we reflect for what a very different purpose this scheme of symbols was contrived, we cannot refrain from expressing our wonder that it should seem, in all respects, as if it had been designed expressly for the purposes of anatomy and physiology.

Another of the uses which the slightest attention to the details of this notation irresistibly forces upon our notice, is to exhibit, in the form of a connected plan or map, the organization of an extensive factory, or any great public institution, in which a vast number of individuals are employed, and their duties regulated (as they generally are or ought to be) by a consistent and well-digested system. The mechanical notation is admirably adapted, not only to express such an organized connexion of human agents, but even to suggest the improvements of which such organization is susceptible—to betray its weak and defective points, and to disclose, at a glance, the origin of any fault which may, from time to time, be observed in the working of the system. Our limits, however, preclude us from pursuing this interesting topic to the extent which its importance would justify. We shall be satisfied if the hints here thrown out should direct to the subject the attention of those who, being most interested in such an enquiry, are likely to prosecute it with greatest success.

One of the consequences which has arisen in the prosecution of the invention of the calculating machinery, has been the discovery of a multitude of mechanical contrivances, which have been elicited by the exigencies of the undertaking, and which are as novel in their nature as the purposes were novel which they were designed to attain. In some cases several different contrivances were devised for the attainment of the same end; and that among them which was best suited for the purpose was finally selected: the rejected expedients—those overflowings or waste of the invention—were not, however, always, found useless. Like the *waste* in various manufactures, they were soon converted to purposes of utility. These rejected contrivances have found their way, in many cases, into the mills of our manufacturers; and we now find them busily effecting purposes, far different from any which their inventor dreamed of, in the spinning-frames of Manchester.*

Another department of mechanical art, which has been enriched by this invention, has been that of *tools*. The great variety of new forms which it was necessary to produce, created the necessity of contriving and constructing a vast number of novel and most valuable tools, by which, with the aid of the lathe, and that alone, the required forms could be given to the different parts of the machinery with all the requisite accuracy.

The idea of calculation by mechanism is not new. Arithmetical instruments, such as the calculating boards of the ancients, on which they made their computations by the aid of counters—the *Abacus*, an instrument for computing by the aid of balls sliding upon parallel rods—the method of calculation invented by Baron Napier, called by him *Rhabdology*. and since called *Napier's bones*—the Swan Pan of the Chinese—and other similar contrivances, among which more particularly may be mentioned the Sliding Rule, of so much use in practical calculations to modern engineers, will occur to every reader: these may more properly be called *arithmetical instruments*, partaking more or less of a mechanical character. But the earliest piece of mechanism to which the name of a 'calculating machine' can fairly be given, appears to have been a machine invented by the celebrated Pascal. This philosopher and mathematician, at a very early age, being engaged with his father, who held an official situation in Upper Normandy, the duties of which required frequent numerical calculations, contrived a piece of mechanism to facilitate the performance of them. This mechanism consisted of a series of wheels, carrying cylindrical barrels, on which were engraved the ten arithmetical characters, in a manner not very dissimilar to that already described. The wheel which expressed each order of units was so connected with the wheel which expressed the superior order, that when the former passed from 9 to 0, the latter was necessarily advanced one figure; and thus the process of carrying was executed by mechanism: when one number was to be added to another by this machine, the addition of each figure to the other was performed by the

---

* An eminent and wealthy retired manufacturer at Manchester assured us, that on the occasion of a visit to London, when he was favoured with a view of the calculating machinery, he found in it mechanical contrivances, which he subsequently introduced with the greatest advantage into his own spinning-machinery.

hand; when it was required to add more than two numbers, the additions were performed in the same manner successively; the second was added to the first, the third to their sum, and so on.

Subtraction was reduced to addition by the method of arithmetical complements; multiplication was performed by a succession of additions; and division by a succession of subtractions. In all cases, however, the operations were executed from wheel to wheel by the hand.*

This mechanism, which was invented about the year 1650, does not appear ever to have been brought into any practical use; and seems to have speedily found its appropriate place in a museum of curiosities. It was capable of performing only particular arithmetical operations, and these subject to all the chances of error in manipulation; attended also with little more expedition (if so much), as would be attained by the pen of an expert computer.

This attempt of Pascal was followed by various others, with very little improvement, and with no additional success. Polenus, a learned and ingenious Italian, invented a machine by which multiplication was performed, but which does not appear to have afforded any material facilities, nor any more security against error than the common process of the pen. A similar attempt was made by Sir Samuel Moreland, who is described as having transferred to wheel-work the figures of *Napier's bones*, and as having made some additions to the machine of Pascal.†

Grillet, a French mechanician, made a like attempt with as little success. Another contrivance for mechanical calculation was made by Saunderson. Mechanical contrivances for performing particular arithmetical processes were also made about a century ago by Depréne and Boitissendeau; but they were merely modifications of Pascal's, without varying or extending its objects. But one of the most remarkable attempts of this kind which has been made since that of Pascal, was a machine invented by Leibnitz, of which we are not aware that any detailed or intelligible description was ever published. Leibnitz described its mode of operation, and its results, in the *Berlin Miscellany*‡ but he appears to have declined any description of its details. In a letter addressed by him to Bernoulli, in answer to a request of the latter that he would afford a description of the machinery, he says, 'Descriptionem ejus dare

* See a description of this machine by Diderot, in the *Encyc. Method.;* also in the works of Pascal, tom. iv., p. 7; Paris, 1819.

† Equidem Morelandus in Anglia, tubae stentoriae author, Rhabdologiam ex baculis in cylindrulos transtulit, et additiones auxiliares peragit in adjuncta machina additionum Pascaliana.

‡ Tom. i., p. 317.

accuratam res non facilis foret. De effectu ex eo judicaveris quod ad multiplicandum numerum sex figurarum, *e. g.* rotam quamdam tantum sexies gyrari necesse est, nulla alia opera mentis, nullis additionibus intervenientibus; quo facto, integrum absolutumque productum oculis objicietur.'* He goes on to say that the process of division is performed independently of a succession of subtractions, such as that used by Pascal.

It appears that this machine was one of an extremely complicated nature, which would be attended with considerable expense of construction, and only fit to be used in cases where numerous and expensive calculations were necessary.† Leibnitz observes to his correspondent, who required whether it might not be brought into common use, 'Non est facta pro his qui olera aut pisculos vendunt, sed pro observatoriis aut cameris computorum, aut aliis, qui sumptus facile ferunt et multo calculo egent.' Nevertheless, it does not appear that this contrivance, of which the inventor states that he caused two models to be made, was ever applied to any useful purpose; nor indeed do the mechanical details of the invention appear ever to have been published.

Even had the mechanism of these machines performed all which their inventors expected from them, they would have been still altogether inapplicable for the purposes to which it is proposed that the calculating machinery of Mr Babbage shall be applied. They were all constructed with a view to perform *particular arithmetical operations*, and in all of them the accuracy of the result depended more or less upon manipulation. The principle of the calculating machinery of Mr Babbage is perfectly general in its nature, not depending on any particular arithmetical operation, and is equally applicable to numerical tables of every kind. This distinguishing characteristic was well expressed by Mr Colebrooke in his address to the Astronomical Society on this invention.

The principle which essentially distinguishes Mr Babbage's invention from all these is, that it proposes to calculate a series of numbers following any law, by the aid of differences, and that by setting a few figures at the outset, a long series of numbers is readily produced by a mechanical operation. The method of differences in a very wide sense is the mathematical principle of the contrivance. A machine to add a number of arbitrary figures together is no economy of time or trouble, since each individual figure must be placed in the machine; but it is

---

* *Com. Epist.* tom. i., p. 289.

† Sed machinam esse sumptuosam et multarum rotarum instar horologii: Huygenius aliquoties admonuit ut absolvi curarem; quod non sine magno sumptu taedioque factum est, dum varie mihi cum opificibus fuit conflictandum.—*Com. Epist.*

otherwise when those figures follow some law. The insertion of a few at first determines the magnitude of the next, and those of the succeeding. It is this constant repetition of similar operations which renders the computation of tables a fit subject for the application of machinery. Mr Babbage's invention puts an engine in the place of the computer; the question is set to the instrument, or the instrument is set to the question, and by simply giving it motion the solution is wrought, and a string of answers is exhibited.

But perhaps the greatest of its advantages is, that it prints what it calculates; and this completely precludes the possibility of error in those numerical results which pass into the hands of the public. 'The usefulness of the instrument,' says Mr Colebrooke, 'is thus more than doubled; for it not only saves time and trouble in transcribing results into a tabular form, and setting types for the printing of the table, but it likewise accomplishes the yet more important object of ensuring accuracy, obviating numerous sources of error through the careless hands of transcribers and compositors.

Life assurance

IN 1824 Babbage was invited to become actuary and manager of a projected new life assurance company, The Protector. The project was stillborn but Babbage had made a detailed investigation of the statistical basis of life premiums, and also of the practice of insurance offices. He published a book, *A Comparative View of the Various Institutions for the Assurance of Lives*, which is in effect a remarkable consumer's guide to life insurance, perhaps the first of the genre. Although Babbage's book was generally more technical than modern consumer's guides, chapter XVII, 'Of Commission to Agents', is down to earth and the problem discussed still of interest. The book also shows the strong moral sentiments of early liberals, which contrast starkly enough with the worldly cynicism of the second half of the twentieth century.

A Comparative View of the Various Institutions for the Assurance of Lives, pp. 132–41

OF COMMISSION TO AGENTS

FEW persons effect assurances on their lives without previously consulting either their solicitor, their agent, their broker, or some other person on whose judgment and integrity they imagine they can rely. It is therefore of the utmost consequence that no motive should be presented to those who are thus confidentially employed, which should induce them, from any prospect of advantage to themselves, to recommend one office in preference to others. That such a motive is constantly held out, and the temptation most frequently accepted, is established by too many proofs to be denied; and the frequency of its occurrence is unfortunately so great, as to cause it in some measure to have lost, in the eyes of those who practise it, the disgrace which, in all other transactions, is attached to the offer or the acceptance of a bribe.

In order to clear the way for the observations which will be offered on this subject, it is necessary to take a short view of the nature of an agent. An individual or a company exercising any trade, may employ a person to travel for them or reside permanently at any place, to procure orders; the persons so employed are known to those with whom they deal as agents, and of course any representations they may make of the merit of the goods they are employed to sell, are duly weighed by the purchasers as coming from persons acknowledged to be interested in the disposal of them. Whatever be the sum paid by the principal to his agent, is of little consequence to the consumer.

Let us now suppose the consumer, doubtful of his judgment, employs an agent of his own; it will never be contended that an individual or a body of men can, with any semblance either of justice or integrity, offer to those agents a premium to buy at their particular establishments the article they are instructed to purchase. If such a principle is once admitted, those who sell the worst goods will both find it necessary, and be able, to offer the highest premium for a breach of trust in the consumer's agent. Yet this is precisely the conduct of almost all the assurance companies: some of them unblushingly offer, even in the statement of their terms, and most of them privately pay, what they call a commission, to those persons who bring assurances to their office.

The following are extracted from some of the printed papers of terms:—

ALBION.—A liberal commission is allowed to solicitors and others who effect life insurance.

DITTO.—A large commission is allowed to solicitors, brokers, and others, who effect life insurances.

BRITISH COMMERCIAL.—A liberal allowance is granted to solicitors and others recommending business to the office.

EAGLE.—Solicitors and others allowed a liberal commission.

EUROPEAN.—A liberal allowance is made to professional and other persons bringing business to the office.

IMPERIAL.—A commission is allowed to solicitors, agents, and brokers, procuring life assurances.

MEDICO-CLERICAL.—Attornies, brokers, and agents, bringing business to this office, will receive a liberal commission.

LAW LIFE ASSOCIATION.—In the circulars forwarded to the solicitors, on the formation of this institution, one of the inducements was a *liberal commission*.

PELICAN.—21st July, 1824, sittings after term, at Westminster, Maynard *versus* Rhode and others, directors of the Pelican Assurance Company.

Mr. Crouch, who had been a clerk to the Pelican for twelve years, examined. 'They gave to annuity agents and others, who brought assurances to their office, five per cent. on the first and every subsequent payment.'

If to this list were added the names of those offices which *privately* follow the same practice, it would be greatly increased; indeed, the exceptions are very few. The consequences of such a system are to corrupt and debase those through whom it is carried on, and frequently to increase a distress which would have been mitigated by a more honourable system. The following is one out of a thousand similar instances.

A clergyman, in order to provide at his death for a numerous family, succeeded, by great economy, in saving from his income sufficient to assure his life for 2000*l.*; being unacquainted with business, he unfortunately trusted the choice of the office at which he assured to the attorney whom he had been in the habit of employing. The attorney effected the policy at one of those offices which make no return of any part of the profits, and which, notwithstanding, charge the same prices as the Equitable. During about twenty years he received a commission* of five per cent. from the office, which was paid out of the annual sum, with difficulty spared from the scanty income of his employer: and on the death of the clergyman his seven surviving orphans received from the office the original sum assured, 2000*l.*, instead of about 3200*l.*, which they might have received from the Equitable, had not the bribe, held out by the other office, been too great for the integrity of their father's solicitor. In contemplating with scorn the mercenary agent who betrayed, for so trifling a sum, the confidence reposed in him by his client, whose distressed family were thus deprived of 1200*l.*, ought not some portion of our indignation to be reserved for those who tempted him to this breach of trust? What would be the sentiments and conduct of the directors of such a company, if, on some other occasion, they were to detect the same attorney offering, to one of their own agents, ten per cent., to induce him to accept for the office a bad life? Yet this would be one of the natural results of that line of conduct, into which they had themselves first introduced him; and whatever difference the law might find between the two cases, the moral turpitude is not very different: the company offer to every agent in the country a temptation to commit a breach of trust; their disciple and humble imitator limits his temptation to an individual.

---

* I do not recollect the precise age in this case, but it may be worth inquiring the profit derived by the agent from the sacrifice of his employer's interest. Supposing the clergyman's age, at the time of assuring, to have been thirty, the annual premium on his life, for an assurance of 2000*l.*, would be, at the Equitable, as well as at the other office, 53*l.* 8*s.* 4*d.*, out of which the agent received annually five per cent., or 2*l.* 13*s.* 5*d.*; so that his whole profit amounted to little more than 50*l.*, whilst the loss to his employer's family was 1200*l.*

The only circumstances urged in extenuation of this practice (for it is looked upon with regret by many of those who think themselves compelled to follow it) are, that all the offices do the same thing, and consequently, if any one were to refuse, that office would lie under a disadvantage in getting business. It has also been contended that, by giving a commission, the person receiving it is *ipso facto* constituted the agent of the giver, and that there is no impropriety in paying your own agent any sum agreed upon. With respect to the first of these arguments, it is not correct to say that all offices adopt the same system; three can certainly be named, the Amicable, the Equitable, and the Economic, which do not;* and it has not been stated that either of those offices complain of a deficiency of business. Besides, the same principle would require all offices to allow an equal commission; (in which case there could be no necessity for any;) and, on referring to the extracts made in pages 134 and 135, it seems evidently to be the intention of some of the offices to raise a belief that they allow a much larger commission than others. With regard to the other argument, (if it deserve the name of one,) it is neither just nor honourable, under whatever name it may be concealed, to offer money to one who is already the agent of another, in order to influence his judgment.

It will naturally be inquired who authorize the practice we have been reprobating, and whether the long lists of respectable names, displayed at the head of many of these institutions, are placed there only to beguile the unwary, and to lead them to suppose that the same honourable principles, which govern the directors in their private capacity, will be adhered to when they act together as a body. There are many persons thus situated, whose known integrity or high rank render it impossible for a moment to suppose, that they are aware of a practice thus carrying on in their name. Who, for instance, in looking over the prospectus of the Medical and Clerical Assurance Society, would not immediately acquit the dignified clergy, who are placed in the first page amongst the list of officers of that institution, of any knowledge of the

---

* I am happy to be able to add to this small but honourable list, the University Life Assurance Society, which, in a recent instance, preferred the risk of losing a large and valuable assurance to the disgrace of bribing an agent. If I have omitted to mention the name of any other society which is free from this reproach, I can only regret my ignorance of any other exception; and add, that the evidence which I possess of the general prevalence of this practice, induces me to believe it to be more for the interest of that office, as well as of the public, thus to call for an explicit denial; which, if properly authenticated, I should gladly introduce into a subsequent edition. It may, however, be remarked, that scarcely any thing less than a distinct denial of the practice in the printed terms of an office, can reasonably satisfy the public.

concluding paragraph of the second? There are, perhaps, amongst the directors of the various companies, many to whom these pages may convey the first information of a practice which secretly prevails in their own offices; and which ought either to be immediately abolished, or else publicly acknowledged, and the arguments which have been brought against it refuted, or proved to be fallacious.

Reform of science

IN 1827 Babbage's life changed dramatically. First his father died and Babbage inherited a fortune, which has been estimated at about £100,000. I have been unable to secure access to banking records which might settle the question, but recently I have come to think that £70,000 might be a more reasonable estimate. This was in any case sufficient to keep Babbage in comfort and finance his research for the rest of his life.

Next Babbage's second son Charles died, followed by Georgiana and a newborn infant. Babbage was devastated and embarked on a long journey, planning to cross Egypt and travel to China. The battle of Navarino closed the routes to the East, and Babbage spent a year in Italy, meeting natural philosophers and spending much time with his Bonaparte friends. He returned through Vienna and Prague to Berlin, where he attended a meeting of the Deutsche Naturforscher Versammlung. Babbage's report of the meeting started the movement which led to formation of the British Association for the Advancement of Science.

On his return to London in the late autumn of 1828 Babbage took steps to get the project for the first Difference Engine on a more secure financial footing. He received much help from Lord Ashley, better known by his later title as the reforming Earl of Shaftesbury, and it would appear that Babbage was instrumental in starting Shaftesbury on his study of the factories. The Duke of Wellington personally inspected the project and became a firm supporter of Babbage's work: Wellington could always be relied on to give a helping hand to the engineers. Work on the Engine continued until 1833, when government support ceased, and then declined into suspended animation until Peel gave the project its quietus in 1842.

In 1829 Babbage leased a larger house with an acre of gardens, a coachhouse and stables, and built extensive workshops. He began to entertain on a large scale and his Saturday evening soirées, attended after a few years by two hundred or three hundred people, become major events of the London season, a meeting place for Europe's liberal intelligentsia.

During the period of the struggle for the Great Reform Bill, Babbage helped to organize the campaigns of Cavendish for the Cambridge parliamentary seat and wrote two major books: *Reflections on the Decline of Science in England and Some of its Causes (1830)*, and *On the Economy of Machinery and Manufactures* (1832). *The Decline of Science* was by far the most polemical of Babbage's books, and I have presented its polemical part in *Charles Babbage, Pioneer of the Computer* (OUP/Princeton, 1982). The first chapter gives a discussion of the relation between education and science which is of remarkably contemporary interest.

Reflections on the Decline of Science in England and Some of its Causes, pp. 3–8

ON THE RECIPROCAL INFLUENCE OF SCIENCE AND EDUCATION

That the state of knowledge in any country will exert a directive influence on the general system of instruction adopted in it, is a principle too obvious to require investigation. And it is equally certain that the tastes and pursuits of our manhood will bear on them the traces of the earlier impressions of our education. It is therefore not unreasonable to suppose that some portion of the neglect of science in England, may be attributed to the system of education we pursue. A young man passes from our public schools to the universities, ignorant almost of the elements of every branch of useful knowledge; and at these latter establishments, formed originally for instructing those who are intended for the clerical profession, classical and mathematical pursuits are nearly the sole objects proposed to the student's ambition.

Much has been done at one of our universities during the last fifteen years, to improve the system of study; and I am confident that there is no one connected with that body, who will not do me the justice to believe that, whatever suggestions I may venture to offer, are prompted by the warmest feelings for the honour and the increasing prosperity of its institutions. The ties which connect me with Cambridge are indeed of no ordinary kind.

Taking it then for granted that our system of academical education ought to be adapted to nearly the whole of the aristocracy of the country, I am inclined to believe that whilst the modifications I should propose would not be great innovations on the spirit of our institutions, they would contribute materially to that important object.

It will be readily admitted, that a degree conferred by an university, ought to be a pledge to the public that he who holds it possesses a certain quantity of knowledge. The progress of society has rendered knowledge far more various in its kinds than it used to be; and to meet this variety in the tastes and inclinations of those who come to us for instruction, we have, besides the regular lectures to which all must attend, other sources of information from whence the students may acquire sound and varied knowledge in the numerous lectures on chemistry, geology, botany, history, &c. It is at present a matter of option with the student, which, and how many of these courses he shall attend, and such it should still remain. All that it would be necessary to add would be, that previously to taking his degree, each person should be examined by those Professors, whose lectures he had attended. The pupils should then be arranged in two classes, according to their merits, and the names included in these classes should be printed. I would then propose that no young man, except his name was found amongst the 'List of Honours,' should be allowed to take his degree, unless he had been placed in the first class of some one at least of the courses given by the professors. But it should still be imperative upon the student to possess such mathematical knowledge as we usually require. If he had attained the first rank in several of these examinations, it is obvious that we should run no hazard in a little relaxing the strictness of his mathematical trial.

If it should be thought preferable, the sciences might be grouped, and the following subjects be taken together:—

{ Modern History
Laws of England
Civil Law

{ Political Economy
Applications of Science to Arts
    and Manufactures

{ Chemistry
Mineralogy
Geology

{ Zoology, including Physiology
    and Comparative Anatomy
Botany, including Vegetable
    Physiology and Anatomy

One of the great advantages of such a system would be, that no young person would have an excuse for not studying, by stating, as is most frequently done, that the only pursuits followed at Cambridge, classics and mathematics, are not adapted either to his taste, or to the wants of his after life. His friends and relatives would then reasonably expect every student to have acquired distinction in *some* pursuit. If it should be feared that this plan would lead to too great a diversity of pursuits in the same individual, a limitation might be placed upon the number of examinations into which the same person might be permitted to enter. It might also be desirable not to restrict the whole of these examinations to the third year, but to allow the student to enter on some portion of them in the first or second year, if he should prefer it.

By such an arrangement, which would scarcely interfere seriously with our other examinations, we should, I think, be enabled effectually to keep pace with the wants of society, and retaining fully our power and our right to direct the studies of those who are intended for the church, as well as of those who aspire to the various offices connected with our academical institutions; we should, at the same time, open a field of honourable ambition to multitudes, who, from the exclusive nature of our present studies, leave us with but a very limited addition to their stock of knowledge.

Much more might be said on a subject so important to the interests of the country, as well as of our university, but my wish is merely to open it for our own consideration and discussion. We have already done so much for the improvement of our system of instruction, that public opinion will not reproach us for any unwillingness to alter. It is our first duty to be well satisfied that we can improve; such alterations ought only to be the result of a most mature consideration, and of a free interchange of sentiments on the subject, in order that we may condense upon the question the accumulated judgment of many minds.

It is in some measure to be attributed to the defects of our system of education, that scientific knowledge scarcely exists amongst the higher classes of society. The discussions in the Houses of Lords or of Commons, which arise on the occurrence of any subjects connected with science, sufficiently prove this fact, which, if I had consulted the extremely limited nature of my personal experience, I should, perhaps, have doubted.

The *Decline of Science* also contains a characteristic discussion of the nature of scientific observation, including hoaxing, forging, and cooking results.

Reflections on the Decline of Science in England, and Some of its Causes, pp. 167–83

OF OBSERVATIONS

THERE are several reflections connected with the art of making observations and experiments, which may be conveniently arranged in this chapter.

Section 1

OF MINUTE PRECISION

No person will deny that the highest degree of attainable accuracy is an object to be desired, and it is generally found that the last advances towards precision require a greater devotion of time, labour, and expense, than those which precede them. The first steps in the path of discovery, and the first approximate measures, are those which add most to the existing knowledge of mankind.

The extreme accuracy required in some of our modern inquiries has, in some respects, had an unfortunate influence, by favouring the opinion, that no experiments are valuable, unless the measures are most minute, and the accordance amongst them most perfect. It may, perhaps, be of some use to show, that even with large instruments, and most practised observers, this is but rarely the case. The following extract is taken from the representation made by the present Astronomer-Royal, to the Council of the Royal Society, on the advantages to be derived from the employment of two mural circles:—

> That by observing, with two instruments, the same objects at the same time, and in the same manner, we should be able to estimate how much of that *occasional discordance from the mean*, which attends *even the most careful observations*, ought to be attributed to irregularity of refraction, and how much to *the imperfections of instruments*.

In confirmation of this may be adduced the opinion of the late M. Delambre, which is the more important, from the statement it contains relative to the necessity of publishing *all* the observations which have been made.

Mais quelque soit le parti que l'on préfère, il me semble qu'on doit tout publier. Ces irrégularités mêmes sont des faits qu'il importe de connoître. *Les soins les plus attentifs n'en sauroient préserver les observateurs les plus exercés*, et celui qui ne produiroit que des angles toujours parfaitment d'accord auroit été singulièrement bien servi par les circonstances ou ne seroit pas bien sincère.—*Base de Système Metrique*, Discours Preliminaire, p. 158.

This desire for extreme accuracy has called away the attention of experimenters from points of far greater importance, and it seems to have been too much overlooked in the present day, that genius marks it tract, not by the observation of quantities inappreciable to any but the acutest senses, but by placing Nature in such circumstances, that she is forced to record her minutest variations on so magnified a scale, that an observer, possessing ordinary faculties, shall find them legibly written. He who can see portions of matter beyond the ken of the rest of his species, confers an obligation on them, by recording what he sees; but their knowledge depends both on his testimony and on his judgment. He contrives a method of rendering such atoms visible to ordinary observers, communicates to mankind an instrument of discovery, and stamps his own observations with a character, alike independent of testimony or of judgment.

Section 2

ON THE ART OF OBSERVING

The remarks in this section are not proposed for the assistance of those who are already observers, but are intended to show to persons not familiar with the subject, that in observations demanding no unrivalled accuracy, the principles of common sense may be safely trusted, and that any gentleman of liberal education may, by perseverance and attention, ascertain the limits within which he may trust both his instrument and himself.

If the instrument is a divided one, the first thing is to learn to read the verniers. If the divisions are so fine that the coincidence is frequently doubtful, the best plan will be for the learner to get some acquaintance who is skilled in the use of instruments, and having set the instrument at hazard, to write down the readings of the verniers, and then request his friend to do the same; whenever there is any difference, he should carefully examine the doubtful one, and ask his friend to point out the minute peculiarities on which he founds his decision. This should be repeated frequently; and after some practice, he should note how many times in a hundred his reading differs from his friend's, and also how many divisions they usually differ.

The next point is, to ascertain the precision with which the learner can bisect an object with the wires of the telescope. This can be done without assistance. It is not necessary even to adjust the instrument, but merely to point it to a distant object. When it bisects any remarkable point, read off the verniers, and write down the result; then displace the telescope a little, and adjust it again. A series of such observations will show the confidence which is due to the observer's eye in bisecting an object, and also in reading the verniers; and as the first direction gave him some measure of the latter, he may, in a great measure, appreciate his skill in the former. He should also, when he finds a deviation in the reading, return to the telescope, and satisfy himself if he has made the bisection as complete as he can. In general, the student should practise each adjustment separately, and write down the results wherever he can measure its deviations.

Having thus practised the adjustments, the next step is to make an observation; but in order to try both himself and the instrument, let him take the altitude of some fixed object, a terrestrial one, and having registered the result, let him derange the adjustment, and repeat the process fifty or a hundred times. This will not merely afford him excellent practice, but enable him to judge of his own skill.

The first step in the use of every instrument, is to find the limits within which its employer can measure the *same object under the same circumstances*. It is only from a knowledge of this, that he can have confidence in his measures of the *same object under different circumstances*, and after that, of *different objects under different circumstances*.

These principles are applicable to almost all instruments. If a person is desirous of ascertaining heights by a mountain barometer, let him begin by adjusting the instrument in his own study; and having made the upper contact, let him write down the reading of the vernier, and then let him derange the *upper* adjustment *only*, re-adjust, and repeat the reading. When he is satisfied about the limits within which he can make that adjustment, let him do the same repeatedly with the lower; but let him not, until he knows his own errors in reading and adjusting, pronounce upon those of the instrument. In the case of a barometer, he must also be assured, that the temperature of the mercury does not change during the interval.

A friend once brought to me a beautifully constructed piece of mechanism, for marking minute portions of time; the three-hundredth parts of a second were indicated by it. It was a kind of watch, with a pin for stopping one of the hands. I proposed that we should each endeavour to stop it twenty times in succession, at the same point. We were both equally unpractised, and our first endeavours showed that we could not be confident of the twentieth part of a second. In fact, both the time occupied in causing the extremities of the fingers to obey the volition, as well as the time employed in compressing the flesh before the fingers acted on the stop, appeared to influence the accuracy of our observations. From some few experiments I made, I thought I perceived that the rapidity of the transmission of the effects of the will, depended on the state of fatigue or health of the body. If any one were to make experiments on this subject, it might be interesting, to compare the rapidity of the transmission of volition in different persons, with the time occupied in obliterating an impression made on one of the senses of the same persons. For example, by having a mechanism to make a piece of ignited charcoal revolve with different degrees of velocity, some persons will perceive a continuous circle of light before others, whose retina does not retain so long impressions that are made upon it.

Section 3

ON THE FRAUDS OF OBSERVERS

Scientific inquiries are more exposed than most others to the inroads of pretenders; and I feel that I shall deserve the thanks of all who really value truth, by stating some of the methods of deceiving practised by unworthy claimants for its honours, whilst the mere circumstance of their arts being known may deter future offenders.

There are several species of impositions that have been practised in science, which are but little known, except to the initiated, and which it may perhaps be possible to render quite intelligible to ordinary understandings. These may be classed under the heads of hoaxing, forging, trimming, and cooking.

Of Hoaxing. This, perhaps, will be better explained by an example. In the year 1788, M. Gioeni, a knight of Malta, published at Naples an account of a new family of Testacea, of which he described, with great minuteness, one species, the specific name of which has been taken from its *habitat*, and the generic he took from his own family, calling it Gioenia Sicula. It consisted of two rounded triangular valves, united by the body of the animal to a smaller valve in front. He gave figures of the animal, and of its parts; described its structure, its mode of advancing along the sand, the figure of the tract it left, and estimated the velocity of its course at about two-thirds of an inch per minute. He then described the structure of the shell, which he treated with nitric acid, and found it approach nearer to the nature of bone than any other shell.

The editors of the *Encyclopédie Methodique*, have copied this description, and have given figures of the Gioenia Sicula. The fact, however, is, that no such animal exists, but that the knight of Malta, finding on the Sicilian shores the three internal bones of one of the species of *Bulla*, of which some are found on the south-western coast of England,* described and figured these bones most accurately, and drew the whole of the rest of the description from the stores of his own imagination.

Such frauds are far from justifiable; the only excuse which has been made for them is, when they have been practised on scientific academies which had reached the period of dotage. It should however be remembered, that the productions of nature are so various, that mere strangeness† is very far from sufficient to render doubtful the existence of any creature for which there is evidence; and that, unless the memoir itself involves principles so contradictory,‡ as to outweigh the evidence of a single witness, it can only be regarded as a deception, without the accompaniment of wit.

Forging differs from hoaxing, inasmuch as in the latter the deceit is intended to last for a time, and then be discovered, to the ridicule of those who have credited it; whereas the forger is one who, wishing to acquire a reputation for science, records observations which he has never made. This is sometimes accomplished in astronomical observations by calculating the time and circumstances of the phenomenon from tables. The observations of the second comet of 1784, which was only seen by the Chevalier D'Angos, were long suspected to be a forgery, and were at length proved to be so by the calculations and reasonings of Encke. The pretended observations did not accord amongst each other in giving any possible orbit. But M. Encke detected an orbit, belonging to some of the observations, from which he found

---

* *Bulla lignaria*.
† The number of vertebrae in the neck of the plesiosaurus is a strange but ascertained fact.
‡ The kind of contradiction which is here alluded to, is that which arises from *well ascertained* final causes; for instance, the ruminating stomach of the hoofed animals, is in no case combined with the claw-shaped form of the extremities, frequent in many of the carniverous animals, and necessary to some of them for the purpose of seizing their prey.

that all the rest might be almost precisely deduced, provided a mistake of a unity in the index of the logarithm of the radius vector were supposed to have been made in all the rest of the calculations. *Zach. Corr. Astron.* Tom. IV. p. 456.

Fortunately instances of the occurrence of forging are rare.

Trimming consists in clipping off little bits here and there from those observations which differ most in excess from the mean, and in sticking them on to those which are too small; a species of 'equitable adjustment,' as a radical would term it, which cannot be admitted in science.

This fraud is not perhaps so injurious (except to the character of the trimmer) as cooking, which the next paragraph will teach, The reason of this is, that the *average* given by the observations of the trimmer is the same, whether they are trimmed or untrimmed. His object is to gain a reputation for extreme accuracy in making observations; but from respect for truth, or from a prudent foresight, he does not distort the position of the fact he gets from nature, and it is usually difficult to detect him. He has more sense or less adventure than the Cook.

Of Cooking. This is an art of various forms, the object of which is to give to ordinary observations the appearance and character of those of the highest degree of accuracy.

One of its numerous processes is to make multitudes of observations, and out of these to select those only which agree, or very nearly agree. If a hundred observations are made, the cook must be very unlucky if he cannot pick out fifteen or twenty which will do for serving up.

Another approved receipt, when the observations to be used will not come within the limit of accuracy, which it has been resolved they shall possess, is to calculate them by two different formulae. The difference in the constants employed in those formulae has sometimes a most happy effect in promoting unanimity amongst discordant measures. If still greater accuracy is required, three or more formulae can be used.

It must be admitted that this receipt is in some instances rather hazardous: but in cases where the positions of stars, as given in different catalogues, occur, or different tables of specific gravities, specific heats, &c. &c., it may safely be employed. As no catalogue contains all stars, the computer must have recourse to several; and if he is obliged to use his judgment in the selection, it would be cruel to deny him any little advantage which might result from it. It may, however, be necessary to guard against one mistake into which persons might fall.

If an observer calculate particular stars from a catalogue which makes them accord precisely with the rest of his results, whereas, had they been computed from other catalogues the difference would have been considerable, it is very unfair to accuse him of *cooking;* for—those catalogues may have been notoriously inaccurate; or—they may have been superseded by others more recent, or made with better instruments; or—the observer may have been totally ignorant of their existence.

It sometimes happens that the constant quantities in formulae given by the highest authorities, although they differ amongst themselves, yet they will not suit the materials. This is precisely the point in which the skill of the artist is shown; and an accomplished cook will carry himself triumphantly through it, provided happily some mean value of such constants will fit his observations. He will discuss the relative merits of formulae he has just knowledge enough to use; and, with admirable candour assigning their proper share of applause to Bessel, to Gauss, and to Laplace, he will take *that* mean value of the constant used by three such philosophers, which will make his own observations accord to a miracle.

There are some few reflections which I would venture to suggest to those who cook, although they may perhaps not receive the attention which, in my opinion, they deserve, from not coming from the pen of an adept.

In the first place, it must require much time to try different formulae. In the next place it may happen that, in the progress of human knowledge, more correct formulae may be discovered, and constants may be determined with far greater precision. Or it may be found that some physical circumstance influences the results, (although unsuspected at the time) the measure of which circumstance may perhaps be recovered from other contemporary registers of facts.* Or if the selection of observations has been made with the view of its agreeing precisely with the latest determination, there is some little danger that the average of the whole may differ from that of the chosen ones, owing to some law of nature, dependent on the interval between the two sets, which law some future philosopher may discover, and thus the very best observations may have been thrown aside.

In all these, and in numerous other cases, it would most probably happen that the cook would procure a temporary reputation for unrivalled accuracy at

---

* Imagine, by way of example, the state of the barometer or thermometer.

the expense of his permanent fame. It might also have the effect of rendering even all his crude observations of no value; for that part of the scientific world whose opinion is of most weight, is generally so unreasonable, as to neglect altogether the observations of those in whom they have, on any occasion, discovered traces of the artist. In fact, the character of an observer, as of a woman, if doubted is destroyed.

The manner in which facts apparently lost are restored to light, even after considerable intervals of time, is sometimes very unexpected, and a few examples may not be without their use. The thermometers employed by the philosophers who composed the Academia Del Cimento, have been lost; and as they did not use the two fixed points of freezing and boiling water, the results of a great mass of observations have remained useless from our ignorance of the value of a degree on their instrument. M. Libri, of Florence, proposed to regain this knowledge by comparing their registers of the temperature of the human body and of that of some warm springs in Tuscany, which have preserved their heat uniform during a century, as well as of other things similarly circumstanced.

Another illustration was pointed out to me by M. Gazzeri, the Professor of Chemistry at Florence. A few years ago an important suit in one of the legal courts of Tuscany depended on ascertaining whether a certain word had been erased by some chemical process from a deed then before the court. The party who insisted that an erasure had been made, availed themselves of the knowledge of M. Gazzeri, who, concluding that those who committed the fraud would be satisfied by the disappearance of the colouring matter of the ink, suspected (either from some colourless matter remaining in the letters, or perhaps from the agency of the solvent having weakened the fabric of the paper itself beneath the supposed letters) that the effect of the slow application of heat would be to render some difference of texture or of applied substance evident, by some variety in the shade of colour which heat in such circumstances might be expected to produce. Permission having been given to try the experiment, on the application of heat the important word reappeared, to the great satisfaction of the court.

Economic theory

On the Economy of Machinery and Manufactures developed out of studies of industry which Babbage made while working on the first Difference Engine. A major contribution to political economy, it had an important influence on John Stuart Mill and Karl Marx. 'When [from the peculiar nature of the produce of each factory] the number of processes into which it is most advantageous to divide it, and the number of individuals to be employed in it, are ascertained, then all factories which do not employ a direct multiple of this latter number, will produce the article at a greater cost . . . . Hence arises one cause of the great size of manufacturing establishments' (*On the Economy of Machinery and Manufactures*, 4th edn, pp. 212–13). Marx gave this quotation in the section on heterogeneous and serial manufacture in *Capital*, and drew the conclusion that the size of factories would continue to increase, bringing increasing numbers of proletarians together to be organized by the very processes of production. This occurred, for example, in Petrograd in 1917 with revolutionary consequences. The same tendency can be observed, albeit in milder form, in Britain until about the 1950s. However, the new technologies have put this whole trend into reverse, with profound political consequences, fundamentally weakening the labour movement. It is amusing to note that, while anti-Marxists have never acknowledged the first part of this argument, Marxists are similarly reluctant to accept the second.

On the Economy of Machinery and Manufactures, pp. 169–267

ON THE DIVISION OF LABOUR

(217.) PERHAPS the most important principle on which the economy of a manufacture depends, is the *division of labour* amongst the persons who perform the work. The first application of this principle must have been made in a very early stage of society; for it must soon have been apparent, that a larger number of comforts and conveniences could be acquired by each individual, if one man restricted his occupation to the art of making bows, another to that of building houses, a third boats, and so on. This division of labour into trades was not, however, the result of an opinion that the general riches of the community would be increased by such an arrangement; but it must have arisen from the circumstance of each individual so employed discovering that he himself could thus make a greater profit of his labour than by pursuing more varied occupations. Society must have made considerable advances before this principle could have been carried into the workshop; for it is only in countries which have attained a high degree of civilization, and in articles in which there is a great competition amongst the producers, that the most perfect system of the division of labour is to be observed. The various principles on which the advantages of this system depend, have been much the subject of discussion amongst writers on Political Economy; but the relative importance of their influence does not appear, in all cases, to have been estimated with sufficient precision. It is my intention, in the first instance, to state shortly those principles, and then to point out what appears to me to have been omitted by those who have previously treated the subject.

(218.) 1. *Of the time require for learning.*—It will readily be admitted, that the portion of time occupied in the acquisition of any art will depend on the difficulty of its execution; and that the greater the number of distinct processes, the longer will be the time which the apprentice must employ in acquiring it. Five or seven years have been adopted, in a great many trades, as the time considered requisite for a lad to acquire a sufficient knowledge of his art, and to enable him to repay by his labour, during the latter portion of his time, the expense incurred by his master at its commencement. If, however, instead of learning *all* the different processes for making a needle, for instance, his attention be confined to one operation, the portion of time consumed unprofitably at the commencement of his apprenticeship will be small, and all the rest of it will be beneficial to his master: and, consequently, if there be any competition amongst the masters, the apprentice will be able to make better terms, and diminish the period of his servitude. Again, the facility of acquiring skill in a single process, and the early period of life at which it can be made a source of profit, will induce a greater number of parents to bring up their children to it; and from this circumstance also, the number of workmen being increased, the wages will soon fall.

(219.) 2. *Of waste of materials in learning.*—A certain quantity of material will, in all cases, be consumed unprofitably, or spoiled by every person who learns an art; and as he applies himself to each new process, he will waste some of the raw material, or of the partly manufactured commodity. But if each man commit this waste in acquiring successively every process, the quantity of waste will be much greater than if each person confine his attention to one process; in this view of the subject, therefore, the division of labour will diminish the price of production.

(220.) 3. Another advantage resulting from the division of labour is, *the saving of that portion of time which is always lost in changing from one occupation to another*. When the human hand, or the human head, has been for some time occupied in any kind of work, it cannot instantly change its employment with full effect. The muscles of the limbs employed have acquired a flexibility during their exertion, and those not in action a stiffness during rest, which renders every change slow and unequal in the commencement. Long habit also produces in the muscles exercised a capacity for enduring fatigue to a much greater degree than they could support under other circumstances. A similar result seems to take place in any change of mental exertion; the attention bestowed on the new subject not being so perfect at first as it becomes after some exercise.

(221.) 4. *Change of tools.*—The employment of different tools in the successive processes is another cause of the loss of time in changing from one operation to another. If these tools are simple, and the change is not frequent, the loss of time is not considerable; but in many processes of the arts the tools are of great delicacy, requiring accurate adjustment every time they are used; and in many cases the time employed in adjusting bears a large proportion to that employed in using the tool. The sliding-rest, the dividing and the drilling-engine, are of this kind; and hence, in manufactories of sufficient extent, it is found to be good economy to keep one machine constantly employed in one kind of work: one lathe, for example, having a screw motion to its sliding-rest along the whole length of its bed, is kept constantly making cylinders; another, having a motion for equalizing the velocity of the work at the point at which it passes the tool, is kept for facing surfaces; whilst a third is constantly employed in cutting wheels.

(222.) 5. *Skill acquired by frequent repetition of the same processes.*—The constant repetition of the same process necessarily produces in the workman a degree of excellence and rapidity in his particular department, which is never possessed by a person who is obliged to execute many different processes. This rapidity is still further increased from the circumstance that most of the operations in factories, where the division of labour is carried to a considerable extent, are paid for as piece-work. It is difficult to estimate in numbers the effect of this cause upon production. In nail-making, Adam Smith has stated, that it is almost three to one; for, he observes, that a smith accustomed to make nails, but whose whole business has not been that of a nailer, can make only from eight hundred to a thousand per day; whilst a lad who had never exercised any other trade, can make upwards of two thousand three hundred a day.

(223.) In different trades, the economy of production arising from the last-mentioned cause will necessarily be different. The case of nail-making is, perhaps, rather an extreme one. It must, however, be observed, that, in one sense, this is not a permanent source of advantage; for, though it acts at the commencement of an establishment, yet every month adds to the skill of the workmen; and at the end of three or four years they will not be very far behind those who have never practised any other branch of their art. Upon an occasion when a large issue of bank-notes was required, a clerk at the Bank of England signed his name, consisting of seven letters, including the initial of his Christian name, five thousand three hundred times during eleven working hours, beside arranging the notes he had signed in parcels of fifty each.

(224.) 6. *The division of labour suggests the contrivance of tools and machinery to execute its processes.*—When each process, by which any article is produced, is the sole occupation of one individual, his whole attention being devoted to a very limited and simple operation, improvements in the form of his tools, or in the mode of using them, are much more likely to occur to his mind, than if it were distracted by a greater variety of circumstances. Such an improvement in the tool is generally the first step towards a machine. If a piece of metal is to be cut in a lathe, for example, there is one particular angle at which the cutting-tool must be held to insure the cleanest cut; and it is quite natural that the idea of fixing the tool at that angle should present itself to an intelligent workman. The necessity of moving the tool slowly, and in a direction parallel to itself, would suggest the use of a screw, and thus arises the sliding-rest. It was probably the idea of mounting a chisel in a frame, to prevent its cutting too deeply, which gave rise to the common carpenter's plane. In cases where a blow from a hammer is employed, experience teaches the proper force required. The transition from the hammer held in the hand to one mounted upon an axis, and lifted regularly to a certain height by some mechanical contrivance, requires perhaps a greater degree of invention than those just instanced; yet it is not difficult to perceive, that, if the hammer always falls from the same height, its effect must be always the same.

(225.) When each process has been reduced to the use of some simple tool, the union of all these tools, actuated by one moving power, constitutes a machine. In contriving tools and simplifying processes, the operative workmen are, perhaps, most successful; but it requires far other habits to combine into one machine these scattered arts. A previous education as a workman in the peculiar trade, is undoubtedly a valuable preliminary; but in order to make such combinations with any reasonable expectation of success, an extensive knowledge of machinery, and the power of making mechanical drawings, are essentially requisite. These accomplishments are now much more common than they were formerly; and their absence was, perhaps, one of the causes of the multitude of failures in the early history of many of our manufactures.

(226.) Such are the principles usually assigned as the causes of the advantage resulting from the division of labour. As in the view I have taken of the question, the most important and influential cause has been altogether unnoticed, I shall re-state those principles in the words of Adam Smith:

> The great increase in the quantity of work, which, in consequence of the division of labour, the same number of people are capable of performing, is owing to three different circumstances: first, to the increase of dexterity in every particular workman; secondly, to the saving of time, which is commonly lost in passing from one species of work to another; and, lastly, to the invention of a great number of machines which facilitate and abridge labour, and enable one man to do the work of many.

Now, although all these are important causes, and each has its influence on the result; yet it appears to me, that any explanation of the cheapness of manufactured articles, as consequent upon the division of labour, would be incomplete if the following principle were omitted to be stated.

That the master manufacturer, by dividing the work to be executed into different processes, each requiring different degrees of skill or of force, can purchase exactly that precise quantity of both which is necessary for each process; whereas, if the whole work were executed by one workman, that person must possess sufficient skill to perform the most difficult, and sufficient strength to execute the most laborious, of the operations into which the art is divided.*

(227.) As the clear apprehension of this principle, upon which a great part of the economy arising from the division of labour depends, is of considerable importance, it may be desirable to point out its precise and numerical application in some specific manufacture. The art of making needles is, perhaps, that which I should have selected for this illustration, as comprehending a very large number of processes remarkably different in their nature;

---

* I have already stated that this principle presented itself to me after a personal examination of a number of manufactories and workshops devoted to different purposes; but I have since found that it had been distinctly pointed out, in the work of Gioja, *Nuovo Prospetto delle Scienze Economiche*, 6 tom. 4to. Milano, 1815, tom. i. capo iv.

but the less difficult art of pin-making, has some claim to attention, from its having been used by Adam Smith; and I am confirmed in the choice of it, by the circumstance of our possessing a very accurate and minute description of that art, as practised in France above half a century ago.

(228.) *Pin-making.*—In the manufacture of pins in England the following processes are employed:—

1. *Wire-drawing.*—(*a.*) The brass wire used for making pins is purchased by the manufacturer in coils of about twenty-two inches in diameter, each weighing about thirty-six pounds. (*b.*) The coils are wound off into smaller ones of about six inches in diameter, and between one and two pounds' weight. (*c.*) The diameter of this wire is now reduced, by drawing it repeatedly through holes in steel plates, until it becomes of the size required for the sort of pins intended to be made. During this process the wire is hardened, and to prevent its breaking, it must be annealed two or three times, according to the diminution of diameter required. (*d.*) The coils are then soaked in sulphuric acid, largely diluted with water, in order to clean them, and are then beaten on stone, for the purpose of removing any oxidated coating which may adhere to them. These operations are usually performed by men, who draw and clean from thirty to thirty-six pounds of wire a day. They are paid at the rate of five farthings per pound, and generally earn about 3s. 6d. per day.

M. Perronnet made some experiments on the extension the wire undergoes in passing through each hole: he took a piece of thick Swedish brass wire, and found

	Feet	In.
Its length to be before drawing	3	8
After passing the first hole	5	5
second hole	7	2
third hole	7	8

It was now annealed, and the length became

	Feet	In.
After passing the fourth hole	10	8
fifth hole	13	1
sixth hole	16	8
And finally, after passing through six other holes	144	0

The holes through which the wire was drawn were not, in this experiment, of regularly decreasing diameter: it is extremely difficult to make such holes, and still more to preserve them in their original dimensions.

(229.) 2. *Straightening the wire.*—The coil of wire now passes into the hands of a woman, assisted by a boy or girl. A few nails, or iron pins, not quite in a line, are fixed into one end of a wooden table about twenty feet in length; the end of the wire is passed alternately between these nails, and is then pulled to the other end of the table. The object of this process is to straighten the wire, which had acquired a considerable curvature in the small coils in which it had been wound. The length thus straightened is cut off, and the remainder of the coil is drawn into similar lengths. About seven nails or pins are employed in straightening the wire, and their adjustment is a matter of some nicety. It seems, that by passing the wire between the first three nails or pins, a bend is produced in an opposite direction to that which the wire had in the coil; this bend, by passing the next two nails, is reduced to another less curved in the first direction, and so on till the curve of the wire may at last be confounded with a straight line.

(230.) 3 *Pointing.*—(*a.*) A man next takes about three hundred of these straightened pieces in a parcel, and putting them into a gauge, cuts off from one end, by means of a pair of shears, moved by his foot, a portion equal in length to rather more than six pins. He continues this operation until the entire parcel is reduced into similar pieces. (*b.*) The next step is to sharpen the ends: for this purpose the operator sits before a *steel mill*, which is kept rapidly revolving: it consists of a cyclinder about six inches diameter, and two and a half inches broad, faced with steel, which is cut in the manner of a file. Another cylinder is fixed on the same axis at a few inches distant; the file on the edge of which is of a finer kind, and is used for finishing off the points. The workman now takes up a parcel of the wires between the finger and thumb of each hand, and presses the ends obliquely on the mill, taking care with his fingers and thumbs to make each wire slowly revolve upon its axis. Having thus pointed all the pieces at one end, he reverses them, and performs the same operation on the other. This process requires considerable skill, but it is not unhealthy; whilst the similar process in needle-making is remarkably destructive of health. (*c.*) The pieces now pointed at both ends, are next placed in gauges, and the pointed ends are cut off, by means of shears, to the proper length of which the pins are to be made. The remaining portions of the wire are now equal to about four pins in length, and are again pointed at each end, and their lengths again cut off. This process is repeated a third time, and the small portion of wire left in the middle is thrown amongst the waste, to be melted along with the dust arising from the sharpening. It is usual for a man, his wife, and a child, to join in performing these processes; and they are paid at the rate of five farthings per pound. They can point from thirty-four to thirty-six and a half pounds per day, and gain from 6s. 6d. to 7s., which may be apportioned thus; 5s. 6d. the man, 1s. the woman, 6d. to the boy or girl.

(231.) 4. *Twisting and cutting the heads.*—The next process is making the heads. For this purpose (*a.*) a boy takes a piece of wire, of the same diameter as the pin to be headed, which he fixes on an axis that can be made to revolve rapidly by means of a wheel and strap connected with it. This wire is called the mould. He then takes a smaller wire, which having passed through an eye in a small tool held in his left hand, he fixes close to the bottom of the mould. The mould is now made to revolve rapidly by means of the right hand, and the smaller wire coils round it until it has covered the whole length of the mould. The boy now cuts the end of the spiral connected with the foot of the mould, and draws it off. (*b.*) When a sufficient quantity of *heading* is thus made, a man takes from thirteen to twenty of these spirals in his left hand, between his thumb and three outer fingers: these he places in such a manner that two turns of the spiral shall be beyond the upper edge of a pair of shears, and with the forefinger of the same hand he feels that only two turns do so project. With his right hand he closes the shears; and the two turns of the spiral being cut off, drop into a basin; the position of the forefinger preventing the heads from flying about when cut off. The workmen who cut the heads are usually paid at the rate of 2½d. to 3d. per pound for large heads, but a higher price is given for the smaller heading. Out of this they pay the boy who spins the spiral; he receives from 4d. to 6d. a day. A good workman can cut from six to about thirty pounds of heading per day, according to its size.

(232.) 5. *Heading.* The process of fixing the head on the body of the pin is usually executed by women and children. Each operator sits before a small steel stake, having a cavity, into which one half of the intended head will fit; immediately above is a steel die, having a corresponding cavity for the other half of the head: this latter die can be raised by a pedal moved by the foot. The weight of the hammer is from seven to ten pounds, and it falls through a very small space, perhaps from one to two inches. The cavities in the centre of these dies are connected with the edge of a small groove, to admit of the body of the pin, which is thus prevented from being flattened by the blow of the die. (*a.*) The operator with his left hand dips the pointed end of the body of a pin into a tray of heads; having passed the point through one of them, he carries it along to the other end with the fore-finger. He now takes the pin in the right hand, and places the head in the cavity of the stake, and, lifting the die with his foot, allows it to fall on the head. This blow tightens the head on the shank, which is then turned round, and the head receives three or four blows on different parts of its circumference. The women and children who fix the heads are paid at the rate of 1s. 6d. for every twenty thousand. A skilful operator can with great exertion do twenty thousand per day; but from ten to fifteen thousand is the usual quantity: children head a much smaller number; varying, of course, with the degree of their skill. About one per cent. of the pins are spoiled in the process; these are picked out afterwards by women, and are reserved, along with the waste from other processes, for the melting-pot. The die in which the heads are struck is varied in form according to the fashion of the time; but the repeated blows to which it is subject render it necessary that it should be repaired after it has been used for about thirty pounds of pins.

(233.) 6. *Tinning*. The pins are now fit to be tinned, a process which is usually executed by a man, assisted by his wife, or by a lad. The quantity of pins operated upon at this stage is usually fifty-six pounds. (*a*.) They are first placed in a pickle, in order to remove any grease or dirt from their surface, and also to render them rough, which facilitates the adherence of the tin with which they are to be covered. (*b*.) They are then placed in a boiler full of a solution of tartar in water, in which they are mixed with a quantity of tin in small grains. In this they are generally kept boiling for about two hours and a half, and are then removed into a tub of water into which some bran has been thrown, for the purpose of washing off the acid liquor. (*c*.) They are then taken out, and, being placed in wooden trays, are well shaken in dry bran: this removes any water adhering to them; and by giving the wooden tray a peculiar kind of motion, the pins are thrown up, and the bran gradually flies off, and leaves them behind in the tray. The man who pickles and tins the pins usually gets one penny per pound for the work, and employs himself, during the boiling of one batch of pins, in drying those previously tinned. He can earn about 9s. per day; but out of this he pays about 3s. for his assistant.

(234.) 7. *Papering*. The pins come from the tinner in wooden bowls, with the points projecting in all directions: the arranging of them side by side in paper is generally performed by women. (*a*.) A woman takes up some, and places them on a comb, and shaking them, some of the pins fall back into the bowl, and the rest, being caught by their heads, are detained between the teeth of the comb. (*b*.) Having thus arranged them in a parallel direction, she fixes the requisite number between two pieces of iron, having twenty-five small grooves, at equal distances; (*c*.) and having previously doubled the paper, she presses it against the points of the pins until they have passed through the two folds which are to retain them. The pins are then relieved from the grasp of the tool, and the process is repeated. A woman gains about 1s. 6d. per day by papering; but children are sometimes employed, who earn from 6d. per day, and upwards.

(235.) Having thus generally described the various processes of pin-making, and having stated the usual cost of each, it will be convenient to present a tabular view of the time occupied by each process, and its cost, as well as the sums which can be earned by the persons who confine themselves solely to each process. As the rate of wages is itself fluctuating, and as the prices paid and quantities executed have been given only between certain limits, it is not to be expected that this table can represent the cost of each part of the work with the minutest accuracy, nor even that it shall accord perfectly with the prices above given: but it has been drawn up with some care, and will be quite sufficient to serve as the basis of those reasonings which it is meant to illustrate. A table nearly similar will be subjoined, which has been deduced from a statement of M. Perronet, respecting the art of pin-making in France, above seventy years ago.

ENGLISH MANUFACTURE

(236.) Pins, '*Elevens*,' 5,546 weigh one pound; '*one dozen*' = 6932 pins weight twenty ounces, and require six ounces of paper.

Name of the process	Workmen	Time for making 1 lb. of pins	Cost of making 1 lb. of pins	Workman earns per day	Price of making each Part of a single Pin, in Millionths of a penny
		Hours.	*Pence.*	*s. d.*	
1. Drawing wire (§ 224.)	Man	.3636	1.2500	3 3	225
2. Straightening wire	Woman	.3000	.2840	1 0	51
(§ 225.)	Girl	.3000	.1420	0 6	26
3. Pointing (§ 226.)	Man	.3000	1.7750	5 3	319
4. Twisting and cutting	Boy	.0400	.0147	0 4½	3
heads (§ 227.)	Man	.0400	.2103	5 4½	38
5. Heading (§ 228.)	Woman	4.0000	5.0000	1 3	901
6. Tinning, or Whiten-	Man	.1071	.6666	6 0	121
ing (§ 229.)	Woman	.1071	.3333	3 0	60
7. Papering (§ 230.)	Woman	2.1314	3.1973	1 6	576
		7.6892	12.8732		2320

Number of Persons employed:—Men, 4; Women, 4; Children, 2, Total, 10.

FRENCH MANUFACTURE

(237.) Cost of 12 000 pins, No. 6, each being eight-tenths of an English inch in length,—as they were manufactured in France about 1760; with the cost of each operation: deduced from the observations and statement of M. Perronet.

Name of the process		Time for making twelve thousand pins	Cost of making twelve thousand pins	Workman usually earns per day	Expense of tools and materials
		Hours.	*Pence.*	*Pence.*	*Pence.*
1.	Wire	—	—	—	24.75
2.	Straightening and cutting	1.2	.5	4.5	—
3.	Coarse pointing	1.2	.625	10.0	—
	Turning wheel*	1.2	.875	7.0	—
	Fine pointing	.8	.5	9.375	—
	Turning wheel	1.2	.5	4.75	—
	Cutting off pointed ends	.6	.375	7.5	—
4.	Turning spiral	.5	.125	3.0	—
	Cutting off heads	.8	.375	5.625	—
	Fuel to anneal ditto	—	—	—	.125
5.	Heading	12.0	.333	4.25	—
6.	Tartar for cleaning	—	—	—	.5
	Tartar for whitening	—	—	—	.5
7.	Papering	4.8	.5	2.0	—
	Paper	—	—	—	1.0
	Wear of Tools	—	—	—	2.0
		24.3	4.708		

(238.) It appears from the analysis we have given of the art of pin-making, that it occupies rather more than seven hours and a half of time, for ten different individuals working in succession on the same material, to convert it into a pound of pins; and that the total expense of their labour, each being paid in the joint ratio of his skill and of the time he is employed, amounts very nearly to 1s. 1d. But from an examination of the first of these tables, it appears that the wages earned by the persons employed vary from 4½d. per day up to 6s., and consequently the skill which is required for their respective employments may be measured by those sums. Now it is evident, that if one

---

* The great expense of turning the wheel appears to have arisen from the person so occupied being unemployed during half his time, whilst the pointer went to another manufactory.

person were required to make the whole pound of pins, he must have skill enough to earn about 5s. 3d. per day, whilst he is pointing the wires or cutting off the heads from the spiral coils,—and 6s. when he is whitening the pins; which three operations together would occupy little more than the seventeenth part of his time. It is also apparent, that during more than one half of his time he must be earning only 1s. 3d. per day, in putting on the heads; although his skill, if properly employed, would, in the same time, produce nearly five times as much. If, therefore, we were to employ, for all the processes, the man who whitens the pins, and who earns 6s. per day, even supposing that he could make the pound of pins in an equally short time, yet we must pay him for his time 46.14 pence, or about 3s. 10d. *The pins would therefore cost, in making, three times and three quarters as much as they now do by the application of the division of labour.*

The higher the skill required of the workman in any one process of a manufacture, and the smaller the time during which it is employed, so much the greater will be the advantage of separating that process from the rest, and devoting one person's attention entirely to it. Had we selected the art of needle-making as our illustration, the economy arising from the division of labour would have been still more striking; for the process of tempering the needles requires great skill, attention, and experience, and although from three to four thousand are tempered at once, the workman is paid a very high rate of wages. In another process of the same manufacture, dry-pointing, which also is executed with great rapidity, the wages earned by the workman reach from 7s. to 12s., 15s., and even, in some instances, to 20s. per day; whilst other processes are carried on by children paid at the rate of 6d. per day.

(239). Some further reflections suggested by the preceding analysis, will be reserved until we have placed before the reader a brief description of a machine for making pins, invented by an American. It is highly ingenious in point of contrivance, and, in respect to its economical principles, will furnish a strong and interesting contrast with the manufacture of pins by the human hand. In this machine a coil of brass wire is placed on an axis; one end of this wire is drawn by a pair of rollers through a small hole in a plate of steel, and is held there by a forceps. As soon as the machine is put in action,—

1. The forceps draws the wire on to a distance equal in length to one pin: a cutting edge of steel then descends close to the hole through which the wire entered, and severs the piece drawn out.
2. The forceps holding the piece thus separated moves on, till it brings the wire to the centre of the *chuck* of a small lathe, which opens to receive it. Whilst the forceps is returning to fetch another piece of wire, the lathe

revolves rapidly, and grinds the projecting end of the wire upon a steel mill, which advances towards it.

3. After this first or coarse pointing, the lathe stops, and another forceps takes hold of the half-pointed pin, (which is instantly released by the opening of the *chuck*,) and conveys it to a similar *chuck* of an adjacent lathe, which receives it, and finishes the pointing on a finer steel mill.

4. This mill again stops, and another forceps removes the pointed pin into a pair of strong steel clams, having a small groove in them by which they hold the pin very firmly. A part of this groove, which terminates at that edge of the steel clams which is intended to form the head of the pin, is made conical. A small round steel punch is now driven forcibly against the end of the wire thus clamped, and the head of the pin is partially formed by compressing the wire into the conical cavity.

5. Another pair of forceps now removes the pin to another pair of clams, and the head of the pin is completed by a blow from a second punch, the end of which is slightly concave. Each pair of forceps returns as soon as it has delivered its burden; and thus there are always five pieces of wire at the same moment in different stages of advance towards finished pin.

The pins so formed are received in a tray, and whitened and papered in the usual manner. About sixty pins can thus be made by this machine in one minute; but each process occupies exactly the same time.

(240.) In order to judge of the value of such a machine, compared with hand-labour, it would be necessary to ascertain:—1. The defects to which pins so made are liable. 2. Their advantages, if any, over those made in the usual way. 3. The prime cost of the machine for making them. 4. The expense of keeping it in repair. 5. The expense of moving the machine and of attending to it.

1. Pins made by the machine are more likely to bend, because the head being 'punched up,' the wire must be in a soft state to admit of that operation. 2. Pins made by the machine are better than common ones, because they are not subject to losing their heads. 3. With respect to the prime cost of a machine, it would be very much reduced if a large number should be required. 4. With regard to its wear and tear, experience only can decide: but it may be remarked, that the steel clams or dies in which the heads are punched up, will wear quickly unless the wire has been softened by annealing; and that if softened, the bodies of the pins will bend too readily. Such an inconvenience might be remedied, either by making the machine spin the heads and fix them on, or by annealing only that end of the wire which is to become the head of the pin: but this would cause a delay between the operations, since the brass is too

brittle, while heated, to bear a blow without crumbling. 5. On comparing the time occupied by the machine with that stated in the analysis, we find that, except in the heading, the human hand is more rapid. Three thousand six hundred pins are pointed by the machine in one hour, whilst a man can point fifteen thousand six hundred in the same time. But in the process of heading, the rapidity of the machine is two and a half times that of the human hand. It must, however, be observed, that the grinding in the machine does not require the application of a force equal to that of one man; for all the processes are executed at once by the machine, and a single labourer can easily work it.

ON THE DIVISION OF MENTAL LABOUR

(241.) We have already mentioned what may, perhaps, appear paradoxical to some of our readers,—that the division of labour can be applied with equal success to mental as to mechanical operations, and that it ensures in both the same economy of time. A short account of its practical applications, in the most extensive series of calculations ever executed, will offer an interesting illustration of this fact, whilst at the same time it will afford an occasion for shewing that the arrangements which ought to regulate the interior economy of a manufactory, are founded on principles of deeper root than may have been supposed, and are capable of being usefully employed in preparing the road to some of the sublimest investigations of the human mind.

(242.) In the midst of that excitement which accompanied the Revolution of France and the succeeding wars, the ambition of the nation, unexhausted by its fatal passion for military renown, was at the same time directed to some of the nobler and more permanent triumphs which mark the era of a people's greatness,—and which receive the applause of posterity long after their conquests have been wrested from them, or even when their existence as a nation may be told only by the page of history. Amongst their enterprises of science, the French government was desirous of producing a series of mathematical tables, to facilitate the application of the decimal system which they had so recently adopted. They directed, therefore, their mathematicians to construct such tables, on the most extensive scale. Their most distinguished philosophers, responding fully to the call of their country, invented new methods for this laborious task; and a work, completely answering the large demands of the government, was produced in a remarkably short period of time. M. Prony, to whom the superintendence of this great undertaking was confided, in speaking of its commencement, observes:

Je m'y livrai avec toute l'ardeur dont j'étois capable, et je m'occupai d'abord du plan général de l'exécution. Toutes les conditions que j'avois à remplir nécessitoient l'emploi d'un grand nombre de calculateurs; et il me vint bientôt à la pensée d'appliquer à la confection de ces Tables la division du travail, dont les Arts de Commerce tirent un parti si avantageux pour réunir à la perfection de main-d'oeuvre l'économie de la dépense et du temps.

The circumstance which gave rise to this singular application of the principle of *the division of labour* is so interesting, that no apology is necessary for introducing it from a small pamphlet printed at Paris a few years since, when a proposition was made by the English to the French government, that the two countries should print these tables at their joint expense.

(243.) The origin of the idea is related in the following extract:—

C'est à un chapitre d'un ouvrage Anglais,* justement célèbre, (I.) qu'est probablement due l'existence de l'ouvrage dont le gouvernement Britannique veut faire jouir le monde savant:—

Voici l'anecdote: M. de Prony s'était engagé, avec les comités de gouvernement, à composer pour *la division centesimale du cercle, des tables logarithmiques et trigonometriques, qui, non seulement ne laissassent rien à desirer quant à l'exactitude, mais qui formassent le monument de calcul le plus vaste et le plus imposant qui eût jamais été exécuté, ou même conçu.* Les logarithmes des nombres de 1 à 200,000 formaient à ce travail un supplement nécessaire et exigé. Il fut aisé à M. de Prony de s'assurer que même en s'associant trois ou quatre habiles co-operateurs, la plus grande durée presumable de sa vie, ne lui suffirai pas pour remplir ses engagements. Il était occupé de cette fâcheuse pensée lorsque, se trouvant devant la boutique d'un marchand de livres, il apperçut la belle edition Anglaise de Smith, donnée a Londres en 1776; il ouvrit le livre au hazard, et tomba sur le premier chapitre, qui traite de *la division du travail,* et où la fabrication des épingles est citée pour exemple. A peine avait-il parcouru les premières pages, que, par une espèce d'inspiration, il conçut l'expédient de mettre ses logarithmes en *manufacture* comme les épingles. Il faisait, en ce moment, à l'école polytechnique, des leçons sur une partie d'analyse liée à ce genre de travail, *la methode des differences,* et ses applications à *l'interpolation.* Il alla passer quelques jours à la campagne, et revint à Paris avec le plan de *fabrication,* qui a été suivi dans l'exécution. Il rassembla deux ateliers, qui faisaient séparément les mêmes calculs, et se servaient de vérification reciproque.†

---

* *An Enquiry into the Nature and Causes of the Wealth of Nations,* by Adam Smith.

† Note sur la publication, proposée par le gouvernement Anglais des grands tables logarithmiques et trigonometriques de M. de Prony.—De l'imprimerie de F. Didot, Dec. 1, 1820, p. 7.

(244.) The ancient methods of computing tables were altogether inapplicable to such a proceeding. M. Prony, therefore, wishing to avail himself of all the talent of his country in devising new methods, formed the first section of those who were to take part in this enterprise out of five or six of the most eminent mathematicians in France.

First Section.—The duty of this first section was to investigate, amongst the various analytical expressions which could be found for the same function, that which was most readily adapted to simple numerical calculation by many individuals employed at the same time. This section had little or nothing to do with the actual numerical work. When its labours were concluded, the formulae on the use of which it had decided, were delivered to the second section.

Second Section.—This section consisted of seven or eight persons of considerable acquaintance with mathematics: and their duty was to convert into numbers the formulae put into their hands by the first section,—an operation of great labour; and then to deliver out these formulae to the members of the third section, and receive from them the finished calculations. The members of this second section had certain means of verifying the calculations without the necessity of repeating, or even of examining, the whole of the work done by the third section.

Third Section.—The members of this section, whose number varied from sixty to eighty, received certain numbers from the second section, and, using nothing more than simple addition and subtraction, they returned to that section the tables in a finished state. It is remarkable that nine-tenths of this class had no knowledge of arithmetic beyond the two first rules which they were thus called upon to exercise, and that these persons were usually found more correct in their calculations, than those who possessed a more extensive knowledge of the subject.

(245.) When it is stated that the tables thus computed occupy seventeen large folio volumes, some idea may perhaps be formed of the labour. From that part executed by the third class, which may almost be termed mechanical, requiring the least knowledge and by far the greatest exertions, the first class were entirely exempt. Such labour can always be purchased at an easy rate. The duties of the second class, although requiring considerable skill in arithmetical operations, were yet in some measure relieved by the higher interest naturally felt in those more difficult operations. The exertions of the first class are not likely to require, upon another occasion, so much skill and labour as they did upon the first attempt to introduce such a method; but when the completion of a calculating-engine shall have produced a substitute for the whole of the third section of computers, the attention of analysts will naturally be directed to simplifying its application, by a new discussion of the methods of converting analytical formulae into numbers.

(246.) The proceeding of M. Prony, in this celebrated system of calculation, much resembles that of a skilful person about to construct a cotton or silk-mill, or any similar establishment. Having, by his own genius, or through the aid of his friends, found that some improved machinery may be successfully applied to his pursuit, he makes drawings of his plans of the machinery, and may himself be considered as constituting the first section. He next requires the assistance of operative engineers capable of executing the machinery he has designed, some of whom should understand the nature of the processes to be carried on; and these constitute his second section. When a sufficient number of machines have been made, a multitude of other persons, possessed of a lower degree of skill, must be employed in using them; these form the third section: but their work, and the just performance of the machines, must be still superintended by the second class.

(247.) As the possibility of performing arithmetical calculations by machinery may appear to non-mathematical readers to be rather too large a postulate, and as it is connected with the subject of the *division of labour*, I shall here endeavour, in a few lines, to give some slight perception of the manner in which this can be done,—and thus to remove a small portion of the veil which covers that apparent mystery.

(248.) *That nearly all tables of numbers which follow any law, however complicated, may be formed, to a greater or less extent, solely by the proper arrangement of the successive addition and subtraction of numbers befitting each table*, is a general principle which can be demonstrated to those only who are well acquainted with mathematics; but the mind, even of the reader who is but very slightly acquainted with that science, will readily conceive that it is not impossible, by attending to the following example.

The subjoined table is the beginning of one in very extensive use, which has been printed and reprinted very frequently in many countries, and is called *a Table of Square Numbers*.

Any number in the table, column A, may be obtained, by multiplying the number which expresses the distance of that term from the commencement of the table by itself; thus, 25 is the fifth term from the beginning of the table, and 5 multiplied by itself, or by 5, is equal to 25. Let us now subtract each term of this table from the next succeeding term, and place the results in another column (B), which may be called first-difference column. If we again subtract each term of this first difference from the succeeding term, we find the result is

Terms of the Table	A. Table	B. First Difference	C. Second Difference
1	1		
		3	
2	4		2
		5	
3	9		2
		7	
4	16		2
		9	
5	25		2
		11	
6	36		2
		13	
7	49		

always the number 2, (column C;) and that the same number will always recur in that column, which may be called the second-difference, will appear to any person who takes the trouble to carry on the table a few terms further. Now when once this is admitted, it is quite clear that, provided the first term (1) of the Table, the first term (3) of the first differences, and the first term (2) of the second or constant difference, are originally given, we can continue the table of square numbers to any extent, merely by addition:—for the series of first differences may be formed by repeatedly adding the constant difference (2) to (3) the first number in column B, and we then have the series of numbers, 3, 5, 6, &c.: and again, by successively adding each of these to the first number (1) of the table, we produce the square numbers.

(249.) Having thus, I hope, thrown some light upon the theoretical part of the question, I shall endeavour to shew that the mechanical execution of such an engine, as would produce this series of numbers, is not so far removed from that of ordinary machinery as might be conceived.* Let the reader imagine three clocks, placed on a table side by side, each having only one hand, and each having a thousand divisions instead of twelve hours marked on the face; and every time a string is pulled, let them strike on a bell the numbers of the divisions to which their hands point. Let him further suppose that two of the clocks, for the sake of distinction called B and C, have some mechanism by

* Since the publication of the Second Edition of this Work, one portion of the engine which I have been constructing for some years past has been put together. It calculates, in three columns, a table with its first and second differences. Each column can be expressed as far as five figures, so that these fifteen figures constitute about one ninth part of the larger engine. The ease and precision with which it works, leave no room to doubt its success in the

which the clock C advances the hand of the clock B one division, for each stroke it makes upon its own bell: and let the clock B by a similar contrivance advance the hand of the clock A one division, for each stroke it makes on its own bell. With such an arrangement, having set the hand of the clock A to the division I., that of B to III., and that of C to II., let the reader imagine the repeating parts of the clocks to be set in motion continually in the following order: viz.—pull the string of clock A; pull the string of clock B; pull the string of clock C.

The table on the following page will then express the series of movements and their results.

If now only those divisions struck or pointed at by the clock A be attended to and written down, it will be found that they produce the series of the squares of the natural numbers. Such a series could, of course, be carried by this mechanism only so far as the numbers which can be expressed by three figures; but this may be sufficient to give some idea of the construction,—and was, in fact, the point to which the first model of the calculating-engine, now in progress, extended.

more extended form. Besides tables of squares, cubes, and portions of logarithmic tables, it possesses the power of calculating certain series whose differences are not constant; and it has already tabulated parts of series formed from the following equations:

$$\Delta^3 u_x = \text{units figure of } \Delta u_x$$

$$\Delta^3 u_x = \text{nearest whole No. to } \left(\frac{1}{10000} \Delta u_x \right)$$

The subjoined is one amongst the series which it has calculated:

0	3486	42972
0	4991	50532
1	6907	58813
14	9295	67826
70	12236	77602
230	15741	88202
495	19861	99627
916	24597	111928
1504	30010	125116
2340	36131	139272

The general term of this is,

$$u_x = \frac{x \cdot x - 1 \cdot x - 2}{1 \cdot 2 \cdot 3} + \text{the whole number in } \frac{x}{10} +$$

$$+ 10 \, \Sigma^3 \left(\text{units figure of } \frac{x \cdot x + 1}{2} \right)$$

Repetitions of Process.	MOVE-MENTS.	CLOCK A. Hand set to I. TABLE.	CLOCK B. Hand set to III. First difference	CLOCK C. Hand set to II. Second difference
1	Pull A.	A. strikes 1	....	....
	——B.	The hand is advanced (by B.) 3 divisions ..	B. strikes 3	....
	——C.	....	The hand is advanced (by C.) 2 divisions ..	C. strikes 2
2	Pull A.	A. strikes 4	....	....
	——B.	The hand is advanced (by B.) 5 divisions ..	B. strikes 5	....
	——C.	....	The hand is advanced (by C.) 2 divisions ..	C. strikes 2
3	Pull A.	A. strikes 9	....	....
	——B.	The hand is advanced (by B.) 7 divisions ..	B. strikes 7	....
	——C.	....	The hand is advanced (by C.) 2 divisions ..	C. strikes 2
4	Pull A.	A. strikes 16	....	....
	——B.	The hand is advanced (by B.) 9 divisions ..	B. strikes 9	
	——C.		The hand is advanced (by C.) 2 divisions ..	C. strikes 2
5	Pull A.	A. strikes 25	....	....
	——B.	The hand is advanced (by B.) 11 divisions ..	B. strikes 11	....
	——C.	....	The hand is advanced (by C.) 2 divisions ..	C. strikes 2
6	Pull A.	A. strikes 36	....	....
	——B.	The hand is advanced (by B.) 13 divisions ..	B. strikes 13	....
	——C.	....	The hand is advanced (by C.) 2 divisions ..	C. strikes 2

(250.) We have seen, then, that the effect of the *division of labour*, both in mechanical and in mental operations, is, that it enables us to purchase and apply to each process precisely that quantity of skill and knowledge which is required for it: we avoid employing any part of the time of a man who can get eight or ten shillings a day by his skill in tempering needles, in turning a wheel, which can be done for sixpence a day; and we equally avoid the loss arising from the employment of an accomplished mathematician in performing the lowest processes of arithmetic.

(251.) The *division of labour* cannot be successfully practised unless there exists a great demand for its produce; and it requires a large capital to be employed in those arts in which it is used. In watchmaking it has been carried, perhaps, to the greatest extent. It was stated in evidence before a committee of the House of Commons, that there are a hundred and two distinct branches of this art, to each of which a boy may be put apprentice: and that he only learns his master's department, and is unable, after his apprenticeship has expired, without subsequent instruction, to work at any other branch. The watch-finisher, whose business is to put together the scattered parts, is the only one, out of the hundred and two persons, who can work in any other department than his own.

(252.) In one of the most difficult arts, that of Mining, great improvements have resulted from the judicious distribution of the duties; and under the arrangements which have gradually been introduced, the whole system of the mine and its government is now placed under the control of the following officers.

1. A Manager, who has the general knowledge of all that is to be done, and who may be assisted by one or more skilful persons.
2. Underground Captains direct the proper mining operations, and govern the working miners.
3. The Purser and Book-keeper manage the accounts.
4. The Engineer erects the engines, and superintends the men who work them.
5. A chief Pitman has charge of the pumps and the apparatus of the shafts.
6. A Surface-captain, with assistants, receives the ores raised, and directs the dressing department, the object of which is to render them marketable.
7. The head Carpenter superintends many constructions.
8. The foreman of the Smiths regulates the ironwork and tools.
9. A Materials-man selects, purchases, receives and delivers all articles required.
10. The Roper has charge of ropes and cordage of all sorts.

ON THE COST OF EACH SEPARATE PROCESS IN A MANUFACTURE

(253.) THE great competition introduced by machinery, and the application of the principle of the subdivision of labour, render it necessary for each producer to be continually on the watch, to discover improved methods by which the cost of the article he manufactures may be reduced; and, with this view, it is of great importance to know the precise expense of every process, as well as of the wear and tear of machinery which is due to it. The same information is desirable for those by whom the manufactured goods are distributed and sold; because it enables them to give reasonable answers or explanations to the objections of inquirers, and also affords them a better chance of suggesting to the manafacturer changes in the fashion of his goods, which may be suitable either to the tastes or to the finances of his customers. To the statesman such knowledge is still more important; for without it he must trust entirely to others, and can form no judgment worthy of confidence, of the effect any tax may produce, or of the injury the manufacturer or the country may suffer by its imposition.

(254.) One of the first advantages which suggests itself as likely to arise from a correct analysis of the expense of the several processes of any manufacture, is the indication which it would furnish of the course in which improvement should be directed. If a method could be contrived of diminishing by one fourth the time required for fixing on the heads of pins, the expense of making them would be reduced about thirteen per cent.; whilst a reduction of one half the time employed in spinning the coil of wire out of which the heads are cut, would scarcely make any sensible difference in the cost of manufacturing of the whole article. It is therefore obvious, that the attention would be much more advantageously directed to shortening the former than the latter process.

(255.) The expense of manufacturing, in a country where machinery is of the rudest kind, and manual labour is very cheap, is curiously exhibited in the price of cotton cloth in the island of Java. The cotton, in the seed, is sold by the Picul, which is a weight of about 133lbs. Not above one fourth or one fifth of this weight, however, is cotton: the natives, by means of rude wooden rollers, can only separate about $1\frac{1}{4}$lb. of cotton from the seed by one day's labour. A Picul of cleansed cotton, therefore, is worth between four and five times the cost of the impure article; and the prices of the same substance, in its different stages of manufacture, are—for one Picul:

	Dollars.
Cotton in the seed	2 to 3
Clean cotton	10–11
Cotton thread	24
Cotton thread died blue	35
Good ordinary cotton cloth	50

Thus it appears that the expense of spinning in Java is 117 per cent. on the value of the raw material; the expense of dying thread blue is 45 per cent. on its value; and that of weaving cotton thread into cloth 117 per cent. on its value. The expense of spinning cotton into a fine thread is, in England, about 33 per cent.

(256.) As an example of the cost of the different processes of a manufacture, perhaps an analytical statement of the expense of the volume now in the reader's hands may not be uninteresting; more especially as it will afford an insight into the nature and extent of the taxes upon literature. It is found economical to print it upon paper of a very large size, so that although thirty-two pages, instead of sixteen, are really contained in each sheet, this work is still called 8vo.

		£	s.	d.
To Printer, for composing (per sheet of 32 pages) 3*l*. 1*s*.	10½ sheets	32	0	6
[This relates to the ordinary size of the type used in the volume.]				
To Printer for composing small type, as in extracts and contents, extra per sheet, 3*s*. 10*d*.		2	0	3
To Printer, for composing table-work, extra per sheet, 5*s*. 6*d*.		2	17	9
Average charge for corrections, per sheet, 3*l*. 2*s*. 10*d*.		33	0	0
Press-work, 3000 being printed off, per sheet, 3*l*. 10*s*.		36	15	0
Paper for 3000, at 1*l*. 11*s*. 6*d*. per ream, weighing 28lbs.: the duty on paper at 3*d*. per lb. amounts to 7*s*. per ream, so that the 63 reams which are required for the work will cost:—				
Paper	77 3 6			
Excise duty	22 1 0			
Total expense of paper		99	4	6
Total expense of printing and paper		205	18	0
Steel-plate for title-page	0 7 6			
Engraving on ditto, Head of Bacon	2 2 0			
Ditto letters	1 1 0			
Total expense of title-page		3	10	6
Printing title-page, at 6*s*. per 100		9	0	0
Paper for ditto, at 1*s*. 9*d*. per 100		2	12	6
Expenses of advertising		40	0	0
Sundries		5	0	0
Total expense in sheets		266	1	0

Cost of a single copy in sheets; 3052 being printed,
including the overplus   o   1   9
Extra boarding   o   o   6
                                                                    _____
Cost of each copy, boarded*   o   2   3

* These charges refer to the edition prepared for the public, and do not relate to the large
paper copies in the hands of some of the author's friends.

(257.) This analysis requires some explanation. The printer usually charges
for composition by the sheet, supposing the type to be all of one kind; and as
this charge is regulated by the size of the letter, on which the quantity in a sheet
depends, little dispute can arise after the price is agreed upon. If there are but
few extracts, or other parts of the work, which require to be printed in smaller
type; or if there are many notes, or several passages in Greek, or in other
languages, requiring a different type, these are considered in the original
contract, and a small additional price per sheet allowed. If there is a large
portion of small type, it is better to have a specific additional charge for it per
sheet. If any work with irregular lines and many figures, and what the printers
call rules, occurs, it is called table-work, and is charged at an advanced price
per sheet. Examples of this are frequent in the present volume. If the page
consists entirely of figures, as in mathematical tables, which require very
careful correction, the charge for composition is usually doubled. A few years
ago I printed a table of logarithms, on a large-sized page, which required great
additional labour and care from the Readers,† in rendering the proofs correct,
and for which, although new punches were not required, several new types
were prepared, and for which stereotype plates were cast, costing about 2*l.* per
sheet. In this case 11*l.* per sheet were charged, although ordinary composition,
with the same sized letter, in demy octavo, could have been executed at thirty-
eight shillings per sheet: but as the expense was ascertained before
commencing the work, it gave rise to no difficulties.
(258.) The charge for *corrections* and *alterations* is one which, from the
difficulty of measuring them, gives rise to the greatest inconvenience, and is as
disagreeable to the publisher (if he be the agent between the author and the
printer), and to the master printer or his foreman, as it is to the author himself.
If the author study economy, he should make the whole of his corrections in

† '*Readers*' are persons employed to correct the press at the printing-office.

the manuscript, and should copy it out fairly: it will then be printed correctly, and he will have little to pay for corrections. But it is scarcely possible to judge of the effect of any passage correctly, without having it set up in type; and there are few subjects, upon which an author does not find he can add some details or explanation, when he sees his views in print. If, therefore, he wish to save his own labour in transcribing, and to give the last polish to the language, he must be content to accomplish these objects at an increased expense. If the printer possess a sufficient stock of type, it will contribute still more to the convenience of the author to have his whole work put up in what are technically called *slips*,* and then to make all the corrections, and to have as few revises as he can. The present work was set up in slips, but the corrections have been unusually large, and the revises frequent.

(259.) The press-work, or *printing off*, is charged at a price agreed upon for each two hundred and fifty sheets; and any broken number is still considered as two hundred and fifty. When a large edition is required, the price for two hundred and fifty is reduced; thus, in the present volume, two hundred and fifty copies, if printed alone, would have been charged eleven shillings per sheet, instead of 5s. 10d., the actual charge. The principle of this mode of charging is good, as it obviates all disputes; but it is to be regretted that the custom of charging the same price for any small number as for two hundred and fifty, is so pertinaciously adhered to, that the workmen will not agree to any other terms when only twenty or thirty copies are required, or even when only three or four are wanted for the sake of some experiment. Perhaps if all numbers above fifty were charged as two hundred and fifty, and all below as for half two hundred and fifty, both parties would derive an advantage.

(260.) The effect of the excise duty is to render the paper thin, in order that it may weigh little; but this is counteracted by the desire of the author to make his book look as thick as possible, in order that he may charge the public as much as he decently can; and so on that ground alone the duty is of no importance. There is, however, another effect of this duty, which both the public and the author feel; for they pay, not merely the duty which is charged, but also the profit on that duty, which the paper-maker requires for the use of additional capital; and also the profit to the publisher and bookseller on the increased price of the volume.

(261.) The estimated charge for advertisements is, in the present case, about the usual allowance for such a volume; and, as it is considered that

---

* *Slips* are long pieces of paper on which sufficient matter is printed to form, when divided, from two to four pages of text.

advertisements in newspapers are the most effectual, where the smallest pays a duty of 3s. 6d., nearly one half of the charge of advertising is a tax.

(262.) It appears then, that, to an expenditure of 224l. necessary to produce the present volume, 42l. are added in the shape of a direct tax. Whether the profits arising from such a mode of manufacturing will justify such a rate of taxation, can only be estimated when the returns from the volume are considered, a subject that will be discussed in a subsequent chapter. It is at present sufficient to observe, that the tax on advertisements is an impolitic tax when contrasted with that upon paper, and on other materials employed. The object of all advertisements is, by making known articles for sale, to procure for them a better price, if the sale is to be by auction; or a larger extent of sale if by retail dealers. Now the more any article is known, the more quickly it is discovered whether it contributes to the comfort or advantage of the public; and the more quickly its consumption is assured if it be found valuable. It would appear, then, that every tax on communicating information respecting articles which are the subjects of taxation in another shape, is one which must reduce the amount that would have been raised, had no impediment been placed in the way of making known to the public their qualities and their price.

ON THE CAUSES AND CONSEQUENCES OF LARGE FACTORIES

(263.) ON examining the analysis which has been given in Chap. XIX. of the operations in the art of pin-making, it will be observed, that ten individuals are employed in it, and also that the time occupied in executing the several processes is very different. In order, however, to render more simple the reasoning which follows, it will be convenient to suppose that each of the seven processes there described requires an equal quantity of time. This being supposed, it is at once apparent, that, to conduct an establishment for pin-making most profitably, the number of persons employed must be a multiple of ten. For if a person with small means has only sufficient capital to enable him to employ half that number of persons, they cannot each of them constantly adhere to the execution of the same process; and if a manufacturer employs any number not a multiple of ten, a similar result must ensue with respect to some portion of them. The same reflection constantly presents itself on examining any well-arranged factory. In that of Mr. Mordan, the patentee of the ever-pointed pencils, one room is devoted to some of the processes by which steel pens are manufactured. Six fly-presses are here constantly at work;—in the first a sheet of thin steel is brought by the workman under the die which at each blow cuts out a flat piece of the metal, having the form intended for the pen. Two other workmen are employed in placing these flat pieces under two other presses, in which a steel chisel cuts the slit. Three other workmen occupy other presses, in which the pieces so prepared receive their semi-cylindrical form. The longer time required for adjusting the small pieces in the two latter operations renders them less rapid in execution than the first; so that two workmen are fully occupied in slitting, and three in bending the flat pieces, which one man can punch out of the sheet of steel. If, therefore, it were necessary to enlarge this factory, it is clear that twelve or eighteen presses would be worked with more economy than any number not a multiple of six.

The same reasoning extends to every manufacture which is conducted upon the principle of the *Division of Labour*, and we arrive at this general conclusion:—*When the number of processes into which it is most advantageous to divide it, and the number of individuals to be employed in it, are ascertained, then all factories which do not employ a direct multiple of this latter number, will produce the article at a greater cost*. This principle ought always to be kept in view in great establishments, although it is quite impossible, even with the best division of the labour, to attend to it rigidly in practice. The proportionate number of the persons who possess the greatest skill, is of course to be first attended to. That exact ratio which is most profitable for a factory employing a hundred workmen, may not be quite the best where there are five hundred; and the arrangements of both may probably admit of variations, without materially increasing the cost of their produce. But it is quite certain that no individual, nor in the case of pin-making could any five individuals, ever hope to compete with an extensive establishment. Hence arises one cause of the great size of manufacturing establishments, which have increased with the progress of civilization. Other circumstances, however, contribute to the same end, and arise also from the same cause—the division of labour.

(264.) The material out of which the manufactured article is produced, must, in the several stages of its progress, be conveyed from one operator to the next in succession; this can be done at least expense when they are all working in the same establishment. If the weight of the material is considerable, this reason acts with additional force; but even where it is light, the danger arising from frequent removal may render it desirable to have all the processes carried on in the same building. In the cutting and polishing of glass this is the case; whilst in the art of needle-making several of the processes are carried on in the cottages of the workmen. It is, however, clear that the latter plan, which is attended with some advantages to the family of the workmen, can be adopted only where there exists a sure and quick method of knowing that the work has been well done, and that the whole of the materials given out have been really employed.

(265.) The inducement to contrive machines for any process of manufacture increases with the demand for the article; and the introduction of machinery, on the other hand, tends to increase the quantity produced, and to lead to the establishment of large factories. An illustration of these principles may be found in the history of the manufacture of patent net.

The first machines for weaving this article were very expensive, costing from a thousand to twelve or thirteen hundred pounds. The possessor of one of these, though it greatly increased the quantity he could produce, was nevertheless unable, when working eight hours a day, to compete with the old methods. This arose from the large capital invested in the machinery; but he quickly perceived that with the same expense of fixed capital, and a small addition to his circulating capital, he could work the machine during the whole twenty-four hours. The profits thus realized soon induced other persons to direct their attention to the improvement of those machines; and the price was greatly reduced, at the same time that the rapidity of production of the patent net was increased. But if machines be kept working through the twenty-four hours, it is necessary that some person shall attend to admit the workmen at the time they relieve each other; and whether the porter or other servant so employed admit one person or twenty, his rest will be equally disturbed. It will also be necessary occasionally to adjust or repair the machine; and this can be done much better by a workman accustomed to machine-making, than by the person who uses it. Now, since the good performance and the duration of machines depend to a very great extent upon correcting every shake or imperfection in their parts as soon as they appear, the prompt attention of a workman resident on the spot will considerably reduce the expenditure arising from the wear and tear of the machinery. But in the case of a single lace-frame, or a single loom, this would be too expensive a plan. Here then arises another circumstance which tends to enlarge the extent of a factory. It ought to consist of such a number of machines as shall occupy the whole time of one workman in keeping them in order: if extended beyond that number, the same principle of economy would point out the necessity of doubling or tripling the number of machines, in order to employ the whole time of two or three skilful workmen.

(266.) Where one portion of the workman's labour consists in the exertion of mere physical force, as in weaving and in many similar arts, it will soon occur to the manufacturer, that if that part were executed by a steam-engine, the same man might, in the case of weaving, attend to two or more looms at once; and, since we already suppose that one or more operative engineers have been employed, the number of his looms may be so arranged that their time shall be fully occupied in keeping the steam-engine and the looms in order. One of the first results will be, that the looms can be driven by the engine nearly twice as fast as before: and as each man, when relieved from bodily labour, can attend to two looms, one workman can now make almost as much cloth as four. This increase of producing power is, however, greater than that which really took place at first; the velocity of some of the parts of the loom being limited by the strength of the thread, and the quickness with which it commences its motion: but an improvement was soon made, by which the motion commenced slowly, and gradually acquired greater velocity than it was safe to give it at once; and the speed was thus increased from 100 to about 120 strokes per minute.

(267.) Pursuing the same principles, the manufactory becomes gradually so enlarged, that the expense of lighting during the night amounts to a considerable sum; and as there are already attached to the establishment persons who are up all night, and can therefore constantly attend to it, and also engineers to make and keep in repair any machinery, the addition of an apparatus for making gas to light the factory leads to a new extension, at the same time that it contributes, by diminishing the expense of lighting, and the risk of accidents from fire, to reduce the cost of manufacturing.

(268.) Long before a factory has reached this extent, it will have been found necessary to establish an accountant's department, with clerks to pay the workmen, and to see that they arrive at their stated times; and this department must be in communication with the agents who purchase the raw produce, and with those who sell the manufactured article.

(269.) We have seen that the application of the *Division of Labour* tends to produce cheaper articles; that it thus increases the demand; and gradually, by the effect of competition, or by the hope of increased gain, that it causes large capitals to be embarked in extensive factories. Let us now examine the influence of this accumulation of capital directed to one object. In the first place, it enables the most important principle on which the advantages of the division of labour depends to be carried almost to its extreme limits: not merely is the precise amount of skill purchased which is necessary for the execution of each process, but throughout every stage,—from that in which the raw material is procured, to that by which the finished produce is conveyed into the hands of the consumer, the same economy of skill prevails. The quantity of work produced by a given number of people is greatly augmented by such an extended arrangement; and the result is necessarily a great reduction in the cost of the article which is brought to market.

(270.) Amongst the causes which tend to the cheap production of any article, and which are connected with the employment of additional capital, may be mentioned, the care which is taken to prevent the absolute waste of any part of the raw material. An attention to this circumstance sometimes causes the union of two trades in one factory, which otherwise might have been separated.

An enumeration of the arts to which the horns of cattle are applicable, will furnish a striking example of this kind of economy. The tanner who has purchased the raw hides, separates the horns, and sells them to the makers of combs and lanterns. The horn consists of two parts, an outward horny case, and an inward conical substance, somewhat intermediate between indurated hair and bone. The first process consists in separating these two parts, by means of a blow against a block of wood. The horny exterior is then cut into three portions with a frame-saw.

1. The lowest of these, next the root of the horn, after undergoing several processes, by which it is flattened, is made into combs.
2. The middle of the horn, after being flattened by heat, and having its transparency improved by oil, is split into thin layers, and forms a substitute for glass, in lanterns of the commonest kind.
3. The tip of the horn is used by the makers of knife-handles, and of the tops of whips, and for other similar purposes.
4. The interior, or core of the horn, is boiled down in water. A large quantity of fat rises to the surface; this is put aside, and sold to the makers of yellow soap.
5. The liquid itself is used as a kind of glue, and is purchased by cloth dressers for stiffening.
6. The insoluble substance, which remains behind, is then sent to the mill, and, being ground down, is sold to the farmers for manure.
7. Besides these various purposes to which the different parts of the horn are applied, the clippings, which arise in comb-making, are sold to the farmer for manure. In the first year after they are spread over the soil they have comparatively little effect, but during the next four or five their efficiency is considerable. The shavings which form the refuse of the lantern-maker, are of a much thinner texture: some of them are cut into various figures and painted, and used as toys; for being hygrometric, they curl up when placed on the palm of a warm hand. But the greater part of these shavings also are

sold for manure, and from their extremely thin and divided form, the full effect is produced upon the first crop.

(271.) Another event which has arisen, in one trade at least, from the employment of large capital, is, that a class of middle-men, formerly interposed between the maker and the merchant, now no longer exist. When calico was woven in the cottages of the workmen, there existed a class of persons who travelled about and purchased the pieces so made, in large numbers, for the purpose of selling them to the exporting merchant. But these middlemen were obliged to examine every piece, in order to know that it was perfect, and of full measure. The greater number of the workmen, it is true, might be depended upon, but the fraud of a few would render this examination indispensable: for any single cottager, though detected by one purchaser, might still hope that the fact would not become known to all the rest.

The value of character, though great in all circumstances of life, can never be so fully experienced by persons possessed of small capital, as by those employing much larger sums: whilst these larger sums of money for which the merchant deals, render his character for punctuality more studied and known by others. Thus it happens that high character supplies the place of an additional portion of capital; and the merchant, in dealing with the great manufacturer, is saved from the expense of verification, by knowing that the loss, or even the impeachment, of the manufacturer's character, would be attended with greater injury to himself than any profit upon a single transaction could compensate.

(272.) The amount of well-grounded confidence, which exists in the character of its merchants and manufacturers, is one of the many advantages that an old manufacturing country always possesses over its rivals. To such an extent is this confidence in character carried in England, that, at one of our largest towns, sales and purchases on a very extensive scale are made daily in the course of business without any of the parties ever exchanging a written document.

(273.) A breach of confidence of this kind, which might have been attended with very serious embarrassment, occurred in the recent expedition to the mouth of the Niger.

'We brought with us from England, Mr. Lander states,

nearly a hundred thousand needles of various sizes, and amongst them was a great quantity of '*Whitechapel Sharps*' warranted '*superfine, and not to cut in the eye.*' Thus highly recommended, we imagined that these needles must have been excellent indeed; but what was our surprise, some time ago, when a number of them which we had disposed of were returned to us, with a complaint that they were all eyeless, thus redeeming with a vengeance the pledge of the manufacturer, 'that they would not cut in the eye.' On an examination afterwards, we found the same fault with the remainder of the 'Whitechapel sharps,' so that to save our credit we have been obliged to throw them away.*

(274.) The influence of established character in producing confidence operated in a very remarkable manner at the time of the exclusion of British manufactures from the Continent during the last war. One of our largest establishments had been in the habit of doing extensive business with a house in the centre of Germany; but, on the closing of the continental ports against our manufactures, heavy penalties were inflicted on all those who contravened the Berlin and Milan decrees. The English manufacturer continued, nevertheless, to receive orders, with directions how to consign them, and appointments for the time and mode of payment, in letters, the handwriting of which was known to him, but which were never signed, except by the Christian name of one of the firm, and even in some instances they were without any signature at all. These orders were executed; and in no instance was there the least irregularity in the payments.

(275.) Another circumstance may be noticed, which to a small extent is more advantageous to large than to small factories. In the export of several articles of manufacture, a drawback is allowed by government, of a portion of the duty paid on the importation of the raw material. In such circumstances, certain forms must be gone through in order to protect the revenue from fraud; and a clerk, or one of the partners, must attend at the custom-house. The agent of the large establishment occupies nearly the same time in receiving a drawback of several thousands, as the smaller exporter does of a few shillings. But if the quantity exported is inconsiderable, the small manufacturer frequently does not find the drawback will repay him for the loss of time.

(276.) In many of the large establishments of our manufacturing districts, substances are employed which are the produce of remote countries, and which are, in several instances, almost peculiar to a few situations. The discovery of any new locality, where such articles exist in abundance, is a matter of great importance to any establishment which consumes them in large quantities; and it has been found, in some instances, that the expense of sending persons to great distances, purposely to discover and to collect such

* Lander's *Journal of an Expedition to the Mouth of the Niger*, vol. ii. p. 42.

produce, has been amply repaid. Thus it has happened, that the snowy mountains of Sweden and Norway, as well as the warmer hills of Corsica, have been almost stripped of one of their vegetable productions, by agents sent expressly from one of our largest establishments for the dying of calicos. Owing to the same command of capital, and to the scale upon which the operations of large factories are carried on, their returns admit of the expense of sending out agents to examine into the wants and tastes of distant countries, as well as of trying experiments, which, although profitable to them, would be ruinous to smaller establishments possessing more limited resources.

These opinions have been so well expressed in the Report of the Committee of the House of Commons on the Woollen Trade, in 1806, that we shall close this chapter with an extract, in which the advantages of great factories are summed up.

Your committee have the satisfaction of seeing, that the apprehensions entertained of factories are not only vicious in principle, but they are practically erroneous; to such a degree, that even the very opposite principles might be reasonably entertained. Nor would it be difficult to prove, that the factories, to a certain extent at least, and in the present day, seem absolutely necessary to the well-being of the domestic system; supplying those very particulars wherein the domestic system must be acknowledged to be inherently defective: for it is obvious, that the little master manufacturers cannot afford, like the man who possesses considerable capital, to try the experiments which are requiste, and incur the risks, and even losses, which almost always occur, in inventing and perfecting new articles of manufacture, or in carrying to a state of greater perfection articles already established. He cannot learn, by personal inspection, the wants and habits, the arts, manufactures, and improvements of foreign countries diligence, economy, and prudence, are the requisites of his character, not invention, taste, and enterprise; nor would he be warranted in hazarding the loss of any part of his small capital. He walks in a sure road as long as he treads in the beaten track; but he must not deviate into the paths of speculation. The owner of a factory, on the contrary, being commonly possessed of a large capital, and having all his workmen employed under his own immediate superintendence, may make experiments, hazard speculation, invent shorter or better modes of performing old processes, may introduce new articles, and improve and perfect old ones, thus giving the range to his taste and fancy, and, thereby alone enabling our manufacturers to stand the competition with their commercial rivals in other countries. Meanwhile, as is well worthy of remark (and experience abundantly warrants the assertion), many of these new fabrics and inventions, when their success is once established, become general among the whole body of manufacturers; the domestic manufacturers themselves thus benefiting, in the end, from those very factories which had been at first the objects of their jealousy. The history of almost all our other manufactures, in which great improvements have been made of late years, in some cases at an immense expense, and after numbers of unsuccessful experiments, strikingly illustrates and enforces the above remarks. It is besides an acknowledged fact, that the owners of factories are often amongst the most extensive purchasers at the halls, where they buy from the domestic clothier the established articles of manufacture, or are able at once to answer a great and sudden order; while, at home, and under their own superintendence, they make their fancy goods, and any articles of a newer, more costly, or more delicate quality, to which they are enabled by the domestic system to apply a much larger proportion of their capital. Thus, the two systems, instead of rivalling, are mutual aids to each other; each supplying the other's defects, and promoting the other's prosperity.

ON THE POSITION OF LARGE FACTORIES

(277.) IT is found in every country, that the situation of large manufacturing establishments is confined to particular districts. In the earlier history of a manufacturing community, before cheap modes of transport have been extensively introduced, it will almost always be found that manufactories are placed near those spots in which nature has produced the raw material: especially in the case of articles of great weight, and in those the value of which depends more upon the material than upon the labour expended on it. Most of the metallic ores being exceedingly heavy, and being mixed up with large quantities of weighty and useless materials, must be smelted at no great distance from the spot which affords them: fuel and power are the requisites for reducing them; and any considerable fall of water in the vicinity will naturally be resorted to for aid in the coarser exertions of physical force; for pounding the ore, for blowing the furnaces, or for hammering and rolling out the iron. There are indeed peculiar circumstances which will modify this. Iron, coal, and limestone, commonly occur in the same tracts; but the union of the fuel in the same locality with the ore does not exist with respect to other metals. The tracts generally the most productive of metallic ores are, geologically speaking, different from those affording coal: thus in Cornwall there are veins of copper and of tin, but no beds of coal. The copper ore, which requires a very large quantity of fuel for its reduction, is sent by sea to the coal-fields of Wales, and is smelted at Swansea; whilst the vessels which convey it, take back coals to work the steam-engines for draining the mines, and to smelt the tin, which requires for that purpose a much smaller quantity of fuel than copper.

(278.) Rivers passing through districts rich in coal and metals, will form the first high roads for the conveyance of weighty produce to stations in which other conveniences present themselves for the further application of human skill. Canals will succeed, or lend their aid to these; and the yet unexhausted applications of steam and of gas, hold out a hope of attaining almost the same advantages for countries to which nature seemed for ever to have denied them. Manufactures, commerce, and civilization, always follow the line of new and cheap communications. Twenty years ago, the Mississippi poured the vast volume of its waters in lavish profusion through thousands of miles of countries, which scarcely supported a few wandering and uncivilized tribes of Indians. The power of the stream seemed to set at defiance the efforts of man to ascend its course; and, as if to render the task still more hopeless, large trees, torn from the surrounding forests, were planted like stakes in its bottom, forming in some places barriers, in others the nucleus of banks; and accumulating in the same spot, which but for accident would have been free from both, the difficulties and dangers of shoals and of rocks. Four months of incessant toil could scarcely convey a small bark with its worn-out crew two thousand miles up this stream. The same voyage is now performed in fifteen days by large vessels impelled by steam, carrying hundreds of passengers enjoying all the comforts and luxuries of civilized life. Instead of the hut of the Indian,—and the far more unfrequent log-house of the thinly scattered settlers,—villages, towns, and cities, have arisen on its banks; and the same engine which stems the force of these powerful waters, will probably tear from their bottom the obstructions which have hitherto impeded and rendered dangerous their navigation.*

* The amount of obstructions arising from the casual fixing of trees in the bottom of the river, may be estimated from the proportion of steam-boats destroyed by running upon them. The subjoined statement is taken from the American Almanack for 1832:—

Between the years 1811 and 1831, three hundred and forty-eight steam-boats were built on the Mississippi and its tributary streams. During that period a hundred and fifty were lost or worn out.

Of this hundred and fifty

worn out	63
lost by snags	36
burnt	14
lost by collision	3
by accidents not ascertained.	34

Thirty-six, or nearly one fourth, being destroyed by accidental obstructions.

Snag is the name given in America to trees which stand nearly upright in the stream, with their roots fixed at the bottom.

It is usual to divide off at the bow of the steam-boats a water-tight chamber, in order that when a hole is made in it by running against the snags, the water may not enter the rest of the vessel and sink it instantly.

(279.) The accumulation of many large manufacturing establishments in the same district has a tendency to bring together purchasers or their agents from great distances, and thus to cause the institution of a public mart or exchange. This contributes to diffuse information relative to the supply of raw materials, and the state of demand for their produce, with which it is necessary manufacturers should be well acquainted. The very circumstance of collecting periodically, at one place, a large number both of those who supply the market and of those who require its produce, tends strongly to check the accidental fluctuations to which a small market is always subject, as well as to render the average of the prices much more uniform.

(280.) When capital has been invested in machinery, and in buildings for its accommodation, and when the inhabitants of the neighbourhood have acquired a knowledge of the modes of working at the machines, reasons of considerable weight are required to cause their removal. Such changes of position do however occur; and they have been alluded to by the Committee on the Fluctuation of Manufactures' Employment, as one of the causes interfering most materially with an uniform rate of wages: it is therefore of particular importance to the workmen to be acquainted with the real causes which have driven manufactures from their ancient seats.

> The migration or change of place of any manufacture has sometimes arisen from improvements of machinery not applicable to the spot where such manufacture was carried on, as appears to have been the case with the woollen manufacture, which has in great measure migrated from Essex, Suffolk, and other southern counties, to the northern districts, where coal for the use of the steam-engine is much cheaper. But this change has, in some instances, been caused or accelerated by the conduct of the workmen, in refusing a reasonable reduction of wages, or opposing the introduction of some kind of improved machinery or process; so that, during the dispute, another spot has in great measure supplied their place in the market. *Any violence used by the workmen against the property of their masters, and any unreasonable combination on their part, is almost sure thus to be injurious to themselves.**

(281.) These removals become of serious consequence when the factories have been long established, because a population commensurate with their wants invariably grows up around them. The combinations in Nottingham-shire, of persons under the name of Luddites, drove a great number of lace-frames from that district, and caused establishments to be formed in

---

* This passage is *not* printed in Italics in the original, but it has been thus marked in the above extract, from its importance, and from the conviction that the most extended discussion will afford additional evidence of its truth.

Devonshire. We ought also to observe, that the effect of driving any establishment into a new district, where similar works have not previously existed, is not merely to place it out of the reach of such combinations; but, after a few years, the example of its success will most probably induce other capitalists in the new district to engage in the same manufacture: and thus, although one establishment only should be driven away, the workmen, through whose combination its removal is effected, will not merely suffer by the loss of that portion of demand for their labour which the factory caused; but the value of that labour will itself be reduced by the competition of a new field of production.

(282.) Another circumstance which has its influence on this question, is the nature of the machinery. Heavy machinery, such as stamping-mills, steam-engines, &c., cannot readily be moved, and must always be taken to pieces for that purpose; but when the machinery of a factory consists of a multitude of separate engines, each complete in itself, and all put in motion by one source of power, such as that of steam, then the removal is much less inconvenient. Thus, stocking-frames, lace-machines, and looms, can be transported to more favourable positions, with but a small separation of their parts.

(283.) It is of great importance that the more intelligent amongst the class of workmen should examine into the correctness of these views; because, without having their attention directed to them, the whole class may, in some instances, be led by designing persons to pursue a course, which, although plausible in appearance, is in reality at variance with their own best interests. I confess I am not without a hope that this volume may fall into the hands of workmen, perhaps better qualified than myself to reason upon a subject which requires only plain common sense, and whose powers are sharpened by its importance to their personal happiness. In asking their attention to the preceding remarks, and to those which I shall offer respecting combinations, I can claim only one advantage over them; namely, that I never have had, and in all human probability never shall have, the slightest pecuniary interest, to influence even remotely, or by anticipation, the judgments I have formed on the facts which have come before me.

ON OVER-MANUFACTURING

(284.) ONE of the natural and almost inevitable consequences of competition is the production of a supply much larger than the demand requires. This result usually arises periodically; and it is equally important, both to the masters and to the workmen, to prevent its occurrence, or to foresee its arrival.

In situations where a great number of very small capitalists exist,—where each master works himself and is assisted by his own family, or by a few journeymen,—and where a variety of different articles is produced, a curious system of compensation has arisen which in some measure diminishes the extent to which fluctuations of wages would otherwise reach. This is accomplished by a species of middle-men or factors, persons possessing some capital, who, whenever the price of any of the articles in which they deal is greatly reduced, purchase it on their own account, in the hopes of selling at a profit when the market is better. These persons, in ordinary times, act as salesmen or agents, and make up assortments of goods at the market price, for the use of the home or foreign dealer. They possess large warehouses in which to make up their orders, or keep in store articles purchased during periods of depression; thus acting as a kind of fly-wheel in equalizing the market price.

(285.) The effect of over-manufacturing upon great establishments is different. When an over supply has reduced prices, one of two events usually occurs: the first is a diminished payment for labour; the other is a diminution of the number of hours during which the labourers work, together with a diminished rate of wages. In the former case production continues to go on at its ordinary rate: in the latter, the production itself being checked, the supply again adjusts itself to the demand as soon as the stock on hand is worked off, and prices then regain their former level. The latter course appears, in the first instance, to be the best both for masters and men; but there seems to be a difficulty in accomplishing this, except where the trade is in few hands. In fact, it is almost necessary, for its success, that there should be a combination amongst the masters or amongst the men; or, what is always far preferable to either, a mutual agreement for their joint interests. Combination among the men is difficult, and is always attended with the evils which arise from the ill-will excited against any persons who, in the perfectly justifiable exercise of their judgement, are disposed not to act with the majority. The combination of the masters, on the other hand, is unavailing, unless the whole body of them agree: for if any one master can procure more labour for his money than the rest, he will be able to undersell them.

(286.) If we look only at the interests of the consumer, the case is different. When too large a supply has produced a great reduction of price, it opens the consumption of the article to a new class, and increases the consumption of those who previously employed it: it is therefore against the interest of both these parties that a return to the former price should occur. It is also certain, that by the diminution of profit which the manufacturer suffers from the diminished price, his ingenuity will be additionally stimulated;—that he will apply himself to discover other and cheaper sources for the supply of his raw material,—that he will endeavour to contrive improved machinery which shall manufacture it at a cheaper rate,—or try to introduce new arrangements into his factory, which shall render the economy of it more perfect. In the event of his success, by any of these courses or by their joint effects, a real and substantial good will be produced. A larger portion of the public will receive advantage from the use of the article, and they will procure it at a lower price; and the manufacturer, though his profit on each operation is reduced, will yet, by the more frequent returns on the larger produce of his factory, find his real gain at the end of the year, nearly the same as it was before; whilst the wages of the workman will return to their level, and both the manufacturer and the workman will find the demand less fluctuating, from its being dependent on a larger number of customers.

(287.) It would be highly interesting, if we could trace, even approximately, through the history of any great manufacture, the effects of gluts in producing improvements in machinery, or in methods of working; and if we could shew what addition to the annual quantity of goods previously manufactured, was produced by each alteration. It would probably be found, that *the increased quantity manufactured by the same capital, when worked with the new improvement, would produce nearly the same rate of profit as other modes of investment.*

Perhaps the manufacture of iron* would furnish the best illustration of this subject; because, by having the actual price of pig and bar iron at the same place and at the same time, the effect of a change in the value of currency, as well as several other sources of irregularity, would be removed.

(288.) At the present moment, whilst the manufacturers of iron are complaining of the ruinously low price of their produce, a new mode of smelting iron is coming into use, which, if it realizes the statement of the patentees, promises to reduce greatly the cost of production. The improvement consists in heating the air previously to employing it for blowing the furnace. One of the results is, that coal may be used instead of coke; and this, in its turn, diminishes the quantity of limestone which is required for the fusion of the iron stone.

The following statement by the proprietors of the patent is extracted from Brewster's Journal, 1832, p. 349:

---

* The average price per ton of pig-iron, bar-iron, and coal, together with the price paid for labour at the works, for a long series of years, would be very valuable, and I shall feel much indebted to any one who will favour me with it for any, even short, period.

Comparative view of the quantity of materials required at the Clyde Iron Works to smelt a ton of foundry pig-iron, and of the quantity of foundry pig-iron smelted from each furnace weekly.

	Fuel in tons of 20 cwt. each cwt. 112 lbs Tons.	Iron-stone	Lime-stone Cwt.	Weekly produce in pig-iron Tons.
1. With air not heated and coke	7	$3\frac{1}{4}$	15	45
2. With air heated and coke	$4\frac{3}{4}$	$3\frac{1}{4}$	10	60
3. With air heated and coals not coked	$2\frac{1}{4}$	$3\frac{1}{4}$	$7\frac{1}{2}$	65

Notes.—1*st*. To the coals stated in the second and third lines, must be added 5 cwt. of small coals, required to heat the air.

2*d*. The expense of the *apparatus* for applying the heated air will be from 200*l*. to 300*l*. per furnace.

3*d*. No coals are now coked at the Clyde Iron Works; at all the three furnaces the iron is smelted with coals.

4*th*. The three furnaces are blown by a double-powered steam engine, with a steam cylinder 40 inches in diameter, and a blowing cylinder 80 inches in diameter, which compresses the air so as to carry $2\frac{1}{2}$ lbs. per square inch. There are two tuyeres to each furnace. The muzzles of the blow-pipes are 3 inches in diameter.

5*th*. The air heated to upwards of 600° of Fahrenheit. It will melt lead at the distance of three inches from the orifice through which it issues from the pipe.

(289.) The increased effect produced by thus heating the air is by no means an obvious result; and an analysis of its action will lead to some curious views respecting the future application of machinery for blowing furnaces.

Every cubic foot of atmospheric air, driven into a furnace, consists of two gases*; about one-fifth being oxygen, and four-fifths azote.

According to the present state of chemical knowlege, the oxygen alone is effective in producing heat; and the operation of blowing a furnace may be thus analyzed.

1. The air is forced into the furnace in a condensed state, and, immediately expanding, abstracts heat from the surrounding bodies.
2. Being itself of moderate temperature, it would, even without expansion,

---

* The accurate proportions are, by measure, oxygen 21, azote 79.

still require heat to raise it to the temperature of the hot substances to which it is to be applied.

3. On coming into contact with the ignited substances in the furnace, the oxygen unites with them, parting at the same moment with a large portion of its latent heat, and forming compounds which have less specific heat than their separate constituents. Some of these pass up the chimney in a gaseous state, whilst others remain in the form of melted slags, floating on the surface of the iron, which is fused by the heat thus set at liberty.

4. The effects of the azote are precisely similar to the first and second of those above described; it seems to form no combinations, and contributes nothing, in any stage, to augment the heat.

The plan, therefore, of heating the air before driving it into the furnace saves, obviously, the whole of that heat which the fuel must have supplied in raising it from the temperature of the external air up to that of 600° Fahrenheit; thus rendering the fire more intense, and the glassy slags more fusible, and perhaps also more effectually decomposing the iron ore. The same quantity of fuel, applied at once to the furnace, would only prolong the duration of its heat, not augment its intensity.

(290.) The circumstance of so large a portion of the air* driven into furnaces being not merely useless but acting really as a cooling, instead of a heating, cause, added to so great a waste of mechanical power in condensing it, amounting, in fact, to four-fifths of the whole, clearly shews the defects of the present method, and the want of some better mode of exciting combustion on a large scale. The following suggestions are thrown out as likely to lead to valuable results, even though they should prove ineffectual for their professed object.

(291.) The great difficulty appears to be to separate the oxygen, which aids combustion, from the azote which impedes it. If either of those gases becomes liquid at a lower pressure than the other, and if those pressures are within the limits of our present powers of compression, the object might be accomplished.

Let us assume, for example, that oxygen becomes liquid under a pressure of 200 atmospheres, whilst azote requires a pressure of 250. Then if atmospheric air be condensed to the two hundredth part of its bulk, the oxygen will be

---

* A similar reasoning may be applied to lamps. An argand burner, whether used for consuming oil or gas, admits almost an unlimited quantity of air. It would deserve inquiry, whether a smaller quantity might not produce greater light; and, possibly, a different supply furnish more heat with the same expenditure of fuel.

found in a liquid state at the bottom of the vessel in which the condensation is effected, and the upper part of the vessel will contain only azote in the state of gas. The oxygen, now liquified, may be drawn off for the supply of the furnace; but as it ought, when used, to have a very moderate degree of condensation, its expansive force may be previously employed in working a small engine. The compressed azote also in the upper part of the vessel, though useless for combustion, may be employed as a source of power, and, by its expansion, work another engine. By these means the mechanical force exerted in the original compression would all be restored, except that small part retained for forcing the pure oxygen into the furnace, and the much larger part lost in the friction of the apparatus.

(292.) The principal difficulty to be apprehended in these operations is that of *packing* a working piston, so as to bear the pressure of 200 or 300 atmospheres: but this does not seem insurmountable. It is possible also that the chemical combination of the two gases which constitute common air may be effected by such pressures: if this should be the case, it might offer a new mode of manufacturing nitrous or nitric acids. The result of such experiments might take another direction: if the condensation were performed over liquids, it is possible that they might enter into new chemical combinations. Thus, if air were highly condensed in a vessel containing water, the latter might unite with an additional dose of oxygen,* which might afterwards be easily disengaged for the use of the furnace.

(293.) A farther cause of the uncertainty of the results of such an experiment arises from the possibility that azote may really contribute to the fusion of the mixed mass in the furnace, though its mode of operating is at present unknown. An examination of the nature of the gases issuing from the chimneys of iron-foundries, might perhaps assist in clearing up this point; and, in fact, if such inquiries were also instituted upon the various products of all furnaces, we might expect the elucidation of many points in the economy of the metallurgic art.

(294.) It is very possible also, that the action of oxygen in a liquid state might be exceedingly corrosive, and that the containing vessels must be lined with platinum or some other substance of very difficult oxydation; and most probably new and unexpected compounds would be formed at such pressures. In some experiments made by Count Rumford in 1797, on the force of fired gunpowder, he noticed a solid compound, which always appeared in the

---

* Deutoxide of hydrogen, the oxygenated water of Thenard.

gunbarrel when the ignited powder had no means of escaping; and, in those cases, the gas which escaped on removing the restraining pressure was usually inconsiderable.

(295.) If the liquefied gases are used, the form of the iron furnace must probably be changed, and perhaps it may be necessary to direct the flame from the ignited fuel upon the ore to be fused, instead of mixing that ore with the fuel itself: by a proper regulation of the blast, an oxygenating or a deoxygenating flame might be procured; and from the intensity of the flame, combined with its chemical agency, we might expect the most refractory ore to be smelted, and that ultimately the metals at present almost infusible, such as platinum, titanium, and others, might be brought into common use, and thus effect a revolution in the arts.

(296.) Supposing, on the occurrence of a glut, that new and cheaper modes of producing are not discovered, and that the production continues to exceed the demand, then it is apparent that too much capital is employed in the trade; and after a time, the diminished rate of profit will drive some of the manufacturers to other occupations. What particular individuals will leave it must depend on a variety of circumstances. Superior industry and attention will enable some factories to make a profit rather beyond the rest; superior capital in others will enable them, without these advantages, to support competition longer, even at a loss, with the hope of driving the smaller capitalists out of the market, and then reimbursing themselves by an advanced price. It is, however, better for all parties, that this contest should not last long; and it is important, that no artificial restraint should interfere to prevent it. An instance of such restriction, and of its injurious effect, occurs at the port of Newcastle, where a particular act of parliament requires that every ship shall be loaded in its turn. The Committee of the House of Commons, in their Report on the Coal Trade, state that,

> Under the regulations contained in this act, if more ships enter into the trade than can be profitably employed in it, the loss produced by detention in port, and waiting for a cargo, which must consequently take place, instead of falling, as it naturally would, upon particular ships, and forcing them from the trade, is now divided evenly amongst them; and the loss thus created is shared by the whole number.—*Report*, p. 6.

(297.) It is not pretended, in this short view, to trace out all the effects or remedies of over-manufacturing; the subject is difficult, and, unlike some of the questions already treated, requires a combined view of the relative influence of many concurring causes.

INQUIRIES PREVIOUS TO COMMENCING
ANY MANUFACTORY

(298.) THERE are many inquiries which ought always to be made previous to the commencement of the manufacture of any new article. These chiefly relate to the expense of tools, machinery, raw materials, and all the outgoings necessary for its production,—to the extent of demand which is likely to arise,—to the time in which the circulating capital will be replaced,—and to the quickness or slowness with which the new article will supersede those already in use.

(299.) The expense of tools and of new machines will be more difficult to ascertain, in proportion as they differ from those already employed; but the variety in constant use in our various manufactories, is such, that few inventions now occur in which considerable resemblance may not be traced to others already constructed. The cost of the raw material is usually less difficult to determine; but cases occasionally arise in which it becomes important to examine whether the supply, at the given price, can be depended upon: for, in the case of a small consumption, the additional demand arising from a factory may produce a considerable temporary rise, though it may ultimately reduce the price.

(300.) The quantity of any new article likely to be consumed is a most important subject for the consideration of the projector of a new manufacture. As these pages are not intended for the instruction of the manufacturer, but rather for the purpose of giving a general view of the subject, an illustration of the way in which such questions are regarded by practical men, will, perhaps, be most instructive. The following extract from the evidence given before a Committee of the House of Commons, in the Report on Artizans and Machinery, shews the extent to which articles apparently the most insignificant, are consumed, and the view which the manufacturer takes of them.

The person examined on this occasion was Mr. Ostler, a manufacturer of glass beads and other toys of the same substance, from Birmingham. Several of the articles made by him were placed upon the table, for the inspection of the Committee of the House of Commons, which held its meetings in one of the committee-rooms.

Question. Is there any thing else you have to state upon this subject?
Answer. Gentlemen may consider the articles on the table as extremely insignificant; but perhaps I may surprise them a little, by mentioning the following fact. Eighteen years ago, on my first journey to London, a respectable-looking man, in

the city, asked me if I could supply him with dolls' eyes; and I was foolish enough to feel half offended; I thought it derogatory to my new dignity as a manufacturer, to make dolls' eyes. He took me into a room quite as wide, and perhaps twice the length of this, and we had just room to walk between stacks, from the floor to the ceiling, of parts of dolls. He said, 'These are only the legs and arms; the trunks are below.' But I saw enough to convince me, that he wanted a great many eyes; and, as the article appeared quite in my own line of business, I said I would take an order by way of experiment; and he shewed me several specimens. I copied the order. He ordered various quantities, and of various sizes and qualities. On returning to the Tavistock hotel, I found that the order amounted to upwards of 500*l*. I went into the country, and endeavoured to make them. I had some of the most ingenious glass toy-makers in the kingdom in my service; but when I shewed it to them, they shook their heads, and said they had often seen the article before, but could not make it. I engaged them by presents to use their best exertions; but after trying and wasting a great deal of time for three or four weeks, I was obliged to relinquish the attempt. Soon afterwards I engaged in another branch of business (chandelier furniture), and took no more notice of it. About eighteen months ago I resumed the trinket trade, and then determined to think of the dolls' eyes; and about eight months since, I accidentally met with a poor fellow who had impoverished himself by drinking, and who was dying in a consumption, in a state of great want. I showed him ten sovereigns; and he said he would instruct me in the process. He was in such a state that he could not bear the effluvia of his own lamp; but though I was very conversant with the manual part of the business, and it related to things I was daily in the habit of seeing, I felt I could do nothing from his description (I mention this to show how difficult it is to convey, by description, the mode of working.) He took me into his garret, where the poor fellow had economized to such a degree, that he actually used the entrails and fat of poultry from Leadenhall market to save oil (the price of the article having been lately so much reduced by competition at home.) In an instant, before I had seen him make three, I felt competent to make a gross; and the difference between his mode and that of my own workmen was so trifling, that I felt the utmost astonishment.

Quest. You can now make dolls' eyes?

Ans. I can. As it was eighteen years ago that I received the order I have mentioned, and feeling doubtful of my own recollection, though very strong, and suspecting that it could [not] have been to the amount stated, I last night took the present very reduced price of that article (less than half now of what it was then), and calculating that every child in this country not using a doll till two years old, and throwing it aside at seven, and having a new one annually, I satisfied myself that the eyes alone would produce a circulation of a great many thousand pounds. I mention this merely to shew the importance of trifles; and to assign one reason, amongst many, for my conviction, that nothing but personal communication can enable our manufactures to be transplanted.

(301.) In many instances it is exceedingly difficult to estimate beforehand the sale of an article, or the effects of a machine; a case, however, occurred during a recent inquiry, which although not quite appropriate as an illustration of probable demand, is highly instructive as to the mode of conducting investigations of this nature. A committee of the House of Commons was appointed to inquire into the tolls proper to be placed on steam-carriages; a question, apparently, of difficult solution, and upon which widely different opinions had been formed, if we may judge by the very different rate of tolls imposed upon such carriages by different 'turnpike trusts.' The principles on which the committee conducted the inquiry were, that 'The only ground on which a fair claim to toll can be made on any public road, is to raise a fund, which, with the strictest economy, shall be just sufficient,—first, to repay the expense of its original formation;—secondly, to maintain it in good and sufficient repair.' They first endeavoured to ascertain, from competent persons, the effect of the atmosphere alone in deteriorating a well-constructed road. The next step was, to determine the proportion in which the road was injured, by the effect of the horses' feet compared with that of the wheels. Mr. Macneill, the superintendent, under Mr. Telford, of the Holyhead roads, was examined, and proposed to estimate the relative injury, from the comparative quantities of iron worn off from the shoes of the horses, and from the tire of the wheels. From the data he possessed, respecting the consumption of iron for the tire of the wheels, and for the shoes of the horses, of one of the Birmingham day-coaches, he estimated the wear and tear of roads, arising from the feet of the horses, to be three times as great as that arising from the wheels. Supposing repairs amounting to a hundred pounds to be required on a road travelled over by a fast coach at the rate of ten miles an hour, and the same amount of injury to occur on another road, used only by waggons, moving at the rate of three miles an hour, Mr. Macneill divides the injuries in the following proportions:—

Injury arising from	Fast Coach.	Heavy Waggon.
Atmospheric changes	20	20
Wheels	20	35.5
Horses' feet drawing	60	44.5
Total injury	100	100

Supposing it, therefore, to be ascertained that the wheels of steam-carriages do no more injury to roads than other carriages of equal weight travelling with the same velocity, the committee now possessed the means of approximating to a just rate of toll for steam-carriages.*

(302.) As connected with this subject, and as affording most valuable information upon points in which, previous to experiment, widely different opinions have been entertained; the following extract is inserted from Mr. Telford's Report on the State of the Holyhead and Liverpool Roads. The instrument employed for the comparison was invented by Mr. Macneill; and the road between London and Shrewsbury was selected for the place of experiment.

The general results, when a waggon weighing 21 cwt. was used on different sorts of roads, are as follows:—

		lbs.
1.	On well-made pavement, the draught is	33
2.	On a broken stone surface, or old flint road	65
3.	On a gravel road	147
4.	On a broken stone road, upon a rough pavement foundation	46
5.	On a broken stone surface, upon a bottoming of concrete, formed of Parker's cement and gravel	46

The following statement relates to the force required to draw a coach weighing 18 cwt., exclusive of seven passengers, up roads of various inclinations:

Inclination.	Force required at six miles per hour	Force at eight miles per hour	Force at ten miles per hour
	lbs.	*lbs.*	*lbs.*
1 in 20	268	296	318
1 in 26	213	219	225
1 in 30	165	196	200
1 in 40	160	166	172
1 in 600	111	120	128

(303.) In establishing a new manufactory, the time in which the goods produced can be brought to market and the returns be realized, should be

---

* One of the results of these inquiries is, that every coach which travels from London to Birmingham distributes about eleven pounds of wrought iron, along the line of road between those two places.

thoroughly considered, as well as the time the new article will take to supersede those already in use. If it is destroyed in using, the new produce will be much more easily introduced. Steel pens readily took the place of quills; and a new form of pen would, if it possessed any advantage, as easily supersede the present one. A new lock, however secure, and however cheap, would not so readily make its way. If less expensive than the old, it would be employed in new work: but old locks would rarely be removed to make way for it; and even if perfectly secure, its advance would be slow.

(304.) Another element in this question which should not be altogether omitted, is the opposition which the new manufacture may create by its real or apparent injury to other interests, and the probable effect of that opposition. This is not always foreseen; and when anticipated is often inaccurately estimated. On the first establishment of steam-boats from London to Margate, the proprietors of the coaches running on that line of road petitioned the House of Commons against them, as likely to lead to the ruin of the coach proprietors. It was, however, found that the fear was imaginary; and in a very few years, the number of coaches on that road was considerably increased, apparently through the very means which were thought to be adverse to it. The fear, which is now entertained, that steam-power and rail-roads may drive out of employment a large proportion of the horses at present in use, is probably not less unfounded. On some particular lines such an effect might be produced; but in all probability the number of horses employed in conveying goods and passengers to the great lines of rail-road, would exceed that which is at present used.

ON A NEW SYSTEM OF MANUFACTURING

(305.) A MOST erroneous and unfortunate opinion prevails amongst work-men in many manufacturing countries, that their own interest and that of their employers are at variance. The consequences are,—that valuable machinery is sometimes neglected, and even privately injured,—that new improve-ments, introduced by the masters, do not receive a fair trial,—and that the talents and observations of the workmen are not directed to the improvement of the processes in which they are employed. This error is, perhaps, most prevalent where the establishment of manufactories has been of recent origin, and where the number of persons employed in them is not very large: thus, in some of the Prussian provinces on the Rhine it prevails to a much greater extent than in Lancashire. Perhaps its diminished prevalence in our own manufacturing districts, arises partly from the superior information spread amongst the workmen; and partly from the frequent example of persons, who by good conduct and an attention to the interests of their employers for a series of years, have become foremen, or who have ultimately been admitted into advantageous partnerships. Convinced as I am, from my own observation, that the prosperity and success of the master manufacturer is essential to the welfare of the workman, I am yet compelled to admit that this connexion is, in many cases, too remote to be always understood by the latter: and whilst it is perfectly true that workmen, as a class, derive advantage from the prosperity of their employers, I do not think that each individual partakes of that advantage exactly in proportion to the extent to which he contributes towards it; nor do I perceive that the resulting advantage is as immediate as it might become under a different system.

(306.) It would be of great importance, if, in every large establishment the mode of payment could be so arranged, that every person employed should derive advantage from the success of the whole; and that the profits of each individual should advance, as the factory itself produced profit, without the necessity of making any change in the wages. This is by no means easy to effect, particularly amongst that class whose daily labour procures for them their daily food. The system which has long been pursued in working the Cornish mines, although not exactly fulfilling these conditions, yet possesses advantages which make it worthy of attention, as having nearly approached towards them, and as tending to render fully effective the faculties of all who are engaged in it. I am the more strongly induced to place before the reader a short sketch of this system, because its similarity to that which I shall afterwards recommend for trial, will perhaps remove some objections to the latter, and may also furnish some valuable hints for conducting any experiment which might be undertaken.

(307.) In the mines of Cornwall, almost the whole of the operations, both above and below ground, are contracted for. The manner of making the contract is nearly as follows. At the end of every two months, the work which it is proposed to carry on during the next period is marked out. It is of three kinds. 1. *Tutwork*, which consists in sinking shafts, driving levels, and making excavations: this is paid for by the fathom in depth, or in length, or by the cubic fathom. 2. *Tribute*, which is payment for raising and dressing the ore, by means of a certain part of its value when rendered merchantable. It is this mode of payment which produces such admirable effects. The miners, who are to be paid in proportion to the richness of the vein, and the quantity of metal extracted from it, naturally become quick-sighted in the discovery of ore, and in estimating its value; and it is their interest to avail themselves of every improvement that can bring it more cheaply to market. 3. *Dressing.* The 'Tributors,' who dig and dress the ore, can seldom afford to dress the coarser parts of what they raise, at their contract price; this portion, therefore, is again let out to other persons, who agree to dress it at an advanced price.

The lots of ore to be dressed, and the works to be carried on, having been marked out some days before, and having been examined by the men, a kind of auction is held by the captains of the mine, in which each lot is put up, and bid for by different gangs of men. The work is then offered, at a price usually below that bid at the auction, to the lowest bidder, who rarely declines it at the rate proposed. The *tribute* is a certain sum out of every twenty shillings' worth of ore raised, and may vary from threepence to fourteen or fifteen shillings. The rate of earnings in tribute is very uncertain: if a vein, which was poor when taken, becomes rich, the men earn money rapidly; and instances have occurred in which each miner of a gang has gained a hundred pounds in the two months. These extraordinary cases, are, perhaps, of more advantage to the owners of the mine than even to the men; for whilst the skill and industry of the workmen are greatly stimulated, the owner himself always derives still greater advantage from the improvement of the vein.* This system has been introduced, by Mr. Taylor, into the lead mines of Flintshire, into those at Skipton in Yorkshire, and into some of the copper mines of Cumberland; and it is desirable that it should become general, because no other mode of payment affords to the workmen a measure of success so directly proportioned to the industry, the integrity, and the talent, which they exert.

(308.) I shall now present the outline of a system which appears to me to be pregnant with the most important results, both to the class of workmen and to the country at large; and which, if acted upon, would, in my opinion, permanently raise the working classes, and greatly extend the manufacturing system.

The general principles on which the proposed system is founded, are—

1st. *That a considerable part of the wages received by each person employed should depend on the profits made by the establishment; and,*

2d. *That every person connected with it should derive more advantage from applying any improvement he might discover, to the factory in which he is employed, than he could by any other course.*

---

* For a detailed account of the method of working the Cornish mines, see a paper of Mr. John Taylor's, *Transactions of the Geological Society*, vol. ii. p. 309.

(309.) It would be difficult to prevail on the large capitalist to enter upon any system, which would change the division of the profits arising from the employment of his capital in setting skill and labour in action; any alteration, therefore, must be expected rather from the small capitalist, or from the higher class of workmen, who combine the two characters; and to these latter classes, whose welfare will be first affected, the change is most important. I shall therefore first point out the course to be pursued in making the experiment; and then, taking a particular branch of trade as an illustration, I shall examine the merits and defects of the proposed system as applied to it.

(310.) Let us suppose, in some large manufacturing town, ten or tewlve of the most intelligent and skilful workmen to unite, whose characters for sobriety and steadiness are good, and are well known among their own class. Such persons will each possess some small portion of capital; and let them join with one or two others who have raised themselves into the class of small master manufacturers, and, therefore possess rather a larger portion of capital. Let these persons, after well considering the subject, agree to establish a manufactory of fire-irons and fenders; and let us suppose that each of the ten workmen can command forty pounds, and each of the small capitalists possesses two hundred pounds: thus they have a capital of 800*l.* with which to commence business; and, for the sake of simplifying, let us further suppose the labour of each of these twelve persons to be worth two pounds a week. One portion of their capital will be expended in procuring the tools necessary for their trade, which we shall take at 400*l.*, and this must be considered as their fixed capital. The remaining 400*l.* must be employed as circulating capital, in purchasing the iron with which their articles are made, in paying the rent of their workshops, and in supporting themselves and their families until some portion of it is replaced by the sale of the goods produced.

(311.) Now the first question to be settled is, what proportion of the profit should be allowed for the use of capital, and what for skill and labour? It does not seem possible to decide this question by any abstract reasoning: if the capital supplied by each partner is equal, all difficulty will be removed; if otherwise, the proportion must be left to find its level, and will be discovered by experience; and it is probable that it will not fluctuate much. Let us suppose it to be agreed that the capital of 800*l.* shall receive the wages of one workman. At the end of each week every workman is to receive one pound as wages, and one pound is to be divided amongst the owners of the capital. After a few weeks the returns will begin to come in; and they will soon become nearly uniform. Accurate accounts should be kept of every expense and of all the sales; and at the end of each week the profit should be divided. A certain portion should be laid aside as a reserved fund, another portion for repair of the tools, and the remainder being divided into thirteen parts, one of these parts would be divided amongst the capitalists and one belong to each workman. Thus each man would, in ordinary circumstances, make up his usual wages of two pounds weekly. If the factory went on prosperously, the wages of the men would increase; if the sales fell off they would be diminished. It is important that every person employed in the establishment, whatever might be the amount paid for his services, whether he act as labourer or porter, as the clerk who keeps the accounts, or as book-keeper employed for a few hours once a week to superintend them, should receive one half of what his service is worth in fixed salary, the other part varying with the success of the undertaking.

(312.) In such a factory, of course, division of labour would be introduced; some of the workmen would be constantly employed in forging the fire-irons, others in polishing them, others in piercing and forming the fenders. It would be essential that the time occupied in each process, and also its expense, should be well ascertained; information which would soon be obtained very precisely. Now, if a workman should find a mode of shortening any of the processes, he would confer a benefit on the whole party, even if they received but a small part of the resulting profit. For the promotion of such discoveries, it would be desirable that those who make them should either receive some reward, to be determined after a sufficient trial by a committee assembling periodically; or if they be of high importance, that the discoverer should receive one-half, or two-thirds, of the profit resulting from them during the next year, or some other determinate period, as might be found expedient. As the advantages of such improvements would be clear gain to the factory, it is obvious that such a share might be allowed to the inventor, that it would be for his interest rather to give the benefit of them to his partners, than to dispose of them in any other way.

(313.) The result of such arrangements in a factory would be,

1. That every person engaged in it would have a *direct* interest in its prosperity; since the effect of any success, or falling off, would almost immediately produce a corresponding change in his own weekly receipts.
2. Every person concerned in the factory would have an immediate interest in preventing any waste or mismanagement in all the departments.
3. The talents of all connected with it would be strongly directed to its improvement in every department.
4. None but workmen of high character and qualifications could obtain admission into such establishments; because when any additional hands

were required, it would be the common interest of all to admit only the most respectable and skilful; and it would be far less easy to impose upon a dozen workmen than upon the single proprietor of a factory.

5.   When any circumstance produced a glut in the market, more skill would be directed to diminishing the cost of production; and a portion of the time of the men might then be occupied in repairing and improving their tools, for which a reserved fund would pay, thus checking present, and at the same time facilitating future production.

6.   Another advantage, of no small importance, would be the total removal of all real or imaginary causes for combinations. The workmen and the capitalist would so shade into each other,—would so *evidently* have a common interest, and their difficulties and distresses would be mutually so well understood, that, instead of combining to oppress one another, the only combination which could exist would be a most powerful union *between* both parties to overcome their common difficulties.

(314.) One of the difficulties attending such a system is, that capitalists would at first fear to embark in it, imagining that the workmen would receive too large a share of the profits: and it is quite true that the workmen would have a larger share than at present: but, at the same time, it is presumed the effect of the whole system would be, that the total profits of the establishment being much increased, the smaller proportion allowed to capital under this system would yet be greater in actual amount, than that which results to it from the larger share in the system now existing.

(315.) It is possible that the present laws relating to partnerships might interfere with factories so conducted. If this interference could not be obviated by confining their purchases under the proposed system to ready money, it would be desirable to consider what changes in the law would be necessary to its existence:—and this furnishes another reason for entering into the question of limited partnerships.

(316.) A difficulty would occur also in discharging workmen who behaved ill, or who were not competent to their work; this would arise from their having a certain interest in the reserved fund, and, perhaps, from their possessing a certain portion of the capital employed; but without entering into detail, it may be observed, that such cases might be determined on by meetings of the whole establishment; and that if the policy of the laws favoured such establishments, it would scarcely be more difficult to enforce just regulations, than it now is to enforce some which are unjust, by means of combinations either amongst the masters or the men.

(317.) Some approach to this system is already practised in several trades:

the mode of conducting the Cornish mines has already been alluded to; the payment to the crew of whaling ships is governed by this principle; the profits arising from fishing with nets on the south coast of England are thus divided: one-half the produce belongs to the owner of the boat and net; the other half is divided in equal portions between the persons using it, who are also bound to assist in repairing the net when injured.

ON CONTRIVING MACHINERY

(318.) THE power of inventing mechanical contrivances, and of combining machinery, does not appear, if we may judge from the frequency of its occurrence, to be a difficult or a rare gift. Of the vast multitude of inventions which have been produced almost daily for a series of years, a large part has failed from the imperfect nature of the first trials; whilst a still larger portion, which had escaped the mechanical difficulties, failed only because the economy of their operations was not sufficiently attended to.

The commissioners appointed to examine into the methods proposed for preventing the forgery of bank notes, state in their report, that out of one hundred and seventy-eight projects communicated to the Bank and to the commissioners, there were only twelve of superior skill, and nine which it was necessary more particularly to examine.

(319.) It is however a curious circumstance, that although the power of combining machinery is so common, yet the more beautiful combinations are exceedingly rare. Those which command our admiration equally by the perfection of their effects and the simplicity of their means, are found only amongst the happiest productions of genius.

To produce movements even of a complicated kind is not difficult. There exist a great multitude of known contrivances for all the more usual purposes, and if the exertion of moderate power is the end of the mechanism to be contrived, it is possible to construct the whole machine upon paper, and to judge of the proper strength to be given to each part as well as to the framework which supports it, and also of its ultimate effect, long before a single part of it has been executed. In fact, all the contrivance, and all the improvements, ought first to be represented in the drawings.

(320.) On the other hand, there are effects dependent upon physical or chemical properties for the determination of which no drawings will be of any use. These are the legitimate objects of direct trial. For example;—if the ultimate result of an engine is to be that it shall impress letters on a copperplate by means of steel punches forced into it, all the mechanism by which the punches and the copper are to be moved at stated intervals, and brought into contact, is within the province of drawing, and the machinery may be arranged entirely upon paper. But a doubt may reasonably spring up, whether the bur that will be raised round the letter, which has been already punched upon the copper, may not interfere with the proper action of the punch for the letter which is to be punched next adjacent to it. It may also be feared that the effect of punching the second letter, if it be sufficiently near to the first, may distort the form of that first figure. If neither of these evils should arise, still the bur produced by the punching might be expected to interfere with the goodness of the impression produced by the copperplate; and the plate itself, after having all but its edge covered with figures, might change its form, from the unequal condensation which it must suffer in this process, so as to render it very difficult to take impressions from it at all. It is impossible by any drawings to solve difficulties such as these, experiment alone can determine their effect. Such experiments having been made, it is found that if the sides of the steel punch are nearly at right angles to the face of the letter, the bur produced is very inconsiderable;—that at the depth which is sufficient for copper-plate printing, no distortion of the adjacent letters takes place, although those letters are placed very close to each other;—that the small bur which arises may easily be scraped off;—and that the copper-plate is not distorted by the condensation of the metal in punching, but is perfectly fit to print from, after it has undergone that process.

(321.) The next stage in the progress of an invention, after the drawings are finished and the preliminary experiments have been made, if any such should be requisite, is the execution of the machine itself. It can never be too strongly impressed upon the minds of those who are devising new machines, that to make the most perfect drawings of every part tends essentially both to the success of the trial, and to economy in arriving at the result. The actual execution from working drawings is comparatively an easy task; provided always that good tools are employed, and that methods of working are adopted, in which the perfection of the part constructed depends less on the personal skill of the workman, than upon the certainty of the method employed.

(322.) The causes of failure in this stage most frequently derive their origin from errors in the preceding one; and it is sufficient merely to indicate a few of their sources. They frequently arise from having neglected to take into consideration that metals are not perfectly rigid but elastic. A steel cylinder of small diameter must not be regarded as an inflexible rod; but in order to ensure its perfect action as an axis, it must be supported at proper intervals. Again, the strength and stiffness of the framing which supports the mechanism must be carefully attended to. It should always be recollected, that the addition of superfluous matter to the immovable parts of a machine produces no additional momentum, and therefore is not accompanied with the same evil that arises when the moving parts are increased in weight. The stiffness of the framing in a machine produces an important advantage. If the bearings of the axis (those places at which they are supported) are once placed in a straight line, they will remain so, if the framing be immovable; whereas if the framework changes its form, though ever so slightly, considerable friction is immediately produced. This effect is so well understood in the districts where spinning factories are numerous, that, in estimating the expense of working a new factory, it is allowed that five per cent. on the power of the steam-engine will be saved if the building is fire-proof: for the greater strength and rigidity of a fire-proof building prevents the movement of the long shafts or axes which drive the machinery, from being impeded by the friction that would arise from the slightest deviation in any of the bearings.

(323.) In conducting experiments upon machinery, it is quite a mistake to suppose that any imperfect mechanical work is good enough for such a purpose. If the experiment is worth making, it ought to be tried with all the advantages of which the state of mechanical art admits; for an imperfect trial may cause an idea to be given up, which better workmanship might have proved to be practicable. On the other hand, when once the efficiency of a contrivance has been established, with good workmanship, it will be easy afterwards to ascertain the degree of perfection which will suffice for its due action.

(324.) It is partly owing to *the imperfection of the original trials*, and partly to the gradual improvements in the art of making machinery, that many inventions which have been tried, and given up in one state of art, have at another period been eminently successful. The idea of printing by means of moveable types had probably suggested itself to the imagination of many persons conversant with impressions taken either from blocks or seals. We find amongst the instruments discovered in the remains of Pompeii and Herculaneum, stamps for words formed out of one piece of metal, and including several letters. The idea of separating these letters, and of recombining them into other words, for the purpose of stamping a book, could scarcely have failed to occur to many: but it would almost certainly have been rejected by those best acquainted with the mechanical arts of that time; for the workmen of those days must have instantly perceived the impossibility of producing many thousand pieces of wood or metal, fitting so perfectly and ranging so uniformly, as the types or blocks of wood now used in the art of printing.

The principle of the press which bears the name of Bramah, was known about a century and a half before the machine, to which it gave rise, existed; but the imperfect state of mechanical art in the time of the discoverer, would have effectually deterred him, if the application of it had occurred to his mind, from attempting to employ it in practice as an instrument for exerting force.

These considerations prove the propriety of repeating, at the termination of intervals during which the art of making machinery has received any great improvement, the trials of methods which, although founded upon just principles, had previously failed.

(325.) When the drawings of a machine have been properly made, and the parts have been well executed, and even when the work it produces possesses all the qualities which were anticipated, still the invention may fail; that is, *it may fail of being brought into general practice*. This will most frequently arise from the circumstance of its producing its work at a greater expense than that at which it can be made by other methods.

(326.) Whenever the new, or improved machine, is intended to become the basis of a manufacture, it is essentially requisite that the *whole* expense attending its operations should be fully considered before its construction is undertaken. It is almost always very difficult to make this estimate of the expense: the more complicated the mechanism, the less easy is the task; and in cases of great complexity and extent of machinery it is almost impossible. It has been estimated roughly, that the first individual of any newly-invented machine, will cost about five times as much as the construction of the second, an estimate which is, perhaps, sufficiently near the truth. If the second machine is to be precisely like the first, the same drawings, and the same patterns will answer for it; but if, as usually happens, some improvements have been suggested by the experience of the first, these must be more or less altered. When, however, two or three machines have been completed, and many more are wanted, they can usually be produced at much less than one-fifth of the expense of the original invention.

(327.) The arts of contriving, of drawing, and of executing, do not usually reside in their greatest perfection in one individual; and in this, as in other arts, *the Division of Labour* must be applied. The best advice which can be offered to a projector of any mechanical invention, is to employ a respectable draughtsman; who, if he has had a large experience in his profession, will assist in finding out whether the contrivance is new, and can then make working drawings of it. The first step, however, the ascertaining whether the contrivance has the merit of novelty, is most important; for it is a maxim equally just in all the arts, and in every science, that *the man who aspires to fortune or to fame by new discoveries, must be content to examine with care the knowledge of his contemporaries, or to exhaust his efforts in inventing again, what he will most probably find has been better executed before.*

(328.) This, nevertheless, is a subject upon which even ingenious men are often singularly negligent. There is, perhaps, no trade or profession existing in which there is so much quackery, so much ignorance of the scientific principles, and of the history of their own art, with respect to its resources and extent, as are to be met with amongst mechanical projectors. The self-constituted engineer, dazzled with the beauty of some, perhaps, really original contrivance, assumes his new profession with as little suspicion that previous instruction, that thought and painful labour, are necessary to its successful exercise, as does the statesman or the senator. Much of this false confidence arises from the improper estimate which is entertained of the difficulty of invention in mechanics. It is, therefore, of great importance to the individuals and to the families of those who are too often led away from more suitable pursuits, the dupes of their own ingenuity and of the popular voice, to convince both them and the public that the power of making new mechanical combinations is a possession common to a multitude of minds, and that the talents which it requires are by no means of the highest order. It is still more important that they should be impressed with the conviction that the great merit, and the great success of those who have attained to eminence in such matters, was almost entirely due to the unremitted perseverance with which they concentrated upon their successful inventions the skill and knowledge which years of study had matured.

On the Economy of Machinery and Manufactures, pp. 364–92

ON THE EXPORTATION OF MACHINERY

(437.) A few years only have elapsed, since our workmen were not merely prohibited by act of Parliament from transporting themselves to countries in which their industry would produce for them higher wages, but were forbidden to export the greater part of the machinery which they were employed to manufacture at home. The reason assigned for this prohibition was, the apprehension that foreigners might avail themselves of our improved machinery, and thus compete with our manufacturers. It was, in fact, a sacrifice of the interests of one class of persons, the makers of machinery, for the imagined benefit of another class, those who use it. Now, independently of the impolicy of interfering, without necessity, between these two classes, it may be observed,—that the first class, or the makers of machinery, are, as a body, far more intelligent than those who only use it; and though, at present, they are not nearly so numerous, yet, when the removal of the prohibition which cramps their ingenuity shall have had time to operate, there appears good reason to believe, that their number will be greatly increased, and may, in time, even surpass that of those who use machinery.

(438.) The advocates of these prohibitions in England seem to rely greatly upon the possibility of preventing the knowledge of new contrivances from being conveyed to other countries; and they take much too limited a view of the possible, and even probable, improvements in mechanics.

(439.) For the purpose of examining this question, let us consider the case of two manufacturers of the same article, one situated in a country in which labour is very cheap, the machinery bad, and the modes of transport slow and expensive; the other engaged in manufacturing in a country in which the price of labour is very high, the machinery excellent, and the means of transport expeditious and economical. Let them both send their produce to the same market, and let each receive such a price as shall give to him the profit ordinarily produced by capital in his own country. It is almost certain that in such circumstances the first improvement in machinery will occur in the country which is most advanced in civilization; because, even admitting that the ingenuity to contrive were the same in the two countries, the means of execution are very different. The effect of improved machinery in the rich country will be perceived in the common market, by a small fall in the price of the manufactured article. This will be the first intimation to the manufacturer of the poor country, who will endeavour to meet the diminution in the selling price of his article by increased industry and economy in his factory; but he will soon find that this remedy is temporary, and that the market-price continues to fall. He will thus be induced to examine the rival fabric, in order to detect, from its structure, any improved mode of making it. If, as would most usually happen, he should be unsuccessful in this attempt, he must endeavour to contrive improvements in his own machinery, or to acquire information respecting those which have been made in the factories of the richer country. Perhaps after an ineffectual attempt to obtain by letters the information he requires, he sets out to visit in person the factories of his competitors. To a foreigner and rival manufacturer such establishments are not easily accessible; and the more recent the improvements, the less likely he will be to gain access to them. His next step, therefore, will be to obtain the knowledge he is in search of from the workmen employed in using or making the machines. Without *drawings*, or an examination of the *machines* themselves, this process will be slow and tedious; and he will be liable, after all, to be deceived by artful and designing workmen, and be exposed to many chances of failure. But suppose he returns to his own country with perfect drawings and instructions, he must then begin to construct his improved machines: and these he cannot execute either so cheaply or so well as his rivals in the richer countries. But after the lapse of some time, we shall suppose the machines thus laboriously improved, to be at last completed, and in working order.

(440.) Let us now consider what will have occurred to the manufacturer in the rich country. He will, in the first instance, have realized a profit by supplying the home market, at the usual price, with an article which it costs him less to produce; he will then reduce the price both in the home and foreign market, in order to produce a more extended sale. It is in this stage that the manufacturer in the poor country first feels the effect of the competition; and if we suppose only two or three years to elapse between the first application of the new improvement in the rich country, and the commencement of its employment in the poor country, yet will the manufacturer who contrived the improvement (even supposing that during the whole of this time he has made only one step) have realized so large a portion of the outlay which it required, that he can afford to make a much greater reduction in the price of his produce, and thus to render the gains of his rivals quite inferior to his own.

(441.) It is contended that by admitting the exportation of machinery, foreign manfacturers will be supplied with machines equal to our own. The first answer which presents itself to this argument is supplied by almost the whole of the present volume; *That in order to succeed in a manufacture, it is necessary not merely to possess good machinery, but that the domestic economy of the factory should be most carefully regulated.*

The truth, as well as the importance of this principle, is so well established in the Report of a Committee of the House of Commons 'On the Export of Tools and Machinery,' that I shall avail myself of the opinions and evidence there stated, before I offer any observations of my own:

> Supposing, indeed, that the same machinery which is used in England could be obtained on the Continent, it is the opinion of some of the most intelligent of the witnesses that a want of arrangement in foreign manufactories, of division of labour in their work, of skill and perseverance in their workmen, and of enterprise in the masters, together with the comparatively low estimation in which the

master-manufacturers are held on the Continent, and with the comparative want of capital, and of many other advantageous circumstances detailed in the evidence, would prevent foreigners from interfering in any great degree by competition with our principal manufactures; on which subject the Committee submit the following evidence as worthy the attention of the House:—

I would ask whether, upon the whole, you consider any danger likely to arise to our manufactures from competition, even if the French were supplied with machinery equally good and cheap as our own?—They will always be behind us until their general habits approximate to ours; and they must be behind us for many reasons that I have before given.

Why must they be behind us?—One other reason is, that a cotton manufacturer who left Manchester seven years ago, would be driven out of the market by the men who are now living in it, provided his knowledge had not kept pace with those who have been during that time constantly profiting by the progressive improvements that have taken place in that period; this progressive knowledge and experience is our great power and advantage.

It should also be observed, that the constant, nay, almost daily, improvements which take place in our machinery itself, as well as in the mode of its application, require that all those means and advantages alluded to above, should be in constant operation; and that, in the opinion of several of the witnesses, although Europe were possessed of every tool now used in the United Kingdom, along with the assistance of English artisans, which she may have in any number, yet, from the natural and acquired advantages possessed by this country, the manufacturers of the United Kingdom would for ages continue to retain the superiority they now enjoy. It is indeed the opinion of many, that if the exportation of machinery were permitted, the exportation would often consist of those tools and machines, which, although already superseded by new inventions, still continue to be employed, from want of opportunity to get rid of them; to the detriment, in many instances, of the trade and manufactures of the country: and it is matter worthy of consideration, and fully borne out by the evidence, that by such increased foreign demand for machinery, the ingenuity and skill of our workmen would have greater scope; and that, important as the improvements in machinery have lately been, they might, under such circumstances, be fairly expected to increase to a degree beyond all precedent.

The many important facilities for the construction of machines and the manufacturing of commodities which we possess, are enjoyed by no other country; nor is it likely that any country can enjoy them to an equal extent for an indefinite period. *It is admitted by every one, that our skill is unrivalled; the industry and power of our people unequalled; their ingenuity, as displayed in the continual improvement in machinery, and production of commodities, without parallel; and apparently, without limit.* The freedom which, under our government, every man has, to use his capital, his labour, and his talents, in the manner most conducive to his interests, is an inestimable advantage; canals are cut, and rail-roads constructed, by the voluntary association of persons whose local knowledge enables them to place them in the most desirable situations; and these great advantages cannot exist under less free governments. These circumstances, when taken together, give such a decided superiority to our people, that no injurious rivalry, either in the construction of machinery or the manufacture of commodities, can reasonably be anticipated.

(442.) But, even if it were desirable to prevent the exportation of a certain class of machinery, it is abundantly evident, that, whilst the exportation of other classes is allowed, it is impossible to prevent the forbidden one from being smuggled out; and that, in point of fact, the additional risk has been well calculated by the smuggler.

(443.) It would appear, also, from various circumstances, that the immediate exportation of improved machinery is not quite so certain as has been assumed; and that the powerful principle of self-interest will urge the makers of it, rather to push the sale in a different direction. When a great maker of machinery has contrived a new machine for any particular process, or has made some great improvement upon those in common use, to whom will he naturally apply for the purpose of selling his new machines? Undoubtedly, in by far the majority of cases, to his nearest and best customers, those to whom he has immediate and personal access, and whose capability to fulfil any contract is best known to him. With these, he will communicate and offer to take their orders for the new machine; nor will he think of writing to foreign customers, so long as he finds the home demand sufficient to employ the whole force of his establishment. Thus, therefore, the machine-maker is himself interested in giving the first advantage of any new improvement to his own countrymen.

(444.) In point of fact, the machine-makers in London greatly prefer home orders, and do usually charge an additional price to their foreign customers. Even the measure of this preference may be found in the evidence before the Committee on the Export of Machinery. It is differently estimated by various engineers; but appears to vary from five up to twenty-five per cent. on the amount of the order. The reasons are:—1. If the machinery be complicated, one of the best workmen, well accustomed to the mode of work in the factory, must be sent out to put it up; and there is always a considerable chance of his having offers that will induce him to remain abroad. 2. If the work be of a more simple kind, and can be put up without the help of an English workman, yet for the credit of the house which supplies it, and to prevent the accidents likely to occur from the want of sufficient instruction in those who use it, the parts are frequently made stronger, and examined more attentively, than they would be for an English purchaser. Any defect or accident also would be attended with more expense to repair, if it occurred abroad, than in England.

(445.) The class of workmen who *make* machinery, possess much more skill, and are paid much more highly than that class who merely *use it;* and, if a free exportation were allowed, the more valuable class would, undoubtedly, be greatly increased; for, notwithstanding the high rate of wages, there is no country in which it can at this moment be made, either so well or so cheaply as in England. We might, therefore, supply the whole world with machinery, at an evident advantage, both to ourselves and our customers. In Manchester, and the surrounding district, many thousand men are wholly occupied in making the machinery, which gives employment to many hundred thousands who use it; but the period is not very remote, when the whole number of those who *used* machines, was not greater than the number of those who at present *manufacture* them. Hence, then, if England should ever become a great exporter of machinery, she would necessarily contain a large class of workmen, to whom skill would be indispensable, and, consequently, to whom high wages would be paid; and although her manufacturers might probably be comparatively fewer in number, yet they would undoubtedly have the advantage of being the first to derive profit from improvement. Under such circumstances, any diminution in the demand for machinery, would, in the first instance, be felt by a class much better able to meet it, than that which now suffers upon every check in the consumption of manufactured goods; and the resulting misery would therefore assume a mitigated character.

(446.) It has been feared, that when other countries have purchased our machines, they will cease to demand new ones: but the statement which has been given of the usual progress in the improvement of the machinery employed in any manufacture, and of the average time which elapses before it is superseded by such improvements, is a complete reply to this objection. If our customers abroad did not adopt the new machinery contrived by us as soon as they could procure it, then our manufacturers would extend their establishments, and undersell their rivals in their own markets.

(447.) It may also be urged, that in each kind of machinery a maximum of perfection may be imagined, beyond which it is impossible to advance; and certainly the last advances are usually the smallest when compared with those which precede them: but it should be observed, that these advances are generally made when the number of machines in employment is already large; and when, consequently, their effects on the power of producing are very considerable. But though it should be admitted that any one species of machinery may, after a long period, arrive at a degree of perfection which would render further improvement nearly hopeless, yet it is impossible to suppose that this can be the case with respect to all kinds of mechanism. In fact the limit of improvement is rarely approached, except in extensive branches of national manufactures; and the number of such branches is, even at present, very small.

(448.) Another argument in favour of the exportation of machinery, is, that *it would facilitate the transfer of capital to any more advantageous mode of employment which might present itself*. If the exportation of machinery were permitted, there would doubtless arise a new and increased demand; and, supposing any particular branch of our manufactures to cease to produce the average rate of profit, the loss to the capitalist would be much less, if a market were open for the sale of his machinery to customers more favourably circumstanced for its employment. If, on the other hand, new improvements in machinery should be imagined, the manufacturer would be more readily enabled to carry them into effect, by having the foreign market opened where he could sell his old machines. The fact, that England can, notwithstanding her taxation and her high rate of wages, actually undersell other nations, seems to be well established: and it appears to depend on the superior goodness and cheapness of those raw materials of machinery the metals,—on the excellence of the tools,—and on the admirable arrangements of the domestic economy of our factories.

(449.) The different degrees of facility with which capital can be transferred from one mode of employment to another, has an important effect on the rate of profits in different trades and in different countries. Supposing all the other causes which influence the rate of profit at any period, to act equally on capital employed in different occupations, yet the real rates of profit would soon alter, on account of the different degrees of loss incurred by removing the capital from one mode of investment to another, or of any variation in the action of those causes.

(450.) This principle will appear more clearly by taking an example. Let two capitalists have embarked 10,000*l*. each, in two trades: A in supplying a district with water, by means of a steam-engine and iron pipes; B in manufacturing bobbin-net. The capital of A will be expended in building a house and erecting a steam-engine, which costs, we shall suppose, 3000*l*.; and in laying down iron pipes to supply his customers, costing 7000*l*. The greatest part of this latter expense is payment for labour: and if the pipes were to be taken up, the damage arising from that operation would render them of little value, except as old metal; whilst the expense of their removal would be considerable. Let us, therefore, suppose, that if A were obliged to give up his trade, he could realize only 4000*l*. by the sale of his stock. Let us suppose again that B, by the sale of his bobbin-net factory and machinery, could realize 8000*l*. and let the usual profit on the capital employed by each party be the same, say 20 per cent: then we have

	Capital invested	Money which would arise from sale of machinery	Annual rate of profit per cent.	Income
Water-works	£10,000	£4,000	£20	£2,000
Bobbin-net factory	10,000	8,000	20	2,000

Now, if, from competition, or any other cause, the rate of profit arising from water-works should fall to 10 per cent., that circumstance would not cause a transfer of capital from the water-works to bobbin-net making; because the reduced income from the water-works, 1000*l*. per annum, would still be greater than that produced by investing 4000*l*., (the whole sum arising from the sale of the materials of the water-works), in a bobbin-net factory, which sum, at 20 per cent., would yield only 800*l*. per annum. In fact, the rate of profit, arising from the water-works, must fall to less than 8 per cent. before the proprietor could increase his income by removing his capital into the bobbin-net trade.

(451.) In any inquiry into the probability of the injury arising to our manufacturers from the competition of foreign countries, particular regard should be had to the facilities of transport, and to the existence in our own country of a mass of capital in roads, canals, machinery, &c., the greater portion of which may fairly be considered as having repaid the expense of its outlay; and also to the cheap rate at which the abundance of our fuel enables us to produce iron, the basis of almost all machinery. It has been justly remarked by M. de Villefosse, in the memoir before alluded to, that '*Ce que l'on nomme en France, la question du prix des fers, est, à proprement parler, la question du prix des bois, et la question, des moyens de communications interieures par les routes, fleuves, rivières et canaux.*'

The price of iron in various countries in Europe has been stated in section 215 of the present volume; and it appears, that in England it is produced at the least expense, and in France at the greatest. The length of the roads which cover England and Wales may be estimated roughly at twenty thousand miles of turnpike, and one hundred thousand miles of road not turnpike. The internal water communication of England and France, as far as I have been able to collect information on the subject, may be stated as follows:

In France

	Miles in length.
Navigable rivers	4668
Navigable canals	915.5
Navigable canals in progress of execution (1824)	1388
	6971.5*

* This table is extracted and reduced from one of *Ravinet, Dictionnaire Hydrographique,* 2 vols. 8vo. Paris, 1824.

But, if we reduce these numbers in the proportion of 3.7 to 1, which is the relative area of France as compared with England and Wales, then we shall have the following comparison:

		England† Miles	Portion of France equal in size to England and Wales Miles
Navigable Rivers		1275.5	1261.6
Tidal Navigation‡		545.9	
Canals, direct	2023.5		
—, branch	150.6		
	2174.1	2174.1	247.4
Canals commenced		—	375.1
Total		3995.5	1884.1
Population in 1831		13894500	8608500

† I am indebted to F. Page, Esq. of Speen, for that portion of this table which relates to the internal navigation of England. Those only who have themselves collected statistical details can be aware of the expense of time and labour, of which the few lines it contains are the result.

‡ The tidal navigation includes—the Thames, from the mouth of the Medway,—the Severn, from the Holmes,—the Trent, from Trent-falls in the Humber,—the Mersey, from Runcorn Gap.

This comparison, between the internal communications of the two countries, is not offered as complete; nor is it a fair view, to contrast the wealthiest portion of one country with the whole of the other: but it is inserted with the hope of inducing those who possess more extensive information on the subject, to supply the facts on which a better comparison may be instituted. The information to be added, would consist of the number of miles in each country,—of sea-coast,—of public roads,—of rail-roads,—of rail-roads on which locomotive engines are used.

(452.) One point of view, in which rapid modes of conveyance increase the power of a country, deserves attention. On the Manchester Rail-road, for example, above half a million of persons travel annually; and supposing each person to save only one hour in the time of transit, between Manchester and Liverpool, saving of five hundred thousand hours, or of fifty thousand working days, of ten hours each, is effected. Now this is equivalent to an addition to the actual power of the country of one hundred and sixty-seven men, without increasing the quantity of food consumed; and it should also be remarked, that the time of the class of men thus supplied, is far more valuable than that of mere labourers.

ON THE FUTURE PROSPECTS OF MANUFACTURES, AS CONNECTED WITH SCIENCE

(453). In reviewing the various processes offered as illustrations of those general principles which it has been the main object of the present volume to support and establish, it is impossible not to perceive that the arts and manufactures of the country are intimately connected with the progress of the severer sciences; and that, as we advance in the career of improvement, every step requires, for its success, that this connexion should be rendered more intimate.

The applied sciences derive their facts from experiment; but the reasonings, on which their chief utility depends, are the province of what is called abstract Science. It has been shown, that the division of labour is no less applicable to mental productions than to those in which material bodies are concerned; and it follows, that the efforts for the improvement of its manufactures which any country can make with the greatest probability of success, must arise from the combined exertions of all those most skilled in the theory, as well as in the practice of the arts; each labouring in that department for which his natural capacity and acquired habits have rendered him most fit.

(454.) The profit arising from the successful application to practice of theoretical principles, will, in most cases, amply reward, in a pecuniary sense, those by whom they are first employed; yet even here, what has been stated with respect to *Patents*, will prove that there is room for considerable amendment in our legislative enactments: but the discovery of the great principles of nature demands a mind almost exclusively devoted to such investigations; and these, in the present state of science, frequently require costly apparatus, and exact an expense of time quite incompatible with professional avocations. It becomes, therefore, a fit subject for consideration, whether it would not be politic in the state to compensate for some of those privations, to which the cultivators of the higher departments of science are exposed; and the best mode of effecting this compensation, is a question which interests both the philosopher and the statesman. Such considerations appear to have had their just influence in other countries, where the pursuit of Science is regarded as a profession, and where those who are successful in its cultivation are not shut out from almost every object of honourable ambition to which their fellow-countrymen may aspire. Having, however, already expressed some opinion upon these subjects in another publication,* I shall here content myself with referring to that work.

(455.) There was, indeed, in our own country, one single position to which science, when concurring with independent fortune, might aspire, as conferring rank and station, an office deriving, in the estimation of the public, more than half its value from the commanding knowledge of its possessor; and it is extraordinary, that even that solitary dignity—that barony by tenure in the world of British science—the chair of the Royal Society, should have been coveted for adventitious rank. It is more extraordinary, that a Prince, distinguished by the liberal views he has invariably taken of public affairs,— and eminent for his patronage of every institution calculated to alleviate those miseries from which, by his rank, he is himself exempted—who is stated by his friends to be the warm admirer of knowledge, and most anxious for its advancement, should have been so imperfectly informed by those friends, as to have wrested from the head of science, the only civic wreath which could adorn its brow.†

---

* *Reflections on the Decline of Science in England, and on some of its Causes.* 8vo. 1830. Fellowes.

† The Duke of Sussex was proposed as President of the Royal Society in opposition to the wish of the Council—in opposition to the public declaration of a body of Fellows, comprising the largest portion of those by whose labours the character of English science had been maintained. The aristocracy of rank and of power, aided by such allies as it can

In the meanwhile the President may learn, through the only medium by which his elevated station admits approach, that those evils which were anticipated from his election, have not proved to be imaginary, and that the advantages by some expected to result from it, have not yet become apparent. It may be right also to state, that whilst many of the inconveniences, which have been experienced by the President of the Royal Society, have resulted from the conduct of his own supporters, those who were compelled to differ from him, have subsequently offered no vexatious opposition:—they wait in patience, convinced that the force of truth must ultimately work its certain, though silent course; not doubting that when His Royal Highness is correctly informed, he will himself be amongst the first to be influenced by its power.

(456.) But younger institutions have arisen to supply the deficiences of the old; and very recently a new combination, differing entirely from the older societies, promises to give additional steadiness to the future march of science. The '*British Association for the Advancement of Science*,' which held its first meeting at York* in the year 1831, would have acted as a powerful ally, even if the Royal Society were all that it might be: but in the present state of that body such an association is almost necessary for the purposes of science. The periodical assemblage of persons, pursuing the same or different branches of knowledge, always produces an excitement which is favourable to the development of new ideas; whilst the long period of repose which succeeds, is advantageous for the prosecution of the reasonings or the experiments then suggested; and the recurrence of the meeting in the succeeding year, will stimulate the activity of the inquirer, by the hope of being then enabled to produce the successful result of his labours. Another advantage is, that such meetings bring together a much larger number of persons actively engaged in

always command, set itself in array against the prouder aristocracy of science. Out of about seven hundred members, only two hundred and thirty balloted; and the Duke of Sussex had a majority of EIGHT. Under such circumstances, it was indeed extraordinary, that His Royal Highness should have condescended to accept the fruits of that doubtful and inauspicious victory.

The circumstances preceding and attending this singular contest have been most ably detailed in a pamphlet, entitled '*A Statement of the Circumstances connected with the late Election for the Presidency of the Royal Society*, 1831, printed by R. Taylor, Red-lion-court, Fleet-street.' The whole tone of the tract is strikingly contrasted with that of the productions of some of those persons by whom it was His Royal Highness's misfortune to be supported.

* The second meeting took place at Oxford in June 1832, and surpassed even the sanguine anticipations of its friends. The third annual meeting will take place at Cambridge in June 1833.

science, or placed in positions in which they can contribute to it, than can ever be found at the ordinary meetings of other institutions, even in the most populous capitals; and combined efforts towards any particular object can thus be more easily arranged.

(457.) But perhaps the greatest benefit which will accrue from these assemblies, is the intercourse which they cannot fail to promote between the different classes of society. The man of science will derive practical information from the great manufacturers;—the chemist will be indebted to the same source for substances which exist in such minute quantity, as only to become visible in most extensive operations;—and persons of wealth and property, resident in each neighbourhood visited by these migratory assemblies, will derive greater advantages than either of those classes, from the real instruction they may procure respecting the produce and manufactures of their country, and the enlightened gratification which is ever attendant on the acquisition of knowledge.*

(458.) Thus it may be hoped that public opinion shall be brought to bear upon the world of science; and that by this intercourse light will be thrown upon the characters of men, and the pretender and the charlatan be driven into merited obscurity. Without the action of public opinion, any administration, however anxious to countenance the pursuits of science, and however ready to reward, by wealth or honours, those whom they might think most eminent, would run the risk of acting like the blind man recently couched, who, having no mode of estimating degrees of distance, mistook the nearest and most insignificant for the largest objects in nature: it becomes, therefore, doubly important, that the man of science should mix with the world.

(459.) It is highly probable that in the next generation, the race of scientific men in England will spring from a class of persons altogether different from that which has hitherto scantily supplied them. Requiring, for the success of their pursuits, previous education, leisure, and fortune, few are so likely to unite these essentials as the sons of our wealthy manufacturers, who, having been enriched by their own exertions, in a field connected with science, will be ambitious of having their children distinguished in its ranks. It must, however, be admitted, that this desire in the parents would acquire great addi-

---

* The advantages likely to arise from such an association, have been so clearly stated in the address delivered by the Rev. Mr. Vernon Harcourt, at its first meeting, that I would strongly recommend its perusal by all those who feel interested in the success of English science.—Vide *First Report of the British Association for the Advancement of Science.* York, 1832.

tional intensity, if worldly honours occasionally followed successful efforts; and that the country would thus gain for science, talents which are frequently rendered useless by the unsuitable situations in which they are placed.

(460.) The discoverers of Iodine and Brome, two substances hitherto undecompounded, were both amongst the class of manufacturers, one being a maker of saltpetre at Paris, the other a manufacturing chemist at Marseilles; and the inventor of balloons filled with rarefied air, was a paper manufacturer near Lyons. The descendants of Mongolfier, the first aërial traveller, still carry on the establishment of their progenitor, and combine great scientific knowledge with skill in various departments of the arts, to which the different branches of the family have applied themselves.

(461.) Chemical science may, in many instances, be of great importance to the manufacturer, as well as to the merchant. The quantity of Peruvian bark which is imported into Europe is very considerable; but chemistry has recently proved that a large portion of the bark itself is useless. The alkali Quinia which has been extracted from it, possesses all the properties for which the bark is valuable, and only forty ounces of this substance, when in combination with sulphuric acid, can be extracted from a hundred pounds of the bark. In this instance then, with every ton of useful matter, thirty-nine tons of rubbish are transported across the Atlantic.

The greatest part of the sulphate of quinia now used in this country is imported from France, where the low price of the alcohol, by which it is extracted from the bark, renders the process cheap; but it cannot be doubted, that when more settled forms of government shall have given security to capital, and when advancing civilization shall have spread itself over the States of Southern America, the alkaline medicine will be extracted from the woody matter by which its efficacy is impaired, and that it will be exported in its most condensed form.

(462.) The aid of chemistry, in extracting and in concentrating substances used for human food, is of great use in distant voyages, where the space occupied by the stores must be economized with the greatest care. Thus the essential oils supply the voyager with flavour; the concentrated and crystalized vegetable acids preserve his health; and alcohol, when sufficiently diluted, supplies the spirit necessary for his daily consumption.

(463.) When we reflect on the very small number of species of plants, compared with the multitude that are known to exist, which have hitherto been cultivated, and rendered useful to man; and when we apply the same observation to the animal world, and even to the mineral kingdom, the field that natural science opens to our view seems to be indeed unlimited. These productions of nature, varied and innumerable as they are, may each, in some future day, become the basis of extensive manufactures, and give life, employment, and wealth, to millions of human beings. But the crude treasures perpetually exposed before our eyes, contain within them other and more valuable principles. All these, likewise, in their numberless combinations, which ages of labour and research can never exhaust, may be destined to furnish, in perpetual succession, new sources of our wealth and of our happiness. Science and knowledge are subject, in their extension and increase, to laws quite opposite to those which regulate the material world. Unlike the forces of molecular attraction, which cease at sensible distances; or that of gravity, which decreases rapidly with the increasing distance from the point of its origin; the further we advance from the origin of our knowledge, the larger it becomes, and the greater power it bestows upon its cultivators, to add new fields to its dominions. Yet, does this continually and rapidly increasing power, instead of giving us any reason to anticipate the exhaustion of so fertile a field, place us at each advance, on some higher eminence, from which the mind contemplates the past, and feels irresistibly convinced, that the whole, already gained, bears a constantly diminishing ratio to that which is contained within the still more rapidly expanding horizon of our knowledge.

(464.) But, if the knowledge of the chemical and physical properties of the bodies which surround us, as well as our imperfect acquaintance with the less tangible elements, light, electricity, and heat, which mysteriously modify or change their combinations, concur to convince us of the same fact; we must remember that another and a higher science, itself still more boundless, is also advancing with a giant's stride, and having grasped the mightier masses of the universe, and reduced their wanderings to laws, has given to us in its own condensed language, expressions, which are to the past as history, to the future as prophecy. It is the same science which is now preparing its fetters for the minutest atoms that nature has created: already it has nearly chained the ethereal fluid, and bound in one harmonious system all the intricate and splendid phenomena of light. It is the science of *calculation*,—which becomes continually more necessary at each step of our progress, and which must ultimately govern the whole of the applications of science to the arts of life.

(465.) But perhaps a doubt may arise in the mind, whilst contemplating the continually increasing field of human knowledge, that the weak arm of man may want the physical force required to render that knowledge available. The experience of the past, has stamped with the indelible character of truth, the maxim, that '*Knowledge is power*.' It not merely gives to its votaries control over the mental faculties of their species, but is itself the generator of physical force. The discovery of the expansive power of steam, its condensation, and the doctrine of latent heat, has already added to the population of this small island, millions of hands. But the source of this power is not without limit, and the coal-mines of the world may ultimately be exhausted. Without adverting to the theory, that new deposites of that mineral are now accumulating under the sea, at the estuaries of some of our larger rivers; without anticipating the application of other fluids requiring a less[er] supply of caloric than [of] water:—we may remark that the sea itself offers a perennial source of power hitherto almost unapplied. The tides, twice in each day, raise a vast mass of water, which might be made available for driving machinery. But supposing heat still to remain necessary, when the exhausted state of our coal-fields renders it expensive: long before that period arrives, other methods will probably have been invented for producing it. In some districts, there are springs of hot water, which have flowed for centuries unchanged in temperature. In many parts of the island of Ischia, by deepening the sources of the hot springs only a few feet, the water boils; and there can be little doubt that, by boring a short distance, steam of high pressure would issue from the orifice.

In Iceland, the sources of heat are still more plentiful; and their proximity to large masses of ice, seems almost to point out the future destiny of that island. The ice of its glaciers may enable its inhabitants to liquefy the gases with the least expenditure of mechanical force; and the heat of its volcanoes may supply the power necessary for their condensation. Thus, in a future age, *power* may become the staple commodity of the Icelanders, and of the inhabitants of other volcanic districts;* and possibly the very process by which they will procure this article of exchange for the luxuries of happier climates may, in some measure, tame the tremendous element which occasionally devastates their provinces.

(466.) Perhaps to the sober eye of inductive philosophy, these anticipations of the future may appear too faintly connected with the history of the past. When time shall have revealed the future progress of our race, those laws which are now obscurely indicated, will then become distinctly apparent; and it may possibly be found that the dominion of mind over the material world advances with an ever-accelerating force.

* In 1828, the author of these pages visited Ischia, with a committee of the Royal Academy of Naples, deputed to examine the temperature and chemical constitution of the springs in that island. During the few first days, several springs which had been represented in the instructions as under the boiling temperature, were found, on deepening the excavations, to rise to the boiling point.

Even now, the imprisoned winds which the earliest poet made the Grecian warrior bear for the protection of his fragile bark; or those which, in more modern times, the Lapland wizards sold to the deluded sailors;—these, the unreal creations of fancy or of fraud, called, at the command of science, from their shadowy existence, obey a holier spell: and the unruly masters of the poet and the seer become the obedient slaves of civilized man.

Nor have the wild imaginings of the satirist been quite unrivalled by the realities of after years: as if in mockery of the College of Laputa, light almost solar has been extracted from the refuse of fish; fire has been sifted by the lamp of Davy; and machinery has been taught arithmetic instead of poetry.

(467.) In whatever light we examine the triumphs and achievements of our species over the creation submitted to its power, we explore new sources of wonder. But if science has called into real existence the visions of the poet—if the accumulating knowledge of ages has blunted the sharpest and distanced the loftiest of the shafts of the satirist, the philosopher has conferred on the moralist an obligation of surpassing weight. In unveiling to him the living miracles which teem in rich exuberance around the minutest atom, as well as throughout the largest masses of ever-active matter, he has placed before him resistless evidence of immeasurable design. Surrounded by every form of animate and inanimate existence, the sun of science has yet penetrated but through the outer fold of Nature's majestic robe; but if the philosopher were required to separate, from amongst those countless evidences of creative power, one being, the masterpiece of its skill; and from that being to select one gift, the choicest of all the attributes of life;—turning within his own breast, and conscious of those powers which have subjugated to his race the external world, and of those higher powers by which he has subjugated to himself that creative faculty which aids his faltering conceptions of a deity,—the humble worshipper at the altar of truth would pronounce that being,—man; that endowment,—human reason.

But however large the interval that separates the lowest from the highest of those sentient beings which inhabit our planet, all the results of observation, enlightened by all the reasonings of the philosopher, combine to render it probable that, in the vast extent of creation, the proudest attribute of our race is but, perchance, the lowest step in the gradation of intellectual existence. For, since every portion of our own material globe, and every animated being it supports, afford, on more scrutinizing inquiry, more perfect evidence of design, it would indeed be most unphilosophical to believe that those sister spheres, obedient to the same law, and glowing with light and heat radiant from the same central source—and that the members of those kindred systems, almost lost in the remoteness of space, and perceptible only from the countless multitude of their congregated globes—should each be no more than a floating chaos of unformed matter;—or, being all the work of the same Almighty architect, that no living eye should be gladdened by their forms of beauty, that no intellectual being should expand its faculties in decyphering their laws.

Life peerage

IN 1833, shortly after the Great Reform Act, Babbage published *A Word to the Wise*, advocating life peerage. Several of Babbage's radical proposals, including peerage for life, abolition of ecclesiastical peerage, and decimal currency, actually date from the period of the English civil war and the interregnum. One imagines that Babbage would have been surprised that a century and a half after the enaction of the Great Reform Bill there are still ecclesiastical peers to obstruct development of scientific education in Britain; though of course the power of the House of Lords has been much reduced since Babbage's time.

A Word to the Wise, pp. 3–16

Precedents are treated by powerful minds as fetters with which to bind down the weak, as reasons with which to mistify the moderately informed, and as reeds which they themselves fearlessly break through whenever new combinations and difficult emergencies demand their highest efforts.

It would be idle to search the past history of this country for a conjunction of circumstances which even its future pages may fail to supply. He is the best friend to that country who, steadily looking on its present position, endeavours to ascertain and make known the direction of public opinion; and who, by stating its origin and probable effects, brings to the test of argument those questions which, if not now calmly discussed, might be settled, at a later period, by passion and violence.

It is impossible to view the present position of the House of Lords, without fearful anticipations that the insanity of one party may render it difficult for the firm but moderate liberals to preserve that institution from destruction by the fury of the opposite extreme of faction. A conviction has arisen throughout the country, that the majority of the Peers are opposed to all those ameliorations in our laws which altered circumstances require: and it is but too apparent, that the widening breach, arising from the want of sympathy of the two Houses of Parliament, requires no ordinary measures for its repair.

During the last half-century the various classes of society in England have made very great, but very unequal, advances in knowledge and in power. Large branches of manufactures have been created, which have given wealth to a body of capitalists possessed not only of great technical skill, but also of general intelligence of an high order; and a population has been called into existence, which depends upon those capitalists for its immediate support. Competition acting upon persons engaged in professions, as well as on those employed in manufactures and commerce, has produced a demand for information upon all subjects, which has given a stimulus to the universal mind of the country.

Amidst this general advance, the aristocracy alone have moved with slow and unequal pace, until at last it is found that amongst all the educated classes of the community, the aristocracy, as a body, are the least enlightened in point of knowledge, and the most separated in point of feeling from any sympathy with the mass of the people.

The splendid exceptions which occasionally present themselves to this too general truth, whilst they command increased respect for the individuals, serve but to render more apparent the general characteristic of the class.

It would, however, be highly unjust to ascribe to want of feeling, that which is mainly due to the position in which the upper classes of the country are placed.

It is pure stupidity to suppose that any class, however ignorant or however enlightened, are uninfluenced by those motives which govern the rest of mankind. The aristocracy of England suffer from having little responsibility; they suffer from the absence of objects of ambition to be acquired only by knowledge or by virtue; and they have suffered still more by that insatiable desire for political power, which has led to acts of the most disgraceful corruption—acts to which public opinion yet hesitates for a while to apply the deserved epithet of infamous.

It is impossible to unteach the mass of a nation; and if its aristocracy get far behind in information, they will inevitably be superseded.

With such dangers in view, no temporizing policy will suffice to avert them: it is necessary to recur to ultimate principles; and, by inquiring into the objects and best construction of an Upper Chamber, to endeavour to point out modes of gradually approximating our institutions to that form which reason indicates as the most useful.

One of the main uses of an Upper Chamber in a popular government, is to give consistency and uniformity to the more fluctuating opinion of the immediate representatives of the people. It is to the political what the fly-wheel is to the mechanical engine. It ought to represent the average, but not the extreme opinions of the people. For this purpose it is necessary that none but persons duly qualified should have seats in the House of Lords: Peers should be elected for life:—the Peerage should *not* be hereditary.

An hereditary Peerage is as much at variance with justice as it is with policy. The conquerors of a distant age might retain by force what they had acquired by violence: the *right* of resistance was never forfeited by their victims, though the prudence of exercising that right might long remain questionable; but no principle of justice or of policy can claim from any people perpetual respect and title for the descendants of the minions of the plunderer who robbed their ancestors. It is equally inconsistent with those principles that the weakness or the folly of a sovereign at one period of society, should be allowed to entail on his people the tax of perpetual respect for the descendants of the objects of his unworthy preference. Power may for a time enforce such absurdities, but reason will ultimately compel their abolition.

Possibly it may be argued, that the institution of hereditary titles is as necessary as that of hereditary property. Let us examine this question.

It is found by experience that more food can be produced on a given plot of ground by securing to the cultivator the possession, during his life, of the fruits arising from his labour on that ground, than can be produced by allowing the same man to cultivate it for the common benefit, paying him for his labour out of the common stock. It is therefore good for the whole community, that property in land should exist during one life. The same principle applies, but with far less force, to the question of allowing the field of the cultivator to descend to his children on his death, or of allowing him to bequeath it to a stranger. The increase of produce in this case is however but small, and at the next step of descent it almost altogether ceases. Experience has indeed fully proved the injury arising from lengthened entails, and the diminished production which results from them. The absurdity of allowing the casual possessors of the soil, at any one period, to devote any portion of that soil, throughout all after time, to the benefit of a peculiar family, or to one particular use, will become more conspicuous, if we suppose it to have been acted upon in a less enlightened age. Had such been the acknowledged nature of landed property, half the lands of England might have been devoted to the support of Druidical rites, and the vested interests of that sect would have been pleaded as an obstacle to the introduction of a purer faith.

The absurdity of maintaining a body of hereditary Legislators is still greater. Talent, respectability, and wealth, are the supposed qualifications for that high station; but who can assure us that these qualities shall always be found united in the descendants of some military chief, of some fawning courtier, or of some political adventurer, who sells himself in order to place his descendants in a position to demand a higher price? Is it wise to hold perpetually before the eyes of a people the living record of despotic caprice, of political intrigue, or of regal profligacy? Let us now, on the other hand, consider what would result from having a House of Peers elected for life.

The advantages to the country of having peerages only for life would be—

That no collision could occur between the two houses; for even though the number of the Upper House were limited, yet a gradual and continual change in its character would arise from the supply of vacancies caused by deaths, and the lapse of a very few years would necessarily bring its opinions into unison with those of the mass of the nation: or if difference of opinion should arise under such a form of the institution, it would be referred not to caprice or corruption, but to the desired and beneficial effects of the *intended* check on hasty legislation.

Thus both the Sovereign and the Ministry would have at their disposal much more extensive and cheaper means of rewarding services, and those means would be applicable to a far larger class of persons, the fortune necessary to support a personal being much less than that which is required for an hereditary honour.

Another advantage, of no ordinary amount, is, that the House of Peers would, in that case, almost necessarily contain talent of various kinds, which, under the hereditary system, is excluded. This would give to its decisions additional character and weight with the nation.

It may possibly be objected to this reasoning, that the same arguments would lead to the establishment of an elective monarchy. There is one great and distinctive difference in the two cases: the immense magnitude of that one great prize, and the small number of competitors for it, would tend to keep the country in a perpetual ferment; whilst the small comparative value of a seat in the Chamber of Peers, the few persons interested in the success of any one individual, and the great multitude of competitors, added to the number of the prizes, would effectually prevent a similar result.

Whether the nominations be made by the Crown, on the responsibility of its Ministers, or by the Lower House jointly with the Sovereign,—whether additional facilities be given for the admission of those whose fathers have sat in it or not;—still a considerable, and that by far the best portion of the nobility of the country, would probably find seats in the Upper House. Wealth, talent, and high character, would possess increased influence. The first and most important effect of such a system would be produced on the aristocracy themselves. They would at once acknowledge a responsibility to public opinion, and they would at the same time have set before them objects to occupy their ambition. If the eldest sons of Peers were admitted to the Upper House, not as a matter of right, but of election, it would give them a powerful stimulus to devote their earlier years to the acquisition of that knowledge, and the cultivation of those qualities which would command for them respect in the eyes of their countrymen. Even if possessed of but moderate abilities, with sufficient application, they could rarely fail; and the few excepted instances in a long hereditary line, by excluding cases of decided unfitness, arising from ignorance, from poverty, or from profligacy, would itself serve to raise the general character of the family. The same motive to virtuous exertion would act on the younger branches of the higher classes, with this additional stimulus to acquire fortune, that other contingencies than that of death might call them to the Upper Chamber. The wish of those already enjoying the Peerage to see their eldest sons sitting in the same House with them during their lives, would act as a moderate, though wholesome check upon the parents, by giving them an additional inducement to preserve in the public eye that high character which placed them there.

What, it may be asked, is the object of the preceding reasoning? It is to state principles which I believe to be sound, but which, at all events, are widely though not ostensibly canvassed at the present time: discussion will advance or refute them. Is it proposed that we should, like our neighbours, at once change our hereditary for a non-heriditary House of Peers? By no means; but we may, by previous discussion, render less abrupt those changes which time, circumstance, passion, and party, will inevitably bring about. All great and sudden changes are to be avoided, except for the prevention of still greater change. The Tory Aristocracy, *perfectly unacquainted* with the wants, the wishes, and the sympathies of the mass of the people of England, had so completely dammed up the current of reform, that it required all the energies of the moderate party to redistribute with safety its accumulated force. No man can doubt, that had the operation of the Reform Bill been spread over the last ten years, its effects would have been more beneficial. The same party, nevertheless, untaught by experience, weakened in numbers, and opposed by far greater power in the hands of the people, are now urging on the House of Peers in a course which, if persevered in, must inevitably lead to its reform or its suppression. It becomes, therefore, the duty of those who, amidst the storms of political controversy, may hope to engage even the smallest share of public attention, to direct that attention to the safest course.

A nation cannot give way to a party; and if the House of Lords sets itself in array against the people, its fate is sealed. A quick and full participation in that people's feelings can alone save it.

The first step of improvement is naturally to sever from the House of Lords a branch, which never ought to have been engrafted on it. The Ecclesiastical Peers, unfitted by education, unfitted by want of time, if they perform the other duties of their stations, occupy a position completely at variance with the religion they profess. They may protest that all their votes are uninfluenced by the hope of translation; and heaven may know that their assertions are true; but man, except it be revealed to him, can acquire little evidence of this. Man can only judge of the actions of his fellow-men by reference to what is known of human motives and of human passions; and if there be a class which claims exemption from these springs of action, it is not from the past history of our species, that evidence on which that claim can be admitted, may be derived.*

This severance will however advance us but a little way in harmonizing the two branches of the legislature. The creation of a sufficient number of hereditary Peers might remedy the evil for the present; but this permanent increase of the number of a body already sufficiently large, ought to be avoided. Circumstances also might arise which, at a future period, might require a renewed application of the same remedy. The smallest deviation from our present system, consistent with the attainment of the desired object, is *an immediate and large creation of* PEERS FOR LIFE.

---

* The Bishop of Exeter is reported to have 'declared it to be his firm conviction, that there was not one Clergyman of the Established Church who was not most anxious to advance its spritual interests, *without any reference to their own temporal advantage.*' It is difficult to perceive who are the dupes of assertions so completely at variance with all we know of human nature. The working Clergy are an educated and highly respectable class. They, like all other men, are acted upon by feelings of personal advantage; and it is precisely from the strong control they usually exercise over those feelings, that they are so generally respected. It is the presence, not the absence of *human feelings*, which excites sympathy.

The praise and the example of a Political Bishop are equally injurious to them.

Babbage's philosophy

In 1837 Babbage wrote *The Ninth Bridgewater Treatise* in which he presented God as Programmer. In chapter II he discussed the relation of the Difference Engine to scientific law in the deterministic Newtonian scheme.

The Ninth Bridgewater Treatise, pp. 30–44

ARGUMENT IN FAVOUR OF DESIGN FROM THE CHANGING OF LAWS IN NATURAL EVENTS

The estimate we form of the intellectual capacity of our race, is founded on an examination of those productions which have resulted from the loftiest flights of individual genius, or from the accumulated labours of generations of men, by whose long-continued exertions a body of science has been raised up, surpassing in its extent the creative powers of any individual, and demanding for its development a length of time, to which no single life extends.

The estimate we form of the Creator of the visible world rests ultimately on the same foundation. Conscious that we each of us employ, in our own productions, *means* intended to accomplish the objects at which we aim, and tracing throughout the actions and inventions of our fellow-creatures the same intention,—judging also of their capacity by the fit selection they make of the means by which they work, we are irresistibly led, when we contemplate the natural world, to attempt to trace each existing fact presented to our senses to some precontrived arrangement, itself perhaps the consequence of a yet more general law; and where the most powerful aids by which we can assist our limited faculties fail in enabling us to detect such connexions, we still, and not the less, believe that a more extended inquiry, or higher powers, would enable us to discover them.

The greater the number of consequences resulting from any law, and the more they are foreseen, the greater the knowledge and intelligence we ascribe to the being by which it was ordained. In the earlier stages of our knowledge, we behold a multitude of distinct laws, all harmonizing to produce results which we deem beneficial to our own species: as science advances, many of these minor laws are found to merge into some more general principles; and with its higher progress these secondary principles appear, in their turn, the mere consequences of some still more general law. Such has been the case in two of the most curious and most elaborately cultivated branches of human knowledge, the sciences of astronomy and optics. All analogy leads us to infer, and new discoveries continually direct our expectation to the idea, that the most extensive laws to which we have hitherto attained, converge to some few simple and general principles, by which the whole of the material universe is sustained, and from which its infinitely varied phenomena emerge as the necessary consequences.

To illustrate the distinction between a system to which the restoring hand of its contriver is applied, either frequently or at distant intervals, and one which had received at its first formation the impress of the will of its author, foreseeing the varied but yet necessary laws of its action throughout the whole of its existence, we must have recourse to some machine, the produce of human skill. But far as all such engines must ever be placed at an immeasurable interval below the simplest of Nature's works, yet, from the vastness of those cycles which even human contrivance in some cases unfolds to our view, we may perhaps be enabled to form a faint estimate of the magnitude of that lowest step in the chain of reasoning, which leads us up to Nature's God.

The illustration which I shall here employ will be derived from the results afforded by the Calculating Engine; and this I am the more disposed to use, because my own views respecting the extent of the laws of Nature were greatly enlarged by considering it, and also because it incidentally presents matter for reflection on the subject of inductive reasoning. Nor will any difficulty arise from the complexity of that engine; no knowledge of its mechanism, nor any acquaintance with mathematical science, are necessary for comprehending the illustration; it being sufficient merely to conceive that computations of great complexity *can* be effected by mechanical means.

Let the reader imagine that such an engine has been adjusted; that it is moved by a weight; and that he sits down before it, and observes a wheel, which moves through a small angle round its axis, at short intervals, presenting to his eye, successively, a series of numbers engraved on its divided circumference.

Let the figures thus seen be the series of natural numbers, 1, 2, 3, 4, 5, &c., each of which exceeds its immediate antecedent by unity.

Now, reader, let me ask how long you will have counted before you are firmly convinced that the engine, supposing its adjustments to remain unaltered, will continue whilst its motion is maintained, to produce the same series of natural numbers? Some minds perhaps are so constituted, that after passing the first hundred terms, they will be satisfied that they are acquainted with the law. After seeing five hundred terms, few will doubt; and after the fifty-thousandth term the propensity to believe that the succeeding term will be fifty thousand and one, will be almost irresistible. That term *will* be fifty thousand and one: the same regular succession will continue; the five-millionth and the fifty-millionth term will still appear in their expected order; and one unbroken chain of natural numbers will pass before your eyes, from *one* up to *one hundred million*.

True to the vast induction which has thus been made, the next succeeding term will be one hundred million and one; but after that the next number presented by the rim of the wheel, instead of being one hundred million and two, is one hundred million *ten thousand* and two. The whole series from the commencement being thus:—

1
2
3
4
5
...
...
......
.....
99 999 999
100 000 000
regularly as far as 100 000 001
100 010 002 :—the law changes
100 030 003
100 060 004
100 100 005
100 150 006
100 210 007
100 280 008
100 360 009
100 450 010
100 550 011
....
....

The law which *seemed* at first to govern this series fails at the hundred million and second term. That term is larger than we expected, by 10 000. The next term is larger than was anticipated, by 30 000, and the excess of each term above what we had expected forms the following table:—

$$10\,000$$
$$30\,000$$
$$60\,000$$
$$100\,000$$
$$150\,000$$
$$\cdots\cdots$$
$$\cdots\cdots$$

being, in fact, the series of *triangular numbers,** each multiplied by 10 000.

If we still continue to observe the numbers presented by the wheel, we shall find, that for a hundred, or even for a thousand terms, they continue to follow the new law relating to the triangular numbers; but after watching them for 2761 terms, we find that *this* law fails in the case of the 2762d term.

If we continue to observe, we shall discover another law then coming into action, which also is dependent, but in a different manner, on triangular numbers. This will continue through about 1430 terms, when a new law is again introduced, which extends over about 950 terms; and this too, like all its predecessors, fails, and gives place to other laws, which appear at different intervals.

Now it must be remarked, that the law *that each number presented by the Engine is greater by unity than the preceding number*, which law the observer had deduced from *an induction of a hundred million instances*, was not the true law that regulated its action; and that the occurrence of the number 100 010 002 at the 100 000 002d term, was *as necessary a consequence* of the original adjust-ment, and might have been as fully foreknown at the commencement, as was

---

* The numbers 1, 3, 6, 10, 15, 21, 28, &c. are formed by adding the successive terms of the series of natural numbers thus;

$$1 = 1.$$
$$1 + 2 = 3.$$
$$1 + 2 + 3 = 6.$$
$$1 + 2 + 3 + 4 = 10, \&c.$$

They are called triangular numbers, because a number of points corresponding to any term can always be placed in the form of a triangle, for instance:—

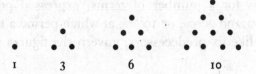

1   3   6   10

the regular succession of any of the intermediate numbers to its immediate antecedent. The same remark applies to the next *apparent* deviation from the new law, which was founded on an induction of 2761 terms, and to all the succeeding laws; with this limitation only—that whilst their consecutive introduction at various definite intervals is a necessary consequence of the mechanical structure of the engine, our knowledge of analysis does not yet enable us to predict the periods at which the more distant laws will be introduced.

Such are some of the facts which, by a certain adjustment of the Calculating Engine, would be presented to the observer. Now, let him imagine another engine, offering to the eye precisely the same figures in the same order of succession; but let it be necessary for the maker of that other engine, previously to each apparent change in the law, to make some new adjustment in the structure of the engine itself, in order to accomplish the ends proposed. The first engine must be susceptible of having embodied in its mechanical structure, that more general law of which all the observed laws were but isolated portions,—a law so complicated, that analysis itself, in its present state, can scarcely grasp the whole question. The second engine might be of far simpler contrivance; it must be capable of receiving the laws impressed upon it from without, but is incapable, by its own intrinsic structure, of changing, at definite periods, and in unlimited succession, those laws by which it acts. Which of these two engines would, in the reader's opinion, give the higher proof of skill in the contriver? He cannot for a moment hesitate in pronouncing that that for which, after its original adjustment, no superintendance is required, displays far greater ingenuity than that which demands, at every change in its law, the direct intervention of its contriver.

The engine we have been considering is but a very small portion (about fifteen figures) of a much larger one, which was preparing, and is partly executed; it was intended, when completed, that it should have presented at once to the eye about one hundred and thirty figures. In that more extended form which recent simplifications have enabled me to give to machinery constructed for the purpose of making calculations, it will be possible, by certain adjustments, to set the engine so that it shall produce the series of natural numbers in regular order, from unity up to a number expressed by more than a thousand places of figures. At the end of that term, another and a different law shall regulate the succeeding terms; this law shall continue in operation perhaps for a number of terms, expressed perhaps by unity, followed by a thousand zeros, or 10^{1000}; at which period a third law shall be introduced, and, like its predecessors, govern the figures produced by the engine during a third of those enormous periods. This change of laws might continue without limit; each individual law being destined to govern for millions of ages the calculations of the engine, and then give way to its successor to pursue a like career.

Thus a series of laws, each simple in itself, successively spring into existence, at distances almost too great for human conception. The full expression of that wider law, which comprehends within it this unlimited sequence of minor consequences, may indeed be beyond the utmost reach of mathematical analysis: but of one remarkable fact, however, we are certain— that the mechanism brought into action for the purpose of changing the nature of the calculation from the production of the merest elementary operations into those highly complicated ones of which we speak, is itself of the simplest kind.

In contemplating the operations of laws so uniform during such immense periods, and then changing so completely their apparent nature, whilst the alterations are in fact only the *necessary* consequences of some far higher law, we can scarcely avoid remarking the analogy which they bear to several of the phenomena of nature.

The laws of animal life which regulate the caterpillar, seem totally distinct from those which, in the subsequent stage of its existence, govern the butterfly. The difference is still more remarkable in the transformations undergone by that class of animals which spend the first portion of their life beneath the surface of the waters, and the latter part as inhabitants of air. It is true that the periods during which these laws continue to act are not, to our senses, enormous, like the mechanical ones above mentioned; but it cannot be doubted that, immeasurably more complex as they are, they were equally foreknown by their Author: and that the first creation of the egg of the moth, or the libellula, involved within its contrivance, as a necessary consequence, the whole of the subsequent transformations of every individual of their respective races.

––––––––––

For Babbage, miracles were, to use a modern term, subroutines called down from the celestial Store.

In chapter X and appendix E Babbage turned aside to demolish Hume's celebrated argument about miracles.

The Ninth Bridgewater Treatise, pp. 120–31

ON HUME'S ARGUMENT AGAINST MIRACLES

Few arguments have excited greater attention, and produced more attempts at refutation, than the celebrated one of David Hume, respecting miracles; and it might be added, that more sophistry has been advanced against it, than its author employed in the whole of his writings.

It must be admitted that in the argument, as originally developed by its author, there exists some confusion between personal experience and that which is derived from testimony; and that there are several other points open to criticism and objection; but the main argument, divested of its less important adjuncts, never has, and never will be refuted. Dr. Johnson seems to have been of this opinion, as the following extract from his life by Boswell proves:—

Talking of Dr. Johnson's unwillingness to believe extraordinary things, I ventured to say—

'Sir, you come near to Hume's argument against miracles—That it is more probable witnesses should lie, or be mistaken, than that they should happen.'

Johnson.—'Why, Sir, Hume, taking the proposition simply, is right. But the Christian revelation is not proved by miracles alone, but as connected with prophecies, and with the doctrines in confirmation of which miracles were wrought.'*

Hume contends that a miracle is a violation of the laws of nature; and as a firm and unalterable experience has established these laws, the proof against a miracle from the very nature of the fact, is as entire as any argument from experience can possibly be imagined.

The plain consequence is (and it is a general maxim worthy of our attention), that no testimony is sufficient to establish a miracle, unless the testimony be of such a kind, that its falsehood would be more miraculous than the fact which it endeavours to establish: and even in that case there is a mutual destruction of arguments, and the superior only gives us an assurance suitable to that degree of force which remains after deducting the inferior.†

The word *miraculous* employed in this passage is evidently equivalent to *improbable*, although the improbability is of a very high degree.

---

* Boswell's *Life of Johnson*. Oxford, 1826. vol. iii. p. 169.
† Hume's *Essays*, Edinburgh, 1817, vol. ii. p. 117.

The condition, therefore, which, it is asserted by the argument of Hume, must be fulfilled with regard to the testimony, is that the *improbability* of its falsehood must be GREATER than the *improbability* of the occurrence of the fact.

This is a condition which, when the terms in which it is expressed are understood, immediately commands our assent. It is in the subsequent stage of the reasoning that the fallacy is introduced. Hume asserts, that this condition cannot be fulfilled by the evidence of *any number* of witnesses, because our experience of the truth of human testimony is not uniform and without any exceptions; whereas, our experience of the course of nature, or our experience against miracles, is uniform and uninterrupted.

The only sound way of trying the validity of this assertion is to *measure* the numerical value of the two improbabilities, one of which it is admitted must be greater than the other; and to ascertain whether, by making any hypothesis respecting the veracity of each witness, it is possible to fulfil that condition by any finite number of such witnesses.

Hume appears to have been but very slightly acquainted with the doctrine of probabilities, and, indeed, at the period when he wrote, the details by which the conclusions he had arrived at could be proved or refuted were yet to be examined and arranged. It is, however, remarkable that the opinion he maintained respecting our knowledge of causation is one which eminently brings the whole question within the province of the calculus of probabilities. In fact, its solution can only be *completely* understood by those who are acquainted with that most difficult branch of science. By those who are not so prepared, certain calculations, which will be found more fully developed in the Note (E), must be taken for granted; and all that can be attempted will be, to convey to them a general outline of the nature of the principles on which these enquiries depend.

A miracle is, according to Hume, an event which has never happened within the experience of the whole human race. Now, the improbability of the future happening of such an occurrence may be calculated according to two different views.

We may conceive an urn, containing *only* black and white balls, from which *m* black balls have been successively drawn and replaced, one by one; and we may calculate the probability of the appearance of a white ball at the next drawing. This would be analogous to the case of one human being raised from the dead after *m* instances to the contrary.

Looking, in another point of view, at a miracle, we may imagine an urn to contain a very large number of tickets, on each of which is written one of the series of natural numbers. These being thoroughly mixed together, a single ticket is drawn: the prediction of the particular number inscribed on the ticket about to be drawn may be assimilated to the occurrence of a miracle.

According to either of these views, the probability of the occurrence of such an event by mere accident may be calculated. Now, the reply to Hume's argument is this: Admitting at once the essential point, viz. that the improbability of the concurrence of the witnesses in falsehood must be *greater* than the improbability of the miracle, it may be denied that this does not take place. Hume has asserted that, in order to prove a miracle, a certain improbability must be *greater* than another; and he has also asserted that this *never* can take place.

Now, as each improbability can be truly measured by number, the *only* way to refute Hume's argument is by examining the *magnitude* of these numbers. This examination depends on known and admitted principles, for which the reader who is prepared by previous study, may refer to the work of Laplace, *Théorie Analytique des Probabilités*; Poisson, *Recherches sur la Probabilité des Jugements*, 1837; or he may consult the article *Probabilities*, by Mr. De Morgan, in the *Encyclopaedia Metropolitana*, in which he will find this subject examined.

One of the most important principles on which the question rests, is the concurrence of the testimony of independent witnesses. This principle has been stated by Campbell, and has been employed by the Archbishop of Dublin,* and also by Dr. Chalmers.† It requires however to be combined with another principle, in order to obtain the numerical values of the quantities spoken of in the argument. The following example may be sufficient for a popular illustration.

Let us suppose that there are witnesses who will speak the truth, and who are not themselves deceived in ninety-nine cases out of a hundred. Now, let us examine what is the probability of the falsehood of a statement about to be made by two such persons absolutely unknown to and unconnected with each other.

Since the order in which independent witnesses give their testimony does not affect their credit, we may suppose that, in a given number of statements, both witnesses tell the truth in the ninety-nine first cases, and the falsehood in the hundredth.

Then the first time the second witness B testifies, he will agree with the testimony of the first witness A, in the ninety-nine first cases, and differ from

* *Elements of Rhetoric*, by R. Whately, D.D. p. 57, 1832.
† *Evidence of Christian Revelation*, vol. i. p. 129.

him in the hundredth. Similarly, in the second testimony of B, he will again agree with A in ninety-nine cases, and differ in the hundredth, and so on for ninety-nine times; so that, after A has testified a hundred, and B ninety-nine times we shall have

99 × 99 cases in which both agree,
      99 cases in which differ, A being wrong.

Now, in the hundredth case in which B testifies, he is wrong; and, if we combine this with the testimony of A, we have ninety-nine cases in which A will be right and B wrong; and one case only in which both A and B will agree in error. The whole number of cases, which amounts to ten thousand, may be thus divided:—

$$99 \times 99 = 9801 \text{ cases in which A and B agree in truth,}$$
$$1 \times 99 = 99 \text{ cases in which B is true and A false,}$$
$$99 \times 1 = 99 \text{ cases in which A is true and B false,}$$
$$1 \times 1 = 1 \text{ case in which both A and B agree in a}$$
$$\text{falsehood.}$$

10 000 cases.

As there is only one case in ten thousand in which two such independent witnesses can agree in error, the probability of their future testimony being false is $\frac{1}{10{,}000}$ or $\frac{1}{(100)^2}$.

The reader will already perceive how great a reliance is due to the *future* concurring testimony of two independent witnesses of tolerably good character and understanding. It appears that, previously to the testimony, the chance of one such witness being in error is $\frac{1}{(100)^1}$; that of two concurring in the same error is $\frac{1}{(100)^2}$; and if the same reasoning be applied to three independent witnesses, it will be found that the probability of their agreeing in error is $\frac{1}{(100)^3}$; or that the odds are 999 999 to 1 against the agreement.

Pursuing the same reasoning, the probability of the falsehood of a fact which six such independent witnesses attest is, previously to the testimony, $\frac{1}{(100)^6}$; or it is, in round numbers,

1 000 000 000 000 to 1 against the falsehood of their testimony.

The improbability of the miracle of a dead man being restored, is, on the principles stated by Hume, $\dfrac{1}{20\,(100)^5}$; or it is—

200 000 000 000 to 1 against its occurrence.

It follows, then, that the chances of accidental or other independent concurrence of only *six* such independent witnesses, is already *five times* as great as the improbability against the miracle of a dead man's being restored to life, deduced from Hume's method of estimating its probability solely from experience.

This illustration shows the great accumulation of probability arising from the concurrence of independent witnesses: we must however combine this principle with another, before we can arrive at the real numerical value of the improbabilities referred to in the argument.

The calculation of the numerical values of these improbabilities I have given in Note (E.)

From this it results that, provided we assume that independent witnesses can be found of whose testimony it can be stated that it is more probable that it is true than that it is false, *we can always assign a number of witnesses which will, according to Hume's argument, prove the truth of a miracle.*

The Ninth Bridgewater Treatise, pp. 192–203

Note E

NOTE TO CHAPTER X. ON HUME'S ARGUMENT AGAINST MIRACLES

The reader will observe, that throughout the chapter to which this note refers, as well as in the note itself, the argument of Hume is taken strictly according to his own interpretation of the terms he uses, and the calculations are founded on them; so that it is from the very argument itself, when fairly pursued to its full extent, that the refutation results.

Both our belief in the truth of human testimony, and our belief in the permanence of the laws of nature, are, according to Hume, founded on experience; we may, therefore, in the complete ignorance in which he assumes we are, with respect to the causes of either, treat the question as one of the probability of an event deduced solely from observations of the past.

The argument of Hume asserts, that one improbability, namely, that of the falsehood of the testimony in favour of a miracle, must always be *greater* than another improbability, namely, that of the occurrence of the miracle itself; and also, that, from the very nature of human experience, this preponderance can *never* take place.

Now the ONLY POSSIBLE mode of disproving the assertion, that one thing cannot, under any circumstances, be greater than another, is to measure, under all circumstances, the numerical value of the two things so compared, and the truth or falsehood of the assertion will then appear. The doctrine of chances, which has been much improved since the time of Hume, now enables us to apply precise measures to this argument; and it is the object of this Note to state the outlines of the calculation, and the results to which it leads. Previously to this, however, it may not be amiss to offer a few remarks on the principles about to be employed.

In the great work of Laplace, *Theorie Analytique des Probabilités*, those principles are established, and they are not merely undisputed, but are admitted by other writers of the highest authority on this subject. They form a part of the received knowledge of the present day, and, as such, they are employed in the present work, in which I propose to use, not to discuss them. I state this, because it has occasionally been asserted by persons unacquainted with the doctrine of chances, that the argument respecting the probability of improbability of miracles does not admit of the application of numbers. The received foundations of science are not to be put aside by such opinions, however highly skilled their authors may be in other branches of knowledge, and however powerful the intellect by which they may have attained those acquirements. The conclusions arrived at by the application of pure analysis must ever rest on the truth of the principles assumed at the commencement of the inquiry; and although a knowledge of mathematics may not appear necessary for forming a right judgement of the accuracy of those principles, yet it is observed, that a clear apprehension of them is not often found in the minds of those who are unacquainted with that science. When, however, the grounds on which the principles employed in the doctrine of chances are called in question by competent authority, it will be time enough to examine the question; and none will more eagerly enter upon that examination than those best versed in it, for none is so well aware of the extreme difficulty and delicacy of the subject.

As confusion sometimes arises from the difference in the meaning of the words *probable* and *improbable* in popular language and in mathematical inquiries, it may be convenient to point it out; and to state, that in this Note it is used in the mathematical sense, unless the reader's attention is directly called to a question relating to its popular sense.

In common language, an event is said to be *probable* when it is more likely to happen than to fail: it is said to be *improbable* when it is more likely to fail than to happen.

Now, an event whose probability is, in mathematical language $\frac{1}{p}$, will be called probable or improbable, in ordinary language, according as p is less or greater than 2.

If, in mathematical language, $\frac{1}{p}$ expresses the *probability* of an event happening,

$1 - \frac{1}{p}$ expresses the probability of its failing, or the *improbability* of its happening.

It has been stated in the text, that two views may be taken of those extraordinary deviations from the usual course of nature, called miracles. According to the first of these, we have to calculate the probability that a *white* ball has been drawn from an urn (containing only white and black balls, out of which m balls have been drawn all black), as deduced from the testimony of witnesses whose probability of speaking truth is known:—or, of the analogous case; it having been observed that m persons have died without any restoration to life, what is the probability that such a resurrection has happened, it having been asserted by n independent witnesses, the probability of each of whose speaking false is $\frac{1}{p}$?

The probability of the death without resurrection of the $(m+1)$th is $\frac{m+1}{m+2}$, and the improbability of such an occurrence, independently of testimony, is $\frac{1}{m+2}$; which is therefore the probability of a contrary occurrence, or that of a person being raised from the dead.

Now only two hypotheses can be formed, collusion being, by hypothesis, out of the question: either the event did happen, and the witnesses agree in speaking the truth, the probability of their concurrence being $\left(1-\frac{1}{p}\right)^n$, and of that of the hypothesis $\frac{1}{m+2}$; or the event did not happen, and the witnesses agree in a falsehood, the probability of their concurrence being $\left(\frac{1}{p}\right)^n$, and that of the hypothesis $\frac{m+1}{m+2}$.

The probability of the witnesses speaking truth, and the event occurring, is therefore,

$$\frac{\left(1-\frac{1}{p}\right)^n \frac{1}{m+2}}{\left(1-\frac{1}{p}\right)^n \frac{1}{m+2}+\left(\frac{1}{p}\right)^n \frac{m+1}{m+2}}=\frac{(p-1)^n}{(p-1)^n+m+1}; \qquad (A.)$$

and the probability of their falsehood is,

$$\frac{\left(\frac{1}{p}\right)^n \frac{m+1}{m+2}}{\left(1-\frac{1}{p}\right)^n \frac{1}{m+2}+\left(\frac{1}{p}\right)^n \frac{m+1}{m+2}}=\frac{m+1}{(p-1)^n+m+1}. \qquad (B.)$$

If we interpret Hume's assertion, 'that the falsehood of the witnesses must be more improbable than the occurrence of the miracle,' according to the mathematical meaning of the word improbable, then we must have,

$$\frac{m+1}{(p-1)^n+m+1}<\frac{1}{m+2};$$

or,

$$(m+1)\cdot(m+2) < (p-1)^n + m + 1;$$

hence,

$$(p-1)^n > (m+1)\cdot(m+2) - (m+1) > (m+1)^2,$$

from which we find,

$$n > \frac{2\log.(m+1)}{\log.(p-1)}.$$

If p is any number greater than two, this equation can always be satisfied.

It follows, therefore, that however large m may be, or however great the quantity of experience against the occurrence of a miracle, (provided only that there are persons whose statements are more frequently correct than incorrect, and who give their testimony in favour of it without collusion,) a certain number n can ALWAYS be found; so that *it shall be a greater improbability that their unanimous statement shall be a falsehood, than that the miracle shall have occurred.*

Let us now suppose each witness to state one falsehood for every ten truths, or $p = 11$, and $m = 1\,000\,000\,000\,000$;

then,

$$n > \frac{2\log.(10^{12}+1)}{\log.10} > 24.$$

or twenty-five such witnesses are sufficient.

If the witnessess only state one falsehood for every hundred truths, then thirteen such witnesses are sufficient.

Another view of the question might be taken; and it might be asserted that, in order to believe in the miracle, the probability of its truth must be greater than the probability of its falsehood; in this case the expression (A) must be greater than (B); or,

$$\frac{(p-1)^n}{(p-1)^n + m + 1} > \frac{m+1}{(p-1)^n + m + 1};$$

hence,

$$(p-1)^n > m+1,$$

and

$$n > \frac{\log.(m+1)}{\log.(p-1)}.$$

In this case also, under the same circumstances, the condition can always be fulfilled of finding a sufficient number of witnesses to render the miracle probable, or even to give to it any required degree of probability.

If $p = 11$, and $m = 1\ 000\ 000\ 000\ 000$, as before, then

$$n > \frac{\log.(10^{12} + 1)}{\log. 10} > 12.$$

According to the second view stated in the text, a miracle may be assimilated to the drawing of a given number i out of an urn, containing all numbers from one to m.

In this case the probability of the occurrence of the event is $\frac{1}{m}$, and the probability of the concurrence of n witnesses in falsehood is $\left(\frac{1}{p}\right)^n$.

Hence the probability that the particular number i was drawn, as deduced from the testimony of n witnesses, each of whose probability of falsehood is $\frac{1}{p}$, is expressed by,

$$\frac{\left(1 - \frac{1}{p}\right)^n \frac{1}{m}}{\left(1 - \frac{1}{p}\right)^n \frac{1}{m} + \left(\frac{1}{p}\right)^n \left(1 - \frac{1}{m}\right)\left(\frac{1}{m-1}\right)^n} =$$

$$\frac{(p-1)^n}{(p-1)^n + (m-1)\left(\frac{1}{m-1}\right)^n} =$$

$$\frac{1}{1 + (m-1)\left(\frac{1}{m-1 \cdot p-1}\right)^n} \qquad (C.)$$

and the probability of the number i not having been drawn, or of their falsehood, is

$$\frac{\left(\frac{1}{p}\right)^n \left(1 - \frac{1}{m}\right)\left(\frac{1}{m-1}\right)^n}{\left(1 - \frac{1}{p}\right)^n \frac{1}{m} + \left(\frac{1}{p}\right)^n \left(1 - \frac{1}{m}\right)\left(\frac{1}{m-1}\right)^n} =$$

$$\frac{(m-1)\left(\frac{1}{m-1}\right)^n}{(p-1)^n + (m-1)\left(\frac{1}{m-1}\right)^n} =$$

$$\frac{m-1}{(p-1 \cdot m-1)^n + (m-1)} \qquad (D.)$$

Hence the improbability of the testimony must, according to Hume, be greater than that of the occurrence of the event; or,

$$\frac{m-1}{(\overline{p-1}\cdot\overline{m-1})^n+m-1} < \frac{1}{m}.$$

Hence, $m\cdot\overline{m-1} < (\overline{p-1}\cdot\overline{m-1})^n + m - 1,$

and, $(\overline{p-1}\cdot\overline{m-1})^n > m\cdot\overline{m-1} - m + 1 > (m-1)^2,$

or, $n > \dfrac{2\log.(m-1)}{\log.(p-1)+\log.(m-1)}.$

If $p = 11$, and $m = 1\ 000\ 000\ 000\ 000$, as above,

$$n > \frac{2\times 12}{1+12} > \frac{24}{13} > 2.$$

If it only required that the probability of the occurrence of the miracle shall be greater than its improbability, then we must make (C) greater than (D); or,

$$\frac{1}{1+(m-1)\left(\dfrac{1}{\overline{m-1}\cdot\overline{p-1}}\right)^n} > \frac{m-1}{(\overline{p-1}\cdot\overline{m-1})^n+m-1};$$

from which,

$$\frac{(\overline{m-1}\cdot\overline{p-1})^n}{(\overline{p-1}\cdot\overline{m-1})^n+m-1} > \frac{m-1}{(\overline{p-1}\cdot\overline{m-1})^n+m-1},$$

or, $(\overline{m-1}\cdot\overline{p-1})^n > m - 1.$

Hence, $n > \dfrac{\log. \overline{m-1}}{\log. \overline{m-1}+\log. \overline{p-1}}.$

If $p = 11$, and $m = 1\ 000\ 000\ 000\ 000$,

$$n > \frac{12}{12+1} > \frac{12}{13}.$$

Hence in this view, also, a sufficient number of witnesses of given veracity may *always* be found to render the improbability of their concurrent independent testimony being false, greater than the improbability of the occurrence of the miracle.
There is, however, one other view, which it seems probable would have been that

taken by Hume himself, had he applied numbers to his own argument. Considering the probability of the coincidence in falsehood of n persons each having the probability $\frac{p-1}{p}$ in favour of his truth, which is $\frac{1}{p^n}$, that probability ought to be less than that of the occurrence of the miracle; or,

$$\frac{1}{p^n} < \frac{1}{m+2};$$

hence, $p^n > m+2$,

$$\text{or, } n > \frac{\log. (m+2)}{\log. p}$$

According to this view also, if $m = 1\ 000\ 000\ 000\ 000$, and $p = 11$, $n > \frac{12}{1.04} > 12$.

This view of the question refers to the probability of the concurrence of the witnesses *before* they have given their testimony. The other four cases relate to the probability of the miracle having happened, as deduced from the fact of the testimony having been given. The last seems to have been that which Hume would have himself arrived at; the others represent the true methods of estimating the probabilities of the various cases: and the important conclusion follows, that, whichever be the interpretation given to the argument of Hume, if independent witnesses can be found, who speak truth more frequently than falsehood, *it is* ALWAYS *possible to assign a number of independent witnesses, the improbability of the falsehood of whose concurring testimony shall be greater than that of the improbability of the miracle itself.*

It is to be observed, that the whole of this argument applies to *independent* witnesses. The possibility of the collusion, and the degree of credit to be assigned to witnesses under any given circumstances, depend on facts which have not yet been sufficiently collected to become the subject of mathematical inquiry. Some of those considerations which bear on this part of the subject, the reader will find treated of in the work of Dr Conyers Middleton, entitled *A Free Inquiry into the Miraculous Powers which are supposed to have subsisted in the Christian Church, from the earliest Ages through several successive Centuries.* London, 1749.

Taxation

In 1848 Babbage wrote a short pamphlet, *Thoughts on the Principles of Taxation with Reference to a Property Tax, and its Exceptions*. Babbage was concerned with the damaging effects of high taxation on the functioning of a capitalist economy, and the ideas are similar to those underlying the present taxation policies of President Reagan of the USA and our Prime Minister, Mrs Thatcher. Babbage was also much concerned that exemptions to taxes would encourage those exempted to vote for ever higher taxes, because they would benefit from the taxes without having to contribute. Babbage returned emphatically to the latter argument in *Thoughts upon an Extension of the Franchise* (1865). These arguments are essentially the same as those underlying the case for a Community Tax, or poll tax, in preference to the old property rating system in Britain.

Thoughts on the Principles of Taxation with Reference to a Property Tax, and its Exceptions, pp. 2–24

Fidèle à son caractère, Napoléon ne craignit pas, le jour même où il briguait le trône, de rétablir, sous le nom des droits-réunis, le plus impopulaire, mais le plus utile des impôts.

Il en fit la première proposition au Conseil d'Etat, et il y soutint avec une sagacité merveilleuse, comme si les finances avaient été l'étude de sa vie, les vrais principes de la matière. A la théorie de l'impôt unique, reposant exclusivement sur la terre, exigeant du fermier et du propriétaire la totalité de la somme nécessaire aux besoins de l'Etat, les obligeant à en faire au moins l'avance dans la supposition la plus favorable pour eux, celle où le renchérissement des produits agricoles les dédommage de cette avance,—à une théorie aussi follement exagérée, il opposa la théorie simple et vraie de l'impôt habilement diversifié, reposant à la fois sur toutes les propriétés et sur toutes les industries, ne demandant à aucune d'elles une portion trop considérable du revenu public, n'amenant par consequent aucun mouvement forcé dans les valeurs, puisant la richesse dans tous les canaux où elle passe abondamment, et puisant dans chacun de ces canaux, de manière à ne pas produire un abaissement trop sensible.

225

Ce système, fruit du temps et de l'expérience, n'est susceptible que d'une seule objection: c'est que la diversité de l'impôt entraine la diversité de la perception, et, des lors, une augmentation des frais: mais il presente tant d'avantages, et le contraire est si violente, que cette légère augmentation de frais ne saurait être une considération sérieuse.—*Histoire du Consulat et de l'Empire, par M. A. Thiers,* vol. v. p. 162. 8vo ed.

THOUGHTS, &C.

There are two essential grounds on which all legitimate taxation ought to rest—

1st. The Protection of property.
2d. The Protection of the person.

I. It is obvious, as a general principle, that all taxation ought to be proportioned to the cost of the service for which the taxes are paid. If it were otherwise, some portion of the people would be compelled to pay for services which they do not receive.

But it is equally obvious, that the cost and difficulty of applying this general principle must put a limit to the extent to which it is politic to carry it into detail. Ships are liable to peculiar dangers from the elements, for protection against one portion of which they pay a special tax to support lighthouses, beacons, and other means of contributing to their security when near the coast. These means are useless for the protection of houses, which, therefore, are not subject to payment for them. But both houses and ships are subject to damage and destruction by fire, and other accidents. Against such misfortunes the owners of both can insure themselves, and will pay to the Insurance Companies in proportion to the nature of their risks.

II. With respect to the expense of protecting personal liberty, some difficulties arise. It may be contended that the personal liberty of a poor man costs as much for its protection as that of a wealthy man or of a peer. If this be admitted, then it follows that a fixed sum might be charged on each individual, as a poll-tax, for his personal protection. But it would be impossible to raise any large sum by such means: because, unless the tax were very small, a considerable portion of the population would be absolutely unable to pay it.

On the other hand, it may with some plausibility be maintained, that the value to any man, of his personal liberty, is in proportion to the amount of property he possesses. It is by no means an uncommon event, that a poor man is convicted of a crime of which he is guiltless, simply from his want of money to pay for legal assistance, and to bring into the presence of his judge the witnesses of his innocence. It is painful to reflect on the instances in which innocent persons have thus suffered even the extreme penalty of the law. Who ever heard of such calamities happening to a rich man?

Amongst those convicted of minor felonies, such instances are more frequent. Even in a newspaper of this morning I observe that the innocence of a man convicted in 1845 of stealing a horse and gig is established, by the confession of another convict in Van Dieman's Land, that he alone was the real thief. In the mean time the unjustly convicted person, who had conducted himself with great propriety during his confinement in Pentonville prison, went as an exile to Australia. This deeply injured and ruined man will now probably receive a *pardon;*—a word, which in the English language always means the forgiveness of an injury done *by* the person to whom it is granted—but which, to the disgrace of English law, implies in such cases an admission that a deep and an unatoned injury has been done by the institutions of the country *to* the person pardoned.*

Undoubtedly ample reparation ought to be made for such sufferings, and as far as money can be a compensation, it ought to be liberally bestowed. The law has already granted compensation to individuals injured by accidents arising from negligence—as in the instance of railroads; and in the case of death, it gives the same relief even to the relatives of the sufferer. Why should not a similar remedy, through the intervention of a jury, be given to men who have been wrongfully injured in their person, their character, and their feelings, by an unjust or a mistaken conviction?

The care taken by the legislature for the protection of property was curiously contrasted in some recent cases, with that which is bestowed on the protection of person.

Not many months ago the public were informed, that a free pardon had been granted to a convict whose innocence had been clearly proved, *after he had suffered some part of his sentence in Van Dieman's Land*, and that on his return to this country the Government had presented him with *ten pounds ! ! !* Much about the same time the public were reminded that certain offices connected with the Court of Chancery, which required but little industry and small talent in their possessors, and had long been greatly overpaid, were to be

---

* 'Forgiveness to the injured does belong,
But they ne'er pardon who commit the wrong.'
DRYDEN

abolished by a decision of the House of Commons, and compensation was of course to be made to the holders. To one of these officers *a pension of six thousand a-year for life was awarded, and an annuity of three thousand a-year to his executors for seven years after his death.**

From such examples it would seem that the estimated value of the personal liberty of the poorer classes is very small. If so, any payment on this ground must also be small, and therefore might be neglected in considering the question of taxation. But if, as appears to be the more reasonable view, the expense of protecting the personal liberty of the wealthy classes bears practically some definite proportion to their means, then the two grounds of taxation follow the same law.

I shall therefore assume in the following pages—

That taxation ought to be proportional to the cost of maintaining those institutions, without which neither property nor industry can be protected, or even exist.

In the application of this general principle, the first question which arises is, whether a portion of the property of the country shall be taken once for all and applied to the support of the Government: or whether certain sums shall be collected at periodical intervals.

The objections to the first of these alternatives are insuperable. It provides only for the protection of that property and those interests which exist at the time of the first arrangement. The amount requisite for protection is also a continually varying sum, which no human foresight can predict. If a portion of land is set apart for this purpose it is usually less improved than that which is in the possession of private owners. All civilized Governments have therefore adopted periodical payments of their revenue, and all their accounts are annual. We may therefore assume that taxation ought to be annual. And if so, its amount must, of course, be regulated by the sum necessary for the protection of property during one year.

It has been objected to this view of the question, that *annual* taxes raised for purposes of protection ought not to be expended on *permanent* structures. The answer is, that a certain portion of the expenditure being so employed, the *average* amount required *annually* will be considerably reduced.

---

* This is unfortunately not a solitary instance of lavish extravagance on the part of a weak government, to conciliate a powerful interest. Can any reasonable being be surprised at the low estimate which is formed of the public integrity of the leaders of party, when such profligate expenditure is contrasted with the paltry funds meted out to science by those who are always ready, when pressed, to deplore the insufficiency, and ever indisposed when urged, to attempt its remedy?

We have now, therefore, arrived at the principle, that each person ought to be taxed *annually* for the protection of his personal liberty and property during that year. It may be observed, that the things to be protected by the taxation during the year, are the income, the advantages, and the enjoyments resulting from property, or from the institutions of the state. This view strictly leads to an income-tax. But without insisting on this inference, there are other reasons which show that the amount of annual taxation for securing the enjoyment of property during each year, ought to be in proportion to its produce.

The power of enjoying property of every kind, depends entirely on the conventions of Society. Whether a man derives an income from an hereditary estate,—from a permanent or a temporary annuity in the funds,—from the produce of his brain in scientific and literary productions,—from the sale of his acquired personal knowledge in medicine and in the law,—or from the same knowledge applied to the employment of capital by the merchant or the shopkeeper,—it is equally essential for the receipt of his annual income that those laws and institutions, without which his profit could not arise, should be maintained during that year.

It is frequently urged that it is unjust to tax a man who has an annuity of 100*l.* for a limited number of years, equally with another person who possesses the same in perpetuity. But the tax is really paid in each case for the annual security of the property; and if he who held the perpetual annuity were taxed more than the other he might justly complain that he is taxed for being richer than the other. Besides, when the annuity to the temporary holder ceases, it reverts to the grantor, who then pays an equal annual sum for its protection.

It is a frequent subject of complaint by professional men, that their uncertain income, dependent on health and other accidental circumstances, pays the same tax as that of the landholder. But it must be observed, that, although the income is precarious to the individual, its protection is not less costly to the State; and in whatever way it may be distributed amongst the members of a profession, the total income of the *whole profession* is often quite as permanent as that of the landlord, and is more certain in its payment. The average annual income received by the whole Bar, and by its various members who have been advanced to the innumerable places to which that profession leads, varies but little, and is certainly much better paid than the rental even of the most fortunate landlord. The income of both depends on the security of property, and the support during the year of the usual institutions of the country. Those who enter the profession of the law are aware of the permanence of its general income, and cannot fairly complain, during the period in which they are profiting by its use, of being obliged to contribute their full share towards the support of those institutions on which the existence of their profession depends.

The same argument applies, more or less in degree, and entirely in principle, to all other professions and trades. It is sometimes urged, that men will be ill and require medical aid, whether any form of government exist or not. Certainly this is so: but unless the medical man can reach his patients and return in safety (for which protection he must sacrifice a portion of his gains) he can neither receive food from his patient to support his own life to-day, nor a fee to supply the wants of himself and his family on the morrow.

In fact, it appears that the cost of protecting the small capitalist is greater in proportion to the amount of his capital than that which is necessary for the larger holders. The Barings and the Rothschilds can with facility transfer their capital, or at least a large portion of it, to the protection of other states, the moment their keen practical eyes perceive the slightest commencing insecurity in the institutions of their own. The helpless vendor of apples at the corner of the street has no such resource. Without the protection of a powerful and expensive police, her humble store would be hopelessly exposed to the plunder of every passing vagabond. When education has fully impressed this fact on the labouring classes of society, one most important step will have been made in the difficult art of government.

Two questions of great importance arise in contemplating a tax upon income:—

1st. As to the amount of the tax on a given amount of income; or, in other words, its rate per cent?

2nd. The amount of income which shall be exempted from taxation.

1st, If the income tax be very high, there is no doubt whatever that it will be considerably evaded in its collection. It may, therefore, become a most unequal tax, and consequently a most unjust one. It would in fact, under such circumstances, be deprived of the whole support of that argument on which its existence has been advocated. Its injustice would be greater, because it would fall with unmitigated force upon the most helpless and the most upright members of the community.

Another evil resulting from a high rate of tax upon income is perhaps of more dangerous consequence, from its being less open to observation. Its necessary effect will be, the *transfer of capital from this to other countries*. No laws however stringent can prevent this consequence, nor follow the transported capital to its adopted home, and there tax its annual produce. The injustice of a government taxing capital which it does not protect, would remove from the minds of its possessors the impediment of moral wrong; and the sagacity of commercial enterprise would soon place it far beyond the grasp of the most rapacious chancellor of the exchequer, even with all the aids which legal ingenuity could devise.

The evil effects of such an abstraction of capital might be at first almost imperceptible; they would be slow, but certain and cumulative. The impost itself would fall more heavily on the capital remaining at home, crippling the manufacturing enterprise of the country, and pressing with weighty force even on that labouring population whose means are so small that they are *nominally* exempt from its infliction. Extended information amongst the masses is the best antidote to these evils, as well as the most faithful trustee of the interests of truth.

2nd, The second question is, *The amount of income which shall be exempt from taxation*. It may be observed, that there are here two limits. It is obviously impolitic to allow any tax to descend below the point at which the cost of collection exceeds the produce. It is also hopeless to attempt to collect it from those whose entire income just enables them to subsist. The remission of the tax might in the latter case be looked upon as an act of charity.

I shall at present refer only to the *economical* ground of national charity, of poor rates, and of other similar institutions, because it is of importance that the operation of the principles of morals and of economy should be investigated separately, before their united action in any system of government is examined. Whenever, for the purposes of government, we arrive, in any state of society, at a class so miserable as to be in want of the common necessaries of life, a new principle must be taken into consideration. The usual restraints which are sufficient for the well-fed are often useless in checking the demands of hungry stomachs. Other and more powerful means must then be employed; a larger array of military or of police force must be maintained. Under such circumstances it may be found considerably cheaper to fill empty stomachs up to the point of ready obedience, than to compel starving wretches to respect the roast-beef of their more industrious neighbours: and it may be expedient, in a mere economical point of view, to supply gratuitously the wants even of able-bodied persons, if it can be done without creating crowds of additional applicants.

In considering the minimum of income on which a tax should be imposed, the *effects of exemption* ought to be thoroughly examined. These effects have hitherto received little attention, although pregnant with danger of the most fatal kind.

The present generation have little notion of the intense feeling of antipathy with which the income tax of ten per cent., existing about a third of a century ago, was then viewed,—nor of the popularity which was acquired by its subsequent abolition, and by the measure which accompanied its extinction, of destroying as far as possible every record tending to an exposure of the circumstances of individuals.

The exemption at that time extended to all incomes under £50; but on its renewal in later times, far more extensive exemptions were admitted: all incomes under £150 were expressly exempted. But even this sacrifice of principle was not thought sufficient; and by the same statute it was enacted, that a farmer who paid £300 a-year rent for his farm should be deemed to make a clear income equal to one-half only of that sum. So that every farmer not possessing other sources of income than a farm, whose rent is less than £300 a-year, is at this moment exempt from this tax.

The machinery for collecting the Income Tax is not expensive; and whether the amount of the tax itself is five or twenty-five per cent., the cost of its collection need not be much augmented. This fact alone is a tempting inducement to a Chancellor of the Exchequer to have recourse to its increase whenever increased expenditure becomes necessary, or whenever a deficiency in the revenue is apprehended.

It is unfortunate, that by the very nature of the exemptions from the income tax, a large number of the electors of this country have a direct pecuniary interest in preferring its augmentation to any other mode of taxation. In consequence of these unjust and unstatesmanlike exemptions, numbers of electors will urge their representatives to pledge themselves to oppose all other taxes:—and the ultimate results might be, that the wealthy would be unjustly plundered,—capital be driven from the land, and at last the ruined fortunes of the rich would be accompanied by the absolute starvation of the poor.

I am not aware of any data or returns by which the number of electors possessing annual incomes of given amounts can be ascertained, nor is anything more than a rough approximation necessary for my argument. It is sufficient for my purpose to show that a very large proportion of the elective body in this country is exposed to an influence tending strongly to mislead its decisions from the path of justice; to corrupt the natural expression of public opinion; and in its endeavour to escape from its own fair share of taxation, to place an undue burden upon other classes.

The total number of electors is about a million, comprising persons of every variety of income, from the mere forty-shilling freeholder up to the millionnaire.

Statistical inquiries have not yet supplied any tables which enable us to ascertain, even approximately, how the population of the country is divided with reference to the income of individual classes;—for example, out of the whole number of inhabitants, what number exist on an income of £20, what number on one of £30, what number on one of £50, and so on. Such a table is of great importance, and its want is continually felt by those who are much engaged on inquiries into economical questions. In the absence of such information I shall avail myself of the valuable returns, published in the Tables of the Board of Trade, of the number of persons receiving certain incomes from funded property.

The following table shows, for the year 1846, the number of persons respectively receiving dividends from the various public funds up to the annual amount of the sums shown in the first column. For example, there were 242 623 persons receiving dividends not exceeding £200 annually.

Annual income up to	Number of persons receiving the same	Proportion of ditto for a million of persons
£		
10	84 613	319 000
20	125 784	472 280
100	218 243	822 900
200	242 623	914 830
400	256 548	967 320
600	260 721	983 040
1000	263 445	993 330
2000	264 671	997 930
4000	265 016	999 250
4000 and upwards.	265 218	1 000 000

Now, if the incomes of the voters follow a similar law, and the number be one million, it would appear that there are above 850 000 electors having an income under £150 yearly.*

It is true that many of the wealthier electors will have more than one vote. But it is also true, especially at general elections, that many votes can very rarely be given by the same individual at different places. A certain deduction must, however, be admitted on this ground. On the other hand, there can be scarcely any doubt that a large number of persons whose farm rents are between £150 and £300 a-year, have really clear incomes above £150, although they are exempted from the income-tax. This number, whatever it may amount to, ought, therefore, to be added to the number of those electors who have a pecuniary interest in the selection of representatives who will vote for the increase of that tax.

However slow the progress of this evil may at first be, the result is inevitable. Public opinion so corrupted, taxation thus unjustly charged, will ultimately work out its natural and necessary consequences. Amidst the political errors of the present century, I know of none possessing so truly revolutionary a character,—none so calculated to accelerate its destructive course by its own accumulated momentum,—none which, although seemingly fatal only to the rich, is in reality more fatal to all industry.

The remedy of these anticipated evils is neither difficult nor obscure. Abolish all exemptions—or else reduce the exemption to the lowest possible point, and disqualify from voting all electors who claim the exemption. Public opinion has already been tampered with;—this change is necessary in order to restore it to a wholesome state on the subject of taxation, and enable it to become the fair representative of the intellect of the country, unbiassed by selfish interests. Such an effort is worthy of a statesman who looks beyond the temporary views and compromises of party. It would possess a character peculiarly its own—for it would be disinterested. Winning for its author no present triumph, it would be duly appreciated only when sounder principles of economy shall have worked their slow progress through the opening mind of the nation.

Few, perhaps, will be inclined to deny the evils which I have pointed out as resulting from exemption, although they may differ from me in the extent of its effects, or doubt the soundness of the principles of taxation on which my reasoning rests. Those who hold the latter opinion, I would request to point

* I am by no means disposed to accept this as the real law, or even as any very near approximation to it. I have employed the only data at present known.

out other principles which they propose as the basis of taxation; and I would further entreat them to unite with me in refuting some common and prevailing errors on this question.

'Tax luxuries,' is the maxim of some. But where is there any consistent and admitted definition of a luxury? The luxuries of one class constitute the necessaries of the class above it. Besides, the desire to possess the luxuries of life and to enjoy them in idleness, is the most active principle of industrial excitement. Fortunately for our happiness, those habits of energetic employment which our minds have acquired in the pursuit, indispose us to enjoy that luxurious inactivity to which we had looked forward as the end of our labour; and thus a double blessing crowns our exertions.

'Tax those who can afford to pay taxes,' is the fallacy of another class of the thoughtless. But who, except the individual himself, can judge what he can afford to spend? All his apparent means, his exact income itself, may be known to his neighbours: but unless all the claims to which he is liable, and all the duties by which he is bound, are equally known, no just opinion can be formed of what he can or cannot afford.

It is not at present my intention to enter on the question of *indirect* taxation. I may, however, be permitted to relate an anecdote which singularly illustrates its effects.

An Irish proprietor, whose country residence was much frequented by beggars, resolved to establish a test for discriminating between the idle and the industrious, and also to obtain some small return for the alms he was in the habit of bestowing. He accordingly added to the pump by which the upper part of his house was supplied with water, a piece of mechanism so contrived, that at the end of a certain number of strokes of the pump handle, a penny fell out from an aperture to repay the labourer for his work. This was so arranged, that labourers who continued at the work, obtained very nearly the usual daily wages of labour in that part of the country. The idlest of the vagabonds of course refused this new labour test: but the greater part of the beggars, whose constant tale was that '*they could not earn a fair day's wages for a fair day's work*,' after earning a few pence, usually went away *cursing* the hardness of their taskmaster.

An Italian gentleman, with greater sagacity, devised a more productive pump, and kept it in action at far less expense. The garden wall of his villa adjoined the great high road leading from one of the capitals of northern Italy, from which it was distant but a few miles. Possessing within his garden a fine spring of water, he erected on the outside of the wall a pump for public use, and chaining to it a small iron ladle, he placed near it some rude seats for the weary traveller, and by a slight roof of climbing plants protected the whole from the mid-day sun. In this delightful shade the tired and thirsty travellers on that well-beaten road ever and anon reposed and refreshed themselves, and did not fail to put in requisition the services which the pump so opportunely presented to them. From morning till night many a dusty and way-worn pilgrim plied its handle and went on his way, *blessing* the liberal proprietor for his kind consideration of the passing stranger.

But the owner of the villa was deeply acquainted with human nature. He knew that in that sultry climate the liquid would be more valued from its scarcity, and from the difficulty of acquiring it. He therefore, in order to enhance the value of the gift, wisely arranged the pump, so that its spout was of rather contracted dimensions, and the handle required a moderate application of force to work it. Under these circumstances the pump raised far more water than could pass through its spout; and, to prevent its being wasted, the surplus was conveyed by an invisible channel to a large reservoir judiciously placed for watering the proprietor's own house, stables, and garden,—into which about five pints were poured for every spoonful passing out of the spout for the benefit of the weary traveller. Even this latter portion was not entirely neglected, for the waste-pipe conveyed the part which ran over from the ladle to some delicious strawberry beds at a lower level. Perhaps, by a small addition to this ingenious arrangement, some kind-hearted travellers might be induced to indulge their mules and asses with a taste of the same cool and refreshing fluid; thus paying an additional tribute to the skill and sagacity of the benevolent proprietor. My accomplished friend would doubtless make a most popular Chancellor of the Exchequer, should his Sardinian majesty require his services in that department of administration.

It has sometimes been objected to indirect taxation, that the people are deceived by it. But to attempt deceit is here quite superfluous; the facility with which such taxes are paid arising in a great measure from the ignorance of those who pay them, and from their conviction that they are, in many cases, at liberty to avoid the payment altogether, as well as from the fact that each pays in proportion to the quantity he consumes.

The result of this inquiry leads to a conclusion perpetually forced upon our conviction in the complicated affairs of human society. No single principle can *alone* explain or be safely applied to all the relations which it influences. In almost all cases, more than one or even than two or three general principles combine to govern important consequences; and the statesman must be ever on the watch to discover those other limiting principles which influence, and sometimes even thrust aside the dominant one. Thus the *principle* of direct taxation by an income-tax has been shown to be consistent with justice; but it has been well remarked, that in order to render it *practically* just, it would require angels for commissioners, and other angels for its collectors.

Amidst conflicting and concurring principles acting upon the welfare of a people, it is the duty of the statesman to choose and to advise, not that combination which is in itself the best, but the best amongst those combinations which the nation can be induced to adopt. This line of policy differs entirely from that of compromise: it needs no concealment; it requires no delusion. Whether opposed by the ignorance of the many, or by the erroneous convictions of the few, it is yet possible for a minister to be honest—for a statesman to be sincere. But to reach this elevation, he must have cast off the conventionalities of party.—Whilst advocating that course which he thinks the best amongst the practicable, he must still boldly proclaim his belief in a better.—Above all things, no temptation must induce him ever to prostitute his talents, by attempting to convince feebler minds of the truth of principles which his own clearer understanding rejects as unsound. This stern moral courage may, perhaps, retard the progress of his earlier reputation, but will add to the solidity of his maturer fame, and contribute to the success of his latest efforts. Misleading no followers,—deceiving no friends,—betraying no party, he will be equally free from the graver charges of sacrificing right to expediency—of apostatizing from the truth for power. With a reputation unimpeached even by suspicion, commanding the admiration of his friends and the confidence of the nation, he will bequeath to his countrymen an example of *intellectual integrity* more valuable to them even than the greatest advantages his political sagacity might have achieved.

Thoughts upon an Extension of the Franchise, pp. 3–8

The approach of a General Election has naturally directed the attention of those members who desire re-election, and of other persons who wish to get into the House of Commons, to the subject of an extension of the franchise. Its imagined popularity with the masses (rather than any examination of its merits and disadvantages) seems to be the chief reason for its present advocacy.

Under these circumstances it may be useful to examine into the principles upon which representative assemblies ought to be constituted. The ultimate object of all government is, in a large sense, to insure—

1. The protection of the person of every individual.
2. The protection of his property.

But these objects can only be accomplished by a considerable expenditure, and by means which are quite different in their nature, and of still greater difference in their extent.

The personal protection of each individual is mainly accomplished directly through the intervention of the police, and indirectly by the aid of certain proceedings of courts of law. The poorer classes of society are much more exposed; and are more subject to attacks, upon their person and their property, than the wealthier classes. A poor woman sitting at her stall at the corner of a street vending a few miserable apples occasionally gets insulted, and perhaps pelted, by young urchins, until, no longer able to endure it, she rushes after the offenders, and having appeased her rage by beating one of them, finds on returning that some older vagabond has plundered her small store, or scattered it over the street.

Looking at the subject in a very general point of view, we may assume that the expense of protecting the person is in every class of society nearly the same. Consequently, each individual ought to pay the same fixed sum for the protection of his person.

With respect to property, nothing can be more fair than to make each kind pay for the cost of its own protection. But in the very general view which alone it is here proposed to take, it will be more convenient to suppose property of every kind to cost for its protection a sum proportional to the value of what it is capable of producing.

Taxes are collected annually, and it therefore appears that they are paid for the purpose of protecting the personal liberty and the annual produce of whatever property a man may possess *during the period of one year*.

It has been maintained that taxation and representation should be coextensive; and certainly it is unreasonable to propose that one class of persons should have the privilege of spending the money which others have contributed. The right of voting for representatives in the House of Commons ought, in accordance with the principle here advocated, to depend on two circumstances—

1. A single vote for the protection of personal liberty.
2. A number of votes in proportion to the annual produce of his property during the year in which it is to be protected.

Such appears to be the most equitable plan if a scheme were to be proposed for a new country. But our own case is that of a country in which the possessor of a freehold, however large its revenue, provided it is all within the same county, is entitled only to a single vote, and is thus placed on the same level, as to his choice of a representative, as his neighbour who only possesses land worth forty shillings a year.

The demands for an enlargement of the franchise will probably be pressed, although not with very great force, at the coming elections. The examples of the United States and of some of our own colonies indicate the danger to be apprehended from the effect of lowering the qualification for the suffrage without some countervailing extension in favour of the influence of property.

If it be maintained that every man capable of supporting himself is entitled to a vote in right of his person, then the difficult question will arise respecting the votes he should enjoy in right of his property.

In the present pamphlet it is not proposed to advocate any one plan, but merely to indicate various systems, some of which will probably be thought to err in excess, and others in defect.

The three following tables will probably be sufficient to explain the proposed system:—

PLAN I.

Annual Income.	Number of Votes to be allowed.
£50	} 1
100	
250	} 2
500	
1,000	} 3
2,000	
4,000	} 4
8,000	
16,000	} 5
32,000	
64,000	} 6

PLAN II.

£50	1
100	2
250	3
500	4
1,000	5
2,000	6
4,000	7
8,000	8
16,000	9
32,000	10
64,000	11

PLAN III.

£50	1
100	2
250	4
500	8
1,000	16
2,000	32
4,000	64
8,000	128
16,000	256
32,000	512
64,000	1,024

The first of these plans gives a very moderate addition to the share of votes of the wealthier classes. The number of persons whose income is more than 50*l*. and less than 1000*l*. probably exceeds that of all those possessing larger incomes. In this case a considerable addition will be made to the influence of the middle classes, and but a very moderate increase will fall to the share of the rich.

The second plan still gives much more influence to the middle classes than to the wealthier.

The third plan gives considerably greater power to the wealthier classes.

These tables have been placed before the reader for the purpose of showing that the proposed system admits of an indefinite variety of applications.

If another column were added, containing the number of persons belonging to each class, the product of the number of votes assigned to each individual of a class by the number contained in that class would express its influence in the legislation of the country.

The question naturally arises, on what principle the adjustment between wealth and the number of votes ought to be regulated. At the present time it is premature to recommend any decisive conclusion; but it may not be without advantage to suggest, that an equitable solution of the question would be, to make the tax such that the number of each class of voters, multiplied by the amount at which they are taxed, should be equal to a constant sum. Under such circumstances, all classes of society would have an equal influence in the legislation of the country.

The amount of income received by each individual ought, in this system of representation, to be the measure of the number of his votes. But at its commencement it might be expedient to adopt it partially, until experience has pointed out those directions in which it might be most advantageous to extend it.

The number of votes to which any individual is entitled may be ascertained by various methods, some of which are indicated below.

1. By the voter's returns to the income tax.
2. By the rent of his landed property.
3. By the dividends on his funded or other government securities.
4. By the dividends on his shares in railroads and other companies.
5. By interest on mortgages on other freehold property.

The time during which such securities have been held ought at the least to exceed twelve months previously to the exercise of the vote.

As a matter of policy, it may be desirable to exclude No. 4, and confine holders of personal property to voters possessing a certain amount in the government funds. The effect of this would be to give greater security to those funds, and therefore to render them more valuable.

Severe penalties, and even disqualification from future voting, ought to be imposed upon a fraudulent exercise of the franchise.

As the object of the present pamphlet is merely to open up for discussion a principle of representation which appears to me to be new, and which is the result of following out certain principles of taxation which I published* many years ago, it is unnecessary at present to enter into the question of the distribution of the votes which such a system would confer.

---

* *Thoughts on the Principles of Taxation with reference to Property Tax and its Exceptions.* 8vo., 1848; 3rd edit., 1852.

The Analytical Engines

AFTER CESSATION of the government-sponsored project to construct a Difference Engine, Babbage embarked on what was to become the great series of Analytical Engines. Beginning in the autumn of 1834 he rapidly increased the potential logical power of the Engines, while at the same time enhancing their efficiency in numerical calculation. A crucial step in the development of the Analytical Engines was the introduction in 1836 of a punched-card input system, which was far more convenient than the 'barrel' he had previously intended for the first stage of input. Programs could also be as long as required. Babbage considered a variety of card control arrangements, but his main method used three types of card: control cards, variable cards, and number cards. The Engines were separated functionally into two principal parts, the store and the mill, much like a modern computer. The control cards controlled operation of the mill. The variable cards defined store addresses and operations in the store, such as read, or write, to use modern terms. The number cards held constants.

The Analytical Engines were decimal machines optimized for rapid numerical calculation. The mill contained ingress and egress axes to hold numbers received from or being sent to the store. There were principal axes for calculation, associated with a carry column. Control was organized by one or more 'barrels', which were microprogram stores. There was also a set of registers, the table axes, associated with special purpose equipment for use in division. Babbage planned a set of output devices, including several types of printer, a curve plotter, and a card punch.

When comparing Babbage's Analytical Engines with modern computers some caution is necessary. A computer is an abstract system mapped on to a physical structure, but the abstract system is not independent of the physical structure in which it is to be realized. Consider for example a set of further registers in the mill, organized, say, as a scratchpad store. The mill of an Analytical Engine did indeed include a set of registers, the table axes. Their utility depended on special-purpose equipment, but any set of registers in an Analytical Engine would have depended on special-purpose equipment. A scratchpad, or a cache store, in a modern computer depends on the possibility of making comparatively small stores with a much shorter access time than the main stores. In Babbage's mechanical constructional system that possibility did not exist. It would have been possible to make the logical connections of a scratchpad but in effect it would have been essentially an extension to the main store. Thus, the correct statement would not be that Babbage did include a scratchpad in his mill, but the stronger statement that a scratchpad would have had little meaning with Babbage's constructional system. A negative statement that Babbage's engines did not include some particular feature is liable to be either trivial or misleading. If it merely states that some feature is absent, then a positive statement of what is included is generally much more useful. If, however, a negative statement carries the further implication that Babbage did not think of the idea, then the statement is liable to be misleading, as the truth is usually that we do not know.

The most detailed published description of the Analytical Engines during Babbage's life was given in a paper by General Menabrea, later prime minister of the newly united Italy. This paper was complemented by detailed notes written by Babbage's close friend, Ada Lovelace, Byron's daughter. Both the Menabrea paper and Ada's notes were prepared under Babbage's detailed guidance. Ada's notes have given rise to the quaint story of Ada as the world's first programmer. The story is nonsense: Babbage was. Indeed, it is not entirely certain that Ada ever wrote a single line of code: the programs, if that is the correct term, were probably all worked out by Babbage. However, Ada's proposal of the Bernoulli numbers for the final program was valuable and well chosen, suggesting that she had a considerably better grasp of mathematics than some commentators have allowed.

'Sketch of the Analytical Engine invented by Charles Babbage, Esq.' [by L. F. Menabrea, with notes by Ada Lovelace], *Scientific Memoirs*, iii, 666-731.

[Before submitting to our readers the translation of M. Menabrea's memoir 'On the Mathematical Principles of the Analytical Engine' invented by Mr. Babbage, we shall present to them a list of the printed papers connected with the subject, and also of those relating to the Difference Engine by which it was preceded.

For information on Mr. Babbage's '*Difference* Engine,' which is but slightly alluded to by M. Menabrea, we refer the reader to the following sources:—

1. *Letter to Sir Humphry Davy, Bart., P.R.S., on the Application of Machinery to Calculate and Print Mathematical Tables.* By Charles Babbage, Esq., F.R.S. London, July 1822. Reprinted, with a Report of the Council of the Royal Society, by order of the House of Commons, May 1823.
2. On the Application of Machinery to the Calculation of Astronomical and Mathematical Tables. By Charles Babbage, Esq.—*Memoirs of the Astronomical Society*, vol. i. part 2. London, 1822.
3. Address to the Astronomical Society by Henry Thomas Colebrooke, Esq., F.R.S., President, in presenting the first Gold Medal of the Society to Charles Babbage, Esq., for the invention of the Calculating Engine.— *Memoirs of the Astronomical Society*. London, 1822.
4. On the Determination of the General Term of a New Class of Infinite Series. By Charles Babbage, Esq.—*Transactions of the Cambridge Philosophical Society*.
5. On Mr. Babbage's New Machine for Calculating and Printing Mathematical Tables.—Letter from Francis Baily, Esq., F.R.S., to M. Schumacher. No. 46, *Astronomische Nachrichten*. Reprinted in the *Philosophical Magazine*, May 1824.
6. On a Method of expressing by Signs the Action of Machinery. By Charles Babbage, Esq.—*Philosophical Transactions*. London, 1826.
7. On Errors common to many Tables of Logarithms. By Charles Babbage, Esq.—*Memoirs of the Astronomical Society*, London, 1827.
8. *Report of the Committee appointed by the Council of the Royal Society to consider the subject referred to in a communication received by them from the Treasury respecting Mr. Babbage's Calculating Engine, and to report thereon.* London, 1829.
9. *Economy of Manufactures*, chap. xx. 8vo. London, 1832.
10. Article on Babbage's Calculating Engine.—*Edinburgh Review*, July 1834. No. 120. vol. lix.

The present state of the Difference Engine, which has always been the property of Government, is as follows:—The drawings are nearly finished, and the mechanical notation of the whole, recording every motion of which it is susceptible, is completed. A part of that Engine, comprising sixteen figures, arranged in three orders of differences, has been put together, and has frequently been used during the last eight years. It performs its work with absolute precision. This portion of the Difference Engine, together with all the drawings, are at present deposited in the Museum of King's College, London.

Of the Analytical Engine, which forms the principal object of the present memoir, we are not aware that any notice has hitherto appeared, except a Letter from the Inventor to M. Quetelet, Secretary to the Royal Academy of Sciences at Brussels, by whom it was communicated to that body. We subjoin a translation of this Letter, which was itself a translation of the original, and was not intended for publication by its author.

ROYAL ACADEMY OF SCIENCES AT BRUSSELS.

GENERAL MEETING OF THE 7TH AND 8TH OF MAY, 1835

A Letter from Mr. Babbage announces that he has for six months been engaged in making the drawings of a new calculating machine of far greater power than the first.

'I am myself astonished,' says Mr. Babbage,

'at the power I have been enabled to give to this machine; a year ago I should not have believed this result possible. This machine is intended to contain a hundred variables (or numbers susceptible of changing); each of these numbers may consist of twenty-five figures, $v_1, v_2, \ldots v_n$ being any numbers whatever, n being less than a hundred; if $f(v_1, v_2, v_3, \ldots v_n)$ be any given function which can be formed by addition, subtraction, multiplication, division, extraction of roots, or elevation to powers, the machine will calculate its numerical value; it will afterwards substitute this value in the place of v, or of any other variable, and will calculate this second function with respect to v. It will reduce to tables almost all equations of finite differences. Let us suppose that we have observed a thousand values of a, b, c, d, and that we wish to calculate them by the formula $p = \sqrt{\dfrac{a+b}{cd}}$, the machine must be set to calculate the formula; the first series of the values of a, b, c, d, must be adjusted to it; it will then calculate them, print them, and reduce them to zero; lastly, it will ring a bell to give notice that a new set of constants must be inserted. When there exists a relation between any number of successive coefficients of a series, provided it can be expressed as has already been said, the machine will calculate them and make their terms known in succession; and it may afterwards be disposed so as to find the value of the series for all the values of the variable.'

Mr. Babbage announces, in conclusion, 'that the greatest difficulties of the invention have already been surmounted, and that the plans will be finished in a few months.'

In the *Ninth Bridgewater Treatise*, Mr. Babbage has employed several arguments deduced from the Analytical Engine, which afford some idea of its powers. See *Ninth Bridgewater Treatise*, 8vo, second edition. London, 1834.

Some of the numerous drawings of the Analytical Engine have been engraved on wooden blocks, and from these (by a mode contrived by

Mr. Babbage) various stereotype plates have been taken. They comprise—

1. Plan of the figure wheels for one method of adding numbers.
2. Elevation of the wheels and axis of ditto.
3. Elevation of framing only of ditto.
4. Section of adding wheels and framing together.
5. Section of the adding wheels, sign wheels and framing complete.
6. Impression from the original wood block.
7. Impressions from a stereotype cast of No. 6, with the letters and signs inserted. Nos. 2, 3, 4 and 5 were stereotypes taken from this.
8. Plan of adding wheels and of long and short pinions, by means of which *stepping* is accomplished.

N.B. This process performs the operation of multiplying or dividing a number by any power of ten.

9. Elevation of long pinions in the position for addition.
10. Elevation of long pinions in the position for stepping.
11. Plan of mechanism for carrying the tens (by anticipation), connected with long pinions.
12. Section of the chain of wires for anticipating carriage.
13. Sections of the elevation of parts of the preceding carriage.

All these were executed about five years ago. At a later period (August 1840) Mr. Babbage caused one of his general plans (No. 25) of the whole Analytical Engine to be lithographed at Paris.

Although these illustrations have not been published, on account of the time which would be required to describe them, and the rapid succession of improvements made subsequently, yet copies have been freely given to many of Mr. Babbage's friends, and were in August 1838 presented at Newcastle to the British Association for the Advancement of Science, and in August 1840 to the Institute of France through M. Arago, as well as to the Royal Academy of Turin through M. Plana.—Editor.]

Those labours which belong to the various branches of the mathematical sciences, although on first consideration they seem to be the exclusive province of intellect, may, nevertheless, be divided into two distinct sections; one of which may be called the mechanical, because it is subjected to precise and invariable laws, that are capable of being expressed by means of the operations of matter; while the other, demanding the intervention of reasoning, belongs more specially to the domain of the understanding. This admitted, we may propose to execute, by means of machinery, the mechanical branch of these labours, reserving for pure intellect that which depends on the reasoning faculties. Thus the rigid exactness of those laws which regulate numerical calculations must frequently have suggested the employment of material instruments, either for executing the whole of such calculations or for abridging them; and thence have arisen several intentions having this object in view, but which have in general but partially attained it. For instance, the much-admired machine of Pascal is now simply an object of curiosity, which, whilst it displays the powerful intellect of its inventor, is yet of little utility in itself. Its powers extended no further than the execution of the four* first operations of arithmetic, and indeed were in reality confined to that of the two first, since multiplication and division were the result of a series of additions and subtractions. The chief drawback hitherto on most of such machines is, that they require the continual intervention of a human agent to regulate their movements, and thence arises a source of errors; so that, if their use has not become general for large numerical calculations, it is because they have not in fact resolved the double problem which the question presents, that of *correctness* in the results, united with *oeconomy* of time.

Struck with similar reflections, Mr. Babbage has devoted some years to the realization of a gigantic idea. He proposed to himself nothing less than the construction of a machine capable of executing not merely arithmetical calculations, but even all those of analysis, if their laws are known. The imagination is at first astounded at the idea of such an undertaking; but the more calm reflection we bestow on it, the less impossible does success appear, and it is felt that it may depend on the discovery of some principle so general, that if applied to machinery, the latter may be capable of mechanically

---

* This remark seems to require further comment, since it is in some degree calculated to strike the mind as being at variance with the subsequent passage (page 675), where it is explained that *an engine which can effect these four operations* can in fact effect *every species of calculation*. The apparent discrepancy is stronger too in the translation than in the original, owing to its being impossible to render precisely into the English tongue all the niceties of distinction which the French idiom happens to admit of in the phrases used for the two passages we refer to. The explanation lies in this: that in the one case the execution of these four operations is the *fundamental starting-point*, and the object proposed for attainment by the machine is the *subsequent combination of these* in every possible variety; whereas in the other case the execution of some *one* of these four operations, selected at pleasure, is the *ultimatum*, the sole and utmost result that can be proposed for attainment by the machine referred to, and which result it cannot any further combine or work upon. The one *begins* where the other *ends*. Should this distinction not now appear perfectly clear, it will become so on perusing the rest of the Memoir, and the Notes that are appended to it.—*Note by translator*.

translating the operations which may be indicated to it by algebraical notation. The illustrious inventor having been kind enough to communicate to me some of his views on this subject during a visit he made at Turin, I have, with his approbation, thrown together the impressions they have left on my mind. But the reader must not expect to find a description of Mr. Babbage's engine; the comprehension of this would entail studies of much length; and I shall endeavour merely to give an insight into the end proposed, and to develope the principles on which its attainment depends.

I must first premise that this engine is entirely different from that of which there is a notice in the 'Treatise on the OEconomy of Machinery,' by the same author. But as the latter gave rise* to the idea of the engine in question, I consider it will be a useful preliminary briefly to recall what were Mr. Babbage's first essays, and also the circumstances in which they originated.

It is well known that the French government, wishing to promote the extension of the decimal system, had ordered the construction of logarithmical and trigonometrical tables of enormous extent. M. de Prony, who had been entrusted with the direction of this undertaking, divided it into three sections, to each of which were appointed a special class of persons. In the first section the formulae were so combined as to render them subservient to the purposes of numerical calculation; in the second, these same formulae were calculated for values of the variable, selected at certain successive distances; and under the third section, comprising about eighty individuals, who were most of them only acquainted with the two first rules of arithmetic, the values which were intermediate to those calculated by the second section were interpolated by means of simple additions and subtractions.

An undertaking similar to that just mentioned having been entered upon in England, Mr. Babbage conceived that the operations performed under the third section might be executed by a machine; and this idea he realized by

---

* The idea that the one engine is the offspring and has grown out of the other, is an exceedingly natural and plausible supposition, until reflection reminds us that no *necessary* sequence and connexion need exist between two such inventions, and that they *may* be wholly independent. M. Menabrea has shared this idea in common with persons who have not his profound and accurate insight into the nature of either engine. In Note A. (see the Notes at the end of the Memoir) it will be found sufficiently explained, however, that this supposition is unfounded. M. Menabrea's opportunities were by no means such as could be adequate to afford him information on a point like this, which would be naturally and almost unconsciously *assumed*, and would scarcely suggest any inquiry with reference to it.—*Note by translator.*

means of mechanism, which has been in part put together, and to which the name Difference Engine is applicable, on account of the principle upon which its construction is founded. To give some notion of this, it will suffice to consider the series of whole square numbers, 1, 4, 9, 16, 25, 36, 49, 64, &c. By subtracting each of these from the succeeding one, we obtain a new series, which we will name the Series of First Differences, consisting of the numbers 3, 5, 7, 9, 11, 13, 15, &c. On subtracting from each of these the preceding one, we obtain the Second Differences, which are all constant and equal to 2. We may represent this succession of operations, and their results, in the following table:—

A. Column of square numbers	B. First differences	C. Second differences
1		
	3	
4		$2\,b$
	5	
9		$2\,d$
	7	
16		2
	9	
25		2
	11	
36		

From the mode in which the two last columns B and C have been formed, it is easy to see that if, for instance, we desire to pass from the number 5 to the succeeding one 7, we must add to the former the constant difference 2; similarly, if from the square number 9 we would pass to the following one 16, we must add to the former the difference 7, which difference is in other words the preceding difference 5, plus the constant difference 2; or again, which comes to the same thing, to obtain 16 we have only to add together the three numbers 2, 5, 9, placed obliquely in the direction ab. Similarly, we obtain the number 25 by summing up the three numbers placed in the oblique direction dc: commencing by the addition 2 + 7, we have the first difference 9 consecutively to 7; adding 16 to the 9 we have the square 25. We see then that the three numbers 2, 5, 9 being given, the whole series of successive square numbers, and that of their first differences likewise, may be obtained by means of simple additions.

Now, to conceive how these operations may be reproduced by a machine, suppose the latter to have three dials, designated as A, B, C, on each of which are traced, say a thousand divisions, by way of example, over which a needle shall pass. The two dials, C, B, shall have in addition a registering hammer, which is to give a number of strokes equal to that of the divisions indicated by the needle. For each stroke of the registering hammer of the dial C, the needle B shall advance one division; similarly, the needle A shall advance one division for every stroke of the registering hammer of the dial B. Such is the general disposition of the mechanism.

This being understood, let us at the beginning of the series of operations we wish to execute, place the needle C on the division 2, the needle B on the division 5, and the needle A on the division 9. Let us allow the hammer of the dial C to strike; it will strike twice, and at the same time the needle B will pass over two divisions. The latter will then indicate the number 7, which succeeds the number 5 in the column of first differences. If we now permit the hammer of the dial B to strike in its turn, it will strike seven times, during which the needle A will advance seven divisions; these added to the nine already marked by it, will give the number 16, which is the square number consecutive to 9. If we now recommence these operations, beginning with the needle C, which is always to be left on the division 2, we shall perceive that by repeating them indefinitely, we may successively reproduce the series of whole square numbers by means of a very simple mechanism.

The theorem on which is based the construction of the machine we have just been describing, is a particular case of the following more general theorem: that if in any polynomial whatever, the highest power of whose variable is m, this same variable be increased by equal degrees; the corresponding values of the polynomial then calculated, and the first, second, third, &c. differences of these be taken (as for the preceding series of squares); the mth differences will all be equal to each other. So that, in order to reproduce the series of values of the polynomial by means of a machine analogous to the one above described, it is sufficient that there be $(m+1)$ dials, having the mutual relations we have indicated. As the differences may be either positive or negative, the machine will have a contrivance for either advancing or retrograding each needle, according as the number to be algebraically added may have the sign *plus* or *minus*.

If from a polynomial we pass to a series having an infinite number of terms, arranged according to the ascending powers of the variable, it would at first appear, that in order to apply the machine to the calculation of the function represented by such a series, the mechanism must include an infinite number of dials, which would in fact render the thing impossible. But in many cases the difficulty will disappear, if we observe that for a great number of functions the series which represent them may be rendered convergent; so that, according to the degree of approximation desired, we may limit ourselves to the calculation of a certain number of terms of the series, neglecting the rest. By this method the question is reduced to the primitive case of a finite polynomial. It is thus that we can calculate the succession of the logarithms of numbers. But since, on this particular instance, the terms which had been originally neglected receive increments in a ratio so continually increasing for equal increments of the variable, that the degree of approximation required would ultimately be affected, it is necessary, at certain intervals, to calculate the value of the function by different methods, and then respectively to use the results thus obtained, as data whence to deduce, by means of the machine, the other intermediate values. We see that the machine here performs the office of the third section of calculators mentioned in describing the tables computed by order of the French government, and that the end originally proposed is thus fulfilled by it.

Such is the nature of the first machine which Mr. Babbage conceived. We see that its use is confined to cases where the numbers required are such as can be obtained by means of simple additions or subtractions; that the machine is, so to speak, merely the expression of one* particular theorem of analysis; and that, in short, its operations cannot be extended so as to embrace the solution of an infinity of other questions included within the domain of mathematical analysis. It was while contemplating the vast field which yet remained to be traversed, that Mr. Babbage, renouncing his original essays, conceived the plan of another system of mechanism whose operations should themselves possess all the generality of algebraical notation, and which, on this account, he denominates the *Analytical Engine*.

Having now explained the state of the question, it is time for me to develope the principle on which is based the construction of this latter machine. When analysis is employed for the solution of any problem, there are usually two classes of operations to execute: firstly, the numerical calculation of the various coefficients; and secondly, their distribution in relation to the quantities affected by them. If, for example, we have to obtain the produce of two binomials $(a + bx) (m + nx)$, the result will be represented by $am + (an + bm)x + bnx^2$, in which expression we must first calculate am, an, bm, bn; then take

* See Note A.

the sum of $an + bm$; and lastly, respectively distribute the coefficients thus obtained, amongst the powers of the variable. In order to reproduce these operations by means of a machine, the latter must therefore possess two distinct sets of powers: first, that of executing numerical calculations; secondly, that of rightly distributing the values so obtained.

But if human intervention were necessary for directing each of these partial operations, nothing would be gained under the heads of correctness and oeconomy of time; the machine must therefore have the additional requisite of executing by itself all the successive operations required for the solution of a problem proposed to it, when once the *primitive numerical data* for this same problem have been introduced. Therefore, since from the moment that the nature of the calculation to be executed or of the problem to be resolved have been indicated to it, the machine is, by its own intrinsic power, of itself to go through all the intermediate operations which lead to the proposed result, it must exclude all methods of trial and guess-work, and can only admit the direct processes of calculation*.

It is necessarily thus; for the machine is not a thinking being, but simply an automaton which acts according to the laws imposed upon it. This being fundamental, one of the earliest researches its author had to undertake, was that of finding means for effecting the division of one number by another without using the method of guessing indicated by the usual rules of arithmetic. The difficulties of effecting this combination were far from being among the least; but upon it depended the success of every other. Under the impossibility of my here explaining the process through which this end is attained, we must limit ourselves to admitting that the four first operations of arithmetic, that is addition, subtraction, multiplication and division, can be performed in a direct manner through the intervention of the machine. This granted, the machine is thence capable of performing every species of numerical calculation, for all such calculations ultimately resolve themselves into the four operations we have just named. To conceive how the machine can now go through its functions according to the laws laid down, we will begin by giving an idea of the manner in which it materially represents numbers.

Let us conceive a pile or vertical column consisting of an indefinite number

---

* This must not be understood in too unqualified a manner. The engine is capable, under certain circumstances, of feeling about to discover which of two or more possible contingencies has occurred, and of then shaping its future course accordingly.—*Note by translator*.

of circular discs, all pierced through their centres by a common axis, around which each of them can take an independent rotatory movement. If round the edge of each of these discs are written the ten figures which constitute our numerical alphabet, we may then, by arranging a series of these figures in the same vertical line, express in this manner any number whatever. It is sufficient for this purpose that the first disc represents units, the second tens, the third hundreds, and so on. When two numbers have been thus written on two distinct columns, we may propose to combine them arithmetically with each other, and to obtain the result on a third column. In general, if we have a series of columns* consisting of discs, which columns we will designate as V_0, V_1, V_2, V_3, V_4, &c., we may require, for instance, to divide the number written on the column V_1 by that on the column V_4, and to obtain the result on the column V_7. To effect this operation, we must impart to the machine two distinct arrangements; through the first it is prepared for executing *a division*, and through the second the columns it is to operate on are indicated to it, and also the column on which the result is to be represented. If this division is to be followed, for example, by the addition of two numbers taken on other columns, the two original arrangements of the machine must be simultaneously altered. If, on the contrary, a series of operations of the same nature is to be gone through, then the first of the original arrangements will remain, and the second alone must be altered. Therefore, the arrangements that may be communicated to the various parts of the machine, may be distinguished into two principal classes:

First, that relative to the *Operations*.
Secondly, that relative to the *Variables*.

By this latter we mean that which indicates the columns to be operated on. As for the operations themselves, they are executed by a special apparatus, which is designated by the name of *mill*, and which itself contains a certain number of columns, similar to those of the Variables. When two numbers are to be combined together, the machine commences by effacing them from the columns where they are written, that is it places *zero*† on every disc of the two vertical lines on which the numbers were represented; and it transfers the numbers to the mill. There, the apparatus having been disposed suitably for

* See Note B.
† Zero is not *always* substituted when a number is transferred to the mill. This is explained further on in the memoir, and still more fully in Note D.—*Note by translator.*

the required operation, this latter is effected, and, when completed, the result itself is transferred to the column of Variables which shall have been indicated. Thus the mill is that portion of the machine which works, and the columns of Variables constitute that where the results are represented and arranged. After the preceding explanations, we may perceive that all fractional and irrational results will be represented in decimal fractions. Supposing each column to have forty discs, this extension will be sufficient for all degrees of approximation generally required.

It will now be inquired how the machine can of itself, and without having recourse to the hand of man, assume the successive dispositions suited to the operations. The solution of this problem has been taken from Jacquard's apparatus*, used for the manufacture of brocaded stuffs, in the following manner:—

Two species of threads are usually distinguished in woven stuffs; one is the *warp* or longitudinal thread, the other the *woof* or transverse thread, which is conveyed by the instrument called the shuttle, and which crosses the longitudinal thread or warp. When a brocaded stuff is required, it is necessary in turn to prevent certain threads from crossing the woof, and this according to a succession which is determined by the nature of the design that is to be reproduced. Formerly this process was lengthy and difficult, and it was requisite that the workman, by attending to the design which he was to copy, should himself regulate the movements the threads were to take. Thence arose the high price of this description of stuffs, especially if threads of various colours entered into the fabric. To simplify this manufacture, Jacquard devised the plan of connecting each group of threads that were to act together, with a distinct lever belonging exclusively to that group. All these levers terminate in rods, which are united together in one bundle, having usually the form of a parallelopiped with a rectangular base. The rods are cylindrical, and are separated from each other by small intervals. The process of raising the threads is thus resolved into that of moving these various lever-arms in the requisite order. To effect this, a rectangular sheet of pasteboard is taken, somewhat larger in size than a section of the bundle of lever-arms. If this sheet be applied to the base of the bundle, and an advancing motion be then communicated to the pasteboard, this latter will move with it all the rods of the bundle, and consequently the threads that are connected with each of them. But if the pasteboard, instead of being plain, were pierced with holes corresponding to the extremities of the levers which meet it, then, since each of the

---

* See Note C.

levers would pass through the pasteboard during the motion of the latter, they would all remain in their places. We thus see that it is easy so to determine the position of the holes in the pasteboard, that, at any given moment, there shall be a certain number of levers, and consequently of parcels of threads, raised, while the rest remain where they were. Supposing this process is successively repeated according to a law indicated by the pattern to be executed, we perceive that this pattern may be reproduced on the stuff. For this purpose we need merely compose a series of cards according to the law required, and arrange them in suitable order one after the other; then, by causing them to pass over a polygonal beam which is so connected as to turn a new face for every stroke of the shuttle, which face shall then be impelled parallelly to itself against the bundle of lever-arms, the operation of raising the threads will be regularly performed. Thus we see that brocaded tissues may be manufactured with a precision and rapidity formerly difficult to obtain.

Arrangements analogous to those just described have been introduced into the Analytical Engine. It contains two principal species of cards: first, Operation cards, by means of which the parts of the machine are so disposed as to execute any determinate series of operations, such as additions, subtractions, multiplications, and divisions; secondly, cards of the Variables, which indicate to the machine the columns on which the results are to be represented. The cards, when put in motion, successively arrange the various portions of the machine according to the nature of the processes that are to be effected, and the machine at the same time executes these processes by means of the various pieces of mechanism of which it is constituted.

In order more perfectly to conceive the thing, let us select as an example the resolution of two equations of the first degree with two unknown quantities. Let the following be the two equations, in which x and y are the unknown quantities:—

$$\begin{cases} m\,x + n\,y = d \\ m'x + n'y = d'. \end{cases}$$

We deduce $x = \dfrac{dn' - d'n}{n'm - nm'}$, and for y an analogous expression. Let us continue to represent by V_0, V_1, V_2, &c. the different columns which contain the numbers, and let us suppose that the first eight columns have been chosen for expressing on them the numbers represented by m, n, d, m', n', d', n and n', which implies that $V_0 = m$, $V_1 = n$, $V_2 = d$, $V_3 = m'$, $V_4 = n'$, $V_5 = d'$, $V_6 = n$, $V_7 = n'$.

The series of operations commanded by the cards, and the results obtained, may be represented in the following table:—

| Number of the operations | Operation-cards | Cards of the variables | | Progress of the operations |
	Symbols indicating the nature of the operations	Columns on which operations are to be performed	Columns which receive results of operations	
1	$\times$	$V_2 \times V_4 =$	V_8	$= dn'$
2	$\times$	$V_5 \times V_1 =$	V_9	$= d'n$
3	$\times$	$V_4 \times V_0 =$	V_{10}	$= n'm$
4	$\times$	$V_1 \times V_3 =$	V_{11}	$= nm'$
5	$-$	$V_8 - V_9 =$	V_{12}	$= dn' - d'n$
6	$-$	$V_{10} - V_{11} =$	V_{13}	$= n'm - nm'$
7	$\div$	$\dfrac{V_{12}}{V_{13}} =$	V_{14}	$= x = \dfrac{dn' - d'n}{n'm - nm'}$

Since the cards do nothing but indicate in what manner and on what columns the machine shall act, it is clear that we must still, in every particular case, introduce the numerical data for the calculation. Thus, in the example we have selected, we must previously inscribe the numerical values of m, n, d, m', n', d', in the order and on the columns indicated, after which the machine when put in action will give the value of the unknown quantity x for this particular case. To obtain the value of y, another series of operations analogous to the preceding must be performed. But we see that they will be only four in number, since the denominator of the expression for y, excepting the sign, is the same as that for x, and equal to $n'm - nm'$. In the preceding table it will be remarked that the column for operations indicates four successive *multiplications*, two *subtractions*, and one *division*. Therefore, if desired, we need only use three operation cards; to manage which, it is sufficient to introduce into the machine an apparatus which shall, after the first multiplication, for instance, retain the card which relates to this operation, and not allow it to advance so as to be replaced by another one, until after the same operation shall have been four times repeated. In the preceding example we have seen, that to find the value of x we must begin by writing the coefficients m, n, d, m', n', d', upon eight columns, thus repeating n and n' twice. According to the same method, if it were required to calculate y likewise, these coefficients must be written on twelve different columns. But it is possible to simplify this process, and thus to diminish the chances of errors, which chances are greater, the larger the number of the quantities that have to be inscribed previous to setting the machine in action. To understand this simplification, we must remember that every number written on a column must, in order to be arithmetically combined with another number, be effaced from the column on which it is, and transferred to the *mill*. Thus, in the example we have discussed, we will take the two coefficients m and n', which are each of them to enter into *two* different products, that is m into mn' and md', n' into mn' and $n'd$. These coefficients will be inscribed on the columns V_0 and V_4. If we commence the series of operations by the product of m into n', these numbers will be effaced from the columns V_0 and V_4, that they may be transferred to the mill, which will multiply them into each other, and will then command the machine to represent the result, say on the column V_6. But as these numbers are each to be used again in another operation, they must again be inscribed somewhere; therefore, while the mill is working out their product, the machine will inscribe them anew on any two columns that may be indicated to it through the cards; and, as in the actual case, there is no reason why they should not resume their former places, we will suppose them again inscribed on V_0 and V_4, whence in short they would not finally disappear, to be reproduced no more, until they should have gone through all the combinations in which they might have to be used.

We see, then, that the whole assemblage of operations requisite for resolving the two* above equations of the first degree, may be definitively represented [see table on p. 258]:

In order to diminish to the utmost the chances of error in inscribing the numerical data of the problem, they are successively placed on one of the columns of the mill; then, by means of cards arranged for this purpose, these same numbers are caused to arrange themselves on the requisite columns, without the operator having to give his attention to it; so that his undivided mind may be applied to the simple inscription of these same numbers.

According to what has now been explained, we see that the collection of columns of Variables may be regarded as a *store* of numbers, accumulated there by the mill, and which, obeying the orders transmitted to the machine by means of the cards, pass alternately from the mill to the store, and from the store to the mill, that they may undergo the transformations demanded by the nature of the calculation to be performed.

Hitherto no mention has been made of the *signs* in the results, and the machine would be far from perfect were it incapable of expressing and combining amongst each other positive and negative quantities. To accom-

* See Note D.

Columns on which are inscribed the primitive data	Number of the operations	Number of the Operation cards	Nature of each operation	Columns acted on by each operation	Columns that receive the result of each operation	Indication of change of value on any column	Statement of results
$^1V_0 = m$	1	1	×	$^1V_0 \times {}^1V_4 =$	$^1V_6$	$\left\{\begin{matrix}^1V_0 = {}^1V_0\\ ^1V_4 = {}^1V_4\end{matrix}\right\}$	$^1V_6 = mn'$
$^1V_1 = n$	2	1	×	$^1V_3 \times {}^1V_1 =$	$^1V_7$	$\left\{\begin{matrix}^1V_3 = {}^1V_3\\ ^1V_1 = {}^1V_1\end{matrix}\right\}$	$^1V_7 = m'n$
$^1V_2 = d$	3	1	×	$^1V_2 \times {}^1V_4 =$	$^1V_8$	$\left\{\begin{matrix}^1V_2 = {}^1V_2\\ ^1V_4 = {}^0V_4\end{matrix}\right\}$	$^1V_8 = dn'$
$^1V_3 = m'$	4	1	×	$^1V_5 \times {}^1V_1 =$	$^1V_9$	$\left\{\begin{matrix}^1V_5 = {}^1V_5\\ ^1V_1 = {}^0V_1\end{matrix}\right\}$	$^1V_9 = d'n$
$^1V_4 = n'$	5	1	×	$^1V_0 \times {}^1V_5 =$	$^1V_{10}$	$\left\{\begin{matrix}^1V_0 = {}^0V_0\\ ^1V_5 = {}^0V_5\end{matrix}\right\}$	$^1V_{10} = d'm$
$^1V_5 = d'$	6	1	×	$^1V_2 \times {}^1V_3 =$	$^1V_{11}$	$\left\{\begin{matrix}^1V_2 = {}^0V_2\\ ^1V_3 = {}^0V_3\end{matrix}\right\}$	$^1V_{11} = dm'$
	7	2	−	$^1V_6 - {}^1V_7 =$	$^1V_{12}$	$\left\{\begin{matrix}^1V_6 = {}^0V_6\\ ^1V_7 = {}^0V_7\end{matrix}\right\}$	$^1V_{12} = mn' - m'n$
	8	2	−	$^1V_8 - {}^1V_9 =$	$^1V_{13}$	$\left\{\begin{matrix}^1V_8 = {}^0V_8\\ ^1V_9 = {}^0V_9\end{matrix}\right\}$	$^1V_{13} = dn' - d'n$
	9	2	−	$^1V_{10} - {}^1V_{11} =$	$^1V_{14}$	$\left\{\begin{matrix}^1V_{10} = {}^0V_{10}\\ ^1V_{11} = {}^0V_{11}\end{matrix}\right\}$	$^1V_{14} = d'm - dm'$
	10	3	÷	$^1V_{13} \div {}^1V_{12} =$	$^1V_{15}$	$\left\{\begin{matrix}^1V_{13} = {}^0V_{13}\\ ^1V_{12} = {}^1V_{12}\end{matrix}\right\}$	$^1V_{15} = \dfrac{dn' - d'n}{mn' - m'n} = x$
	11	3	÷	$^1V_{14} \div {}^1V_{12} =$	$^1V_{16}$	$\left\{\begin{matrix}^1V_{14} = {}^0V_{14}\\ ^1V_{12} = {}^0V_{12}\end{matrix}\right\}$	$^1V_{16} = \dfrac{d'm - dm'}{mn' - m'n} = y$
1	2	3	4	5	6	7	8

plish this end, there is, above every column, both of the mill and of the store, a disc, similar to the discs of which the columns themselves consist. According as the digit on this disc is even or uneven, the number inscribed on the corresponding column below it will be considered as positive or negative. This granted, we may, in the following manner, conceive how the signs can be algebraically combined in the machine. When a number is to be transferred from the store to the mill, and *vice versa*, it will always be transferred with its sign, which will be effected by means of the cards, as has been explained in what precedes. Let any two numbers then, on which we are to operate arithmetically, be placed in the mill with their respective signs. Suppose that we are first to add them together; the operation-cards will command the addition: if the two numbers be of the same sign, one of the two will be entirely effaced from where it was inscribed, and will go to add itself on the column which contains the other number; the machine will, during this operation, be able, by means of a certain apparatus, to prevent any movement in the disc of signs which belongs to the column on which the addition is made, and thus the result will remain with the sign which the two given numbers originally had. When two numbers have two different signs, the addition commanded by the card will be changed into a subtraction through the intervention of mechanisms which are brought into play by this very difference of sign. Since the subtraction can only be effected on the larger of the two numbers, it must be arranged that the disc of signs of the larger number shall not move while the smaller of the two numbers is being effaced from its column and subtracted from the other, whence the result will have the sign of this latter, just as in fact it ought to be. The combinations to which algebraical subtraction give rise, are analogous to the preceding. Let us pass on to multiplication. When two numbers to be multiplied are of the same sign, the result is positive; if the signs are different, the product must be negative. In order that the machine may act conformably to this law, we have but to conceive that on the column containing the product of the two given numbers, the digit which indicates the sign of that product, has been formed by the mutual addition of the two digits that respectively indicated the signs of the two numbers; it is then obvious that if the digits of the signs are both even, or both odd, their sum will be an even number, and consequently will express a positive number; but that if, on the contrary, the two digits of the signs are one even and the other odd, their sum will be an odd number, and will consequently express a negative number. In the case of division, instead of adding the digits of the discs, they must be subtracted one from the other, which will produce results analogous to the preceding; that is to say, that if these figures are both even or both uneven, the remainder of this subtraction will be even; and it will be uneven in the contrary case. When I speak of mutually adding or subtracting the numbers expressed by the digits of the signs, I merely mean that one of the sign-discs is made to advance or retrograde a number of divisions equal to that which is expressed by the digit on the other sign-disc. We see, then, from the preceding explanation, that it is possible mechanically to combine the signs of quantities so as to obtain results comformable to those indicated by algebra*.

The machine is not only capable of executing those numerical calculations which depend on a given algebraical formula, but it is also fitted for analytical calculations in which there are one or several variables to be considered. It

---

* Not having had leisure to discuss with Mr. Babbage the manner of introducing into his machine the combination of algebraical signs, I do not pretend here to expose the method he uses for this purpose; but I considered that I ought myself to supply the deficiency, conceiving that this paper would have been imperfect if I had omitted to point out one means that might be employed for resolving this essential part of the problem in question.

must be assumed that the analytical expression to be operated on can be developed according to powers of the variable, or according to determinate functions of this same variable, such as circular functions, for instance; and similarly for the result that is to be attained. If we then suppose that above the columns of the store, we have inscribed the powers or the functions of the variable, arranged according to whatever is the prescribed law of development, the coefficients of these several terms may be respectively placed on the corresponding column below each. In this manner we shall have a representation of an analytical development; and, supposing the position of the several terms composing it to be invariable, the problem will be reduced to that of calculating their co-efficients according to the laws demanded by the nature of the question. In order to make this more clear, we shall take the following* very simple example, in which we are to multiply $(a + bx^1)$ by $(A + B\cos^1 x)$. We shall begin by writing x^0, x^1, $\cos^0 x$, $\cos^1 x$, above the columns V_0, V_1, V_2, V_3; then, since from the form of the two functions to be combined, the terms which are to compose the products will be of the following nature, $x^0 \cdot \cos^0 x$, $x^0 \cdot \cos^1 x$, $x^1 \cdot \cos^0 x$, $x^1 \cdot \cos^0 x$; these will be inscribed above the columns V_4, V_5, V_6, V_7. The coefficients of x^0, x^1, $\cos^0 x$, $\cos^1 x$ being given, they will, by means of the mill, be passed to the columns V_0, V_1, V_2 and V_3. Such are the primitive data of the problem. It is now the business of the machine to work out its solution, that is to find the coefficients which are to be inscribed on V_4, V_5, V_6, V_7. To attain this object, the law of formation of these same coefficients being known, the machine will act through the intervention of the cards, in the manner [opposite].

It will now be perceived that a general application may be made of the principle developed in the preceding example, to every species of process which it may be proposed to effect on series submitted to calculation. It is sufficient that the law of formation of the coefficients be known, and that this law be inscribed on the cards of the machine, which will then of itself execute all the calculations requisite for arriving at the proposed result. If, for instance, a recurring series were proposed, the law of formation of the coefficients being here uniform, the same operations which must be performed for one of them will be repeated for all the others; there will merely be a change in the locality of the operation, that is it will be performed with different columns. Generally, since every analytical expression is susceptible of being expressed in a series ordered according to certain functions of the variable, we perceive that the machine will include all analytical calculations

* See Note E.

*Columns above which are written the functions of the variable	Coefficients — Given	Coefficients — To be formed	Cards of the operations — Number of the operations	Cards of the operations — Nature of the operation	Cards of the variables — Columns on which operations are to be performed	Cards of the variables — Columns on which are to be inscribed the results of the operations	Cards of the variables — Indication of change of value on any column submitted to an operation	Cards of the variables — Results of the operations
$x^0 \ldots\, {}^{I}V_0$	a	—	—	—	—	—	—	x^0
$x^1 \ldots\, {}^{I}V_1$	b	—	—	—	—	—	—	x^1
$\cos^0 x \ldots\, {}^{I}V_2$	A	—	—	—	—	—	—	$\cos^0 x$
$\cos^1 x \ldots\, {}^{I}V_3$	B	—	—	—	—	—	—	$\cos^1 x$
		aA	1	$\times$	${}^{I}V_0 \times {}^{I}V_2 = {}^{I}V_4$	${}^{I}V_4$	$\left\{ {}^{I}V_0 = {}^{I}V_0 \atop {}^{I}V_2 = {}^{I}V_2 \right\}$	${}^{I}V_4 = aA$ coefficients of $x^0 \cos^0 x$
		aB	2	$\times$	${}^{I}V_0 \times {}^{I}V_3 = {}^{I}V_5$	${}^{I}V_5$	$\left\{ {}^{I}V_0 = {}^{0}V_0 \atop {}^{I}V_3 = {}^{I}V_3 \right\}$	${}^{I}V_5 = aB$ coefficients of $x^0 \cos^1 x$
		bA	3	$\times$	${}^{I}V_1 \times {}^{I}V_2 = {}^{I}V_6$	${}^{I}V_6$	$\left\{ {}^{I}V_1 = {}^{I}V_1 \atop {}^{I}V_2 = {}^{0}V_2 \right\}$	${}^{I}V_6 = bA$ coefficients of $x^1 \cos^0 x$
		bB	4	$\times$	${}^{I}V_1 \times {}^{I}V_3 = {}^{I}V_7$	${}^{I}V_7$	$\left\{ {}^{I}V_1 = {}^{0}V_1 \atop {}^{I}V_3 = {}^{0}V_3 \right\}$	${}^{I}V_7 = bB$ coefficients of $x^1 \cos^1 x$

* For an explanation of the upper left-hand indices attached to the V's in this and in the preceding Table, we must refer the reader to Note D, amongst those appended to the memoir.—*Note by translator.*

which can be definitively reduced to the formation of coefficients according to certain laws, and to the distribution of these with respect to the variables.

We may deduce the following important consequence from these explanations, viz. that since the cards only indicate the nature of the operations to be performed, and the columns of Variables with which they are to be executed, these cards will themselves possess all the generality of analysis, of which they are in fact merely a translation. We shall now further examine some of the difficulties which the machine must surmount, if its assimilation to analysis is to be complete. There are certain functions which necessarily change in nature when they pass through zero or infinity, or whose values cannot be admitted when they pass these limits. When such cases present themselves, the machine is able, by means of a bell, to give notice that the passage through zero or infinity is taking place, and it then stops until the attendant has again set it in action for whatever process it may next be desired that it shall perform. If this process has been foreseen, then the machine, instead of ringing, will so dispose itself as to present the new cards which have relation to the operation that is to succeed the passage through zero and infinity. These new cards may follow the first, but may only come into play contingently upon one or other of the two circumstances just mentioned taking place.

Let us consider a term of the form ab^n; since the cards are but a translation of the analytical formula, their number in this particular case must be the same, whatever be the value of n; that is to say, whatever be the number of multiplications required for elevating b to the nth power (we are supposing for the moment that n is a whole number). Now, since the exponent n indicates that b is to be multiplied n times by itself, and all these operations are of the same nature, it will be sufficient to employ one single operation-card, viz. that which orders the multiplication.

But when n is given for the particular case to be calculated, it will be further requisite that the machine limit the number of its multiplications according to the given values. The process may be thus arranged. The three numbers a, b and n will be written on as many distinct columns of the store; we shall designate them V_0, V_1, V_2; the result ab^n will place itself on the column V_3. When the number n has been introduced into the machine, a card will order a certain registering-apparatus to mark $(n-1)$, and will at the same time execute the multiplication of b by b. When this is completed, it will be found that the registering-apparatus has effaced a unit, and that it only marks $(n-2)$; while the machine will now again order the number b written on the column V_1 to multiply itself with the product b^2 written on the column V_3, which will give b^3. Another unit is then effaced from the registering-apparatus, and the same processes are continually repeated until it only marks zero. Thus the number

b^n will be found inscribed on V_3, when the machine, pursuing its course of operations, will order the product of b^n by a; and the required calculation will have been completed without there being any necessity that the number of operation-cards used should vary with the value of n. If n were negative, the cards, instead of ordering the multiplication of a by b^n, would order its division; this we can easily conceive, since every number, being inscribed with its respective sign, is consequently capable of reacting on the nature of the operations to be executed. Finally, if n were fractional, of the form p/q, an additional column would be used for the inscription of q, and the machine would bring into action two sets of processes, one for raising b to the power p, the other for extracting the qth root of the number so obtained.

Again, it may be required, for example, to multiply an expression of the form $ax^m + bx^n$ by another $Ax^p + Bx^q$, and then to reduce the product to the least number of terms, if any of the indices are equal. The two factors being ordered with respect to x, the general result of the multiplication would be $Aax^{m+p} + Abx^{n+p} + Bax^{m+q} + Bbx^{n+q}$. Up to this point the process presents no difficulties; but suppose that we have $m = p$ and $n = q$, and that we wish to reduce the two middle terms to a single one $(Ab + Ba)x^{m+q}$. For this purpose, the cards may order $m + q$ and $n + p$ to be transferred into the mill, and there subtracted one from the other; if the remainder is nothing, as would be the case on the present hypothesis, the mill will order other cards to bring to it the coefficients Ab and Ba, that it may add them together and give them in this state as a coefficient for the single term $x^{n+p} = x^{m+q}$.

This example illustrates how the cards are able to reproduce all the operations which intellect performs in order to attain a determinate result, if these operations are themselves capable of being precisely defined.

Let us now examine the following expression:—

$$2 \cdot \frac{2^2 \cdot 4^2 \cdot 6^2 \cdot 8^2 \cdot 10^2 \ldots (2n)^2}{1^2 \cdot 3^2 \cdot 5^2 \cdot 7^2 \cdot 9^2 \ldots (2n-1)^2 \cdot (2n+1)^2},$$

which we know becomes equal to the ratio of the circumference to the diameter, when n is infinite. We may require the machine not only to perform the calculation of this fractional expression, but further to give indication as soon as the value becomes identical with that of the ratio of the circumference to the diameter when n is infinite, a case in which the computation would be impossible. Observe that we should thus require of the machine to interpret a result not of itself evident, and that this is not amongst its attributes, since it is no thinking being. Neverthelss, when $n = \infty$ has been foreseen [In the original paper by Menabrea, 'le cas $n = \infty$' was misprinted as 'le cos $n = \infty$.' Ada translated this as 'when the cos of $n = \infty$,' which is nonsense.], a card may immediately order the substitution of the value of π (π being the ratio of the circumference to the diameter), without going through the series of calculations indicated. This would merely require that the machine contain a special card, whose office it should be to place the number π in a direct and independent manner on the column indicated to it. And here we should introduce the mention of a third species of cards, which may be called *cards of numbers*. There are certain numbers, such as those expressing the ratio of the circumference to the diameter, the Numbers of Bernoulli, &c., which frequently present themselves in calculations. To avoid the necessity for computing them every time they have to be used, certain cards may be combined specially in order to give these numbers ready made into the mill, whence they afterwards go and place themselves on those columns of the store that are destined for them. Through this means the machine will be susceptible of those simplifications afforded by the use of numerical tables. It would be equally possible to introduce, by means of these cards, the logarithms of numbers; but perhaps it might not be in this case either the shortest or the most appropriate method; for the machine might be able to perform the same calculations by other more expeditious combinations, founded on the rapidity with which it executes the four first operations of arithmetic. To give an idea of this rapidity, we need only mention that Mr Babbage believes he can, by his engine, form the product of two numbers, each containing twenty figures, in *three minutes*.

Perhaps the immense number of cards required for the solution of any rather complicated problem may appear to be an obstacle; but this does not seem to be the case. There is no limit to the number of cards that can be used. Certain stuffs require for their fabrication not less than *twenty thousand* cards, and we may unquestionably far exceed even this quantity*.

Resuming what we have explained concerning the Analytical Engine, we may conclude that it is based on two principles: the first, consisting in the fact that every arithmetical calculation ultimately depends on four principal operations—addition, subtraction, multiplication, and division; the second, in the possibility of reducing every analytical calculation to that of the coefficients for the several terms of a series. If this last principle be true, all the operations of analysis come within the domain of the engine. To take another point of view: the use of the cards offers a generality equal to that of algebraical formulae, since such a formula simply indicates the nature and order of the operations requisite for arriving at a certain definite result, and similarly the cards merely command the engine to perform these same operations; but in

* See Note F.

order that the mechanisms may be able to act to any purpose, the numerical data of the problem must in every particular case be introduced. Thus the same series of cards will serve for all questions whose sameness of nature is such as to require nothing altered excepting the numerical data. In this light the cards are merely a translation of algebraical formulae, or, to express it better, another form of analytical notation.

Since the engine has a mode of acting peculiar to itself, it will in every particular case be necessary to arrange the series of calculations conformably to the means which the machine possesses; for such or such a process which might be very easy for a calculator, may be long and complicated for the engine, and *vice versa*.

Considered under the most general point of view, the essential object of the machine being to calculate, according to the law dictated to it, the values of numerical coefficients which it is then to distribute appropriately on the columns which represent the variables, it follows that the interpretation of formulae and of results is beyond its province, unless indeed this very interpretation be itself susceptible of expression by means of the symbols which the machine employs. Thus, although it is not itself the being that reflects, it may yet be considered as the being which executes the conceptions of intelligence*. The cards receive the impress of these conceptions, and transmit to the various trains of mechanism composing the engine the orders necessary for their action. When once the engine shall have been constructed, the difficulty will be reduced to the making out of the cards; but as these are merely the translation of algebraical formulae, it will, by means of some simple notations, be easy to consign the execution of them to a workman. Thus the whole intellectual labour will be limited to the preparation of the formulae, which must be adapted for calculation by the engine.

Now, admitting that such an engine can be constructed, it may be inquired: what will be its utility? To recapitulate; it will afford the following advantages:—First, rigid accuracy. We know that numerical calculations are generally the stumbling-block to the solution of problems, since errors easily creep into them, and it is by no means always easy to detect these errors. Now the engine, by the very nature of its mode of acting, which requires no human intervention during the course of its operations, presents every species of security under the head of correctness; besides, it carries with it its own check; for at the end of every operation it prints off, not only the results, but likewise the numerical data of the question; so that it is easy to verify whether the question has been correctly proposed. Secondly, economy of time: to convince

* See Note G.

ourselves of this, we need only recollect that the multiplication of two numbers, consisting each of twenty figures, requires at the very utmost three minutes. Likewise, when a long series of identical computations is to be performed, such as those required for the formation of numerical tables, the machine can be brought into play so as to give several results at the same time, which will greatly abridge the whole amount of the processes. Thirdly, economy of intelligence: a simple arithmetical computation requires to be performed by a person possessing some capacity; and when we pass to more complicated calculations, and wish to use algebraical formulae in particular cases, knowledge must be possessed which pre-supposes preliminary mathematical studies of some extent. Now the engine, from its capability of performing by itself all these purely material operations, spares intellectual labour, which may be more profitably employed. Thus the engine may be considered as a real manufactory of figures, which will lend its aid to those many useful sciences and arts that depend on numbers. Again, who can foresee the consequences of such an invention? In truth, how many precious observations remain practically barren for the progress of the sciences, because there are not powers sufficient for computing the results! And what discouragement does the perspective of a long and arid computation cast into the mind of a man of genius, who demands time exclusively for meditation, and who beholds it snatched from him by the material routine of operations! Yet it is by the laborious route of analysis that he must reach truth; but he cannot pursue this unless guided by numbers; for without numbers it is not given us to raise the veil which envelopes the mysteries of nature. Thus the idea of constructing an appparatus capable of aiding human weakness in such researches, is a conception which, being realized, would mark a glorious epoch in the history of the sciences. The plans have been arranged for all the various parts, and for all the wheel-work, which compose this immense apparatus, and their action studied; but these have not yet been fully combined together in the drawings* and mechanical notation†. The confidence which the genius of Mr. Babbage must inspire, affords legitimate ground for hope that this enterprise will be crowned with success; and while we render homage to the intelligence which directs it, let us breathe aspirations for the accomplishment of such an undertaking.

---

* This sentence has been slightly altered in the translation in order to express more exactly the present state of the engine.—*Note by translator.*

† The notation here alluded to is a most interesting and important subject, and would have well deserved a separate and detailed Note upon it, amongst those appended to the Memoir. It has, however, been impossible, within the space allotted, even to touch upon so wide a field.—*Note by translator.*

NOTES BY THE TRANSLATOR

NOTE A.—PAGE 251

The particular function whose integral the Difference Engine was constructed to tabulate, is

$$\Delta^7 u_z = 0.$$

The purpose which that engine has been specially intended and adapted to fulfil, is the computation of nautical and astronomical tables. The integral of

$$\Delta^7 u_z = 0$$

being

$$u_z = a + bx + cx^2 + dx^3 + ex^4 + fx^5 + gx^6,$$

the constants, a, b, c, &c. are represented on the seven columns of discs, of which the engine consists. It can therefore tabulate *accurately* and to an *unlimited extent*, all series whose general term is comprised in the above formula; and it can also tabulate *approximately* between *intervals of greater or less extent*, all other series which are capable of tabulation by the method of Differences.

The Analytical Engine, on the contrary, is not merely adapted for *tabulating* the results of one particular function and of no other, but for *developing and tabulating* any function whatever. In fact the engine may be described as being the material expression of any indefinite function of any degree of generality and complexity, such as for instance,

$$\mathrm{F}\,(x, y, z, \log x, \sin y, x^p, \&c.),$$

which is, it will be observed, a function of all other possible functions of any number of quantities.

In this, which we may call the *neutral* or *zero* state of the engine, it is ready to receive at any moment, by means of cards constituting a portion of its mechanism (and applied on the principle of those used in the Jacquard-loom), the impress of whatever *special* function we may desire to develope or to tabulate. These cards contain within themselves (in a manner explained in the Memoir itself, pages 677 and 678) the law of development of the particular function that may be under consideration, and they compel the mechanism to act accordingly in a certain corresponding order. One of the simplest cases would be, for example, to suppose that

$$\mathrm{F}\,(x, y, z, \&c.\ \&c.)$$

is the particular function

$$\Delta^n u_z = 0$$

which the Difference Engine tabulates for values of n only up to 7. In this case the cards would order the mechanism to go through that succession of operations which would tabulate

$$u^z = a + bx + cx^2 + \ldots mx^{n+1},$$

Where n might be any number whatever.

These cards, however, have nothing to do with the regulation of the particular *numerical* data. They merely determine the *operations** to be effected, which operations may of course be performed on an infinite variety of particular numerical values, and do not bring out any definite numerical results unless the numerical data of the problem have been impressed on the requisite portions of the train of mechanism. In the above examples, the first essential step towards an arithmetical result, would be the substitution of specific numbers for n, and for the other primitive quantities which enter into the function.

Again, let us suppose that for F we put two complete equations of the fourth degree between x and y. We must then express on the cards the law of elimination for such equations. The engine would follow out those laws, and would ultimately give the equation of one variable which results from such elimination. Various *modes* of elimination might be selected; and of course the cards must be made out accordingly. The following is one mode that might be adopted. The engine is able to multiply together any two functions of the form

$$a + bx + cx^2 + \ldots px^n.$$

This granted, the two equations may be arranged according to the powers of y, and the coefficients of the powers of y may be arranged according to powers of x. The elimination of y will result from the successive multiplications and subtractions of several such functions. In this, and in all other instances, as was explained above, the particular *numerical* data and the *numerical* results are determined by means and by portions of the mechanism which act quite independently of those that regulate the *operations*.

In studying the action of the Analytical Engine, we find that the peculiar and independent nature of the considerations which in all mathematical analysis belong to *operations*, as distinguished from *the objects operated upon* and from the *results* of the operations performed upon these objects, is very strikingly defined and separated.

It is well to draw attention to this point, not only because its full appreciation is

---

* We do not mean to imply that the *only* use made of the Jacquard cards is that of regulating the algebraical *operations*. But we mean to explain that *those* cards and portions of mechanism which regulate these *operations*, are wholly independent of those which are used for other purposes. M. Menabrea explains that there are *three* classes of cards used in the engine for three distinct sets of objects, viz. *Cards of the Operations, Cards of the Variables*, and certain *Cards of Numbers*. (See pages 678 and 687.)

essential to the attainment of any very just and adequate general comprehension of the powers and mode of action of the Analytical Engine, but also because it is one which is perhaps too little kept in view in the study of mathematical science in general. It is, however, impossible to confound it with other considerations, either when we trace the manner in which that engine attains its results, or when we prepare the data for its attainment of those results. It were much to be desired, that when mathematical processes pass through the human brain instead of through the medium of inanimate mechanism, it were equally a necessity of things that the reasonings connected with *operations* should hold the same just place as a clear and well-defined branch of the subject of analysis, a fundamental but yet independent ingredient in the science, which they must do in studying the engine. The confusion, the difficulties, the contradictions which, in consequence of a want of accurate distinctions in this particular, have up to even a recent period encumbered mathematics in all those branches involving the consideration of negative and impossible quantities, will at once occur to the reader who is at all versed in this science, and would alone suffice to justify dwelling somewhat on the point, in connexion with any subject so peculiarly fitted to give forcible illustration of it, as the Analytical Engine. It may be desirable to explain, that by the word *operation*, we mean *any process which alters the mutual relation of two or more things*, be this relation of what kind it may. This is the most general definition, and would include all subjects in the universe. In abstract mathematics, of course operations alter those particular relations which are involved in the considerations of number and space, and the *results* of operations are those peculiar results which correspond to the nature of the subjects of operation. But the science of operations, as derived from mathematics more especially, is a science of itself, and has its own abstract truth and value; just as logic has its own peculiar truth and value, independently of the subjects to which we may apply its reasonings and processes. Those who are accustomed to some of the more modern views of the above subject, will know that a few fundamental relations being true, certain other combinations of relations must of necessity follow; combinations unlimited in variety and extent if the deductions from the primary relations be carried on far enough. They will also be aware that one main reason why the separate nature of the science of operations has been little felt, and in general little dwelt on, is the *shifting* meaning of many of the symbols used in mathematical notation. First, the symbols of *operation* are frequently *also* the symbols of the *results* of operations. We may say that these symbols are apt to have both a *retrospective* and a *prospective* signification. They may signify either relations that are the consequence of a series of processes already performed, or relations that are yet to be effected through certain processes. Secondly, figures, the symbols of *numerical magnitude*, are frequently *also* the symbols of *operations*, as when they are the indices of powers. Wherever terms have a shifting meaning, independent sets of considerations are liable to become complicated together, and reasonings and results are frequently falsified. Now in the Analytical Engine the operations which come under the first of the above heads, are ordered and combined by means of a notation and of a train of mechanism which belong exclusively to themselves; and with respect to the second head, whenever numbers meaning *operations* and not *quantities* (such as the indices of powers), are inscribed on any column or set of columns, those columns immediately act in a wholly separate and independent manner, becoming connected with the *operating mechanism* exclusively, and re-acting upon this. They never come into combination with numbers upon any other columns meaning *quantities*; though, of course, if there are numbers meaning *operations* upon *n* columns, these may *combine amongst each other*, and will often be required to do so, just as numbers meaning *quantities* combine with each other in any variety. It might have been arranged that all numbers meaning *operations* should have appeared on some separate portion of the engine from that which presents numerical *quantities*; but the present mode is in some cases more simple, and offers in reality quite as much distinctness when understood.

The operating mechanism can even be thrown into action independently of any object to operate upon (although of course no *result* could then be developed). Again, it might act upon other things besides *number*, were objects found whose mutual fundamental relations could be expressed by those of the abstract science of operations, and which should be also susceptible of adaptations to the action of the operating notation and mechanism of the engine. Supposing, for instance, that the fundamental relations of pitched sounds in the science of harmony and of musical composition were susceptible of such expression and adaptations, the engine might compose elaborate and scientific pieces of music of any degree of complexity or extent.

The Analytical Engine is an *embodying of the science of operations*, constructed with peculiar reference to abstract number as the subject of those operations. The Difference Engine is the embodying of *one particular and very limited set of operations*, which (see the notation used in note B) may be expressed thus, $(+, +, +, +, +, +)$, or thus, $6(+)$. Six repetitions of the one operation, $+$, is, in fact, the whole sum and object of that engine. It has seven columns, and a number on any column can add itself to a number on the next column to its *right-hand*. So that, beginning with the column furthest to the left, six additions can be effected, and the result appears on the seventh column, which is the last on the right-hand. The *operating* mechanism of this engine acts in as separate and independent a manner as that of the Analytical Engine; but being susceptible of only one unvarying and restricted combination, it has little force or interest in illustration of the distinct nature of the *science of operations*. The importance of regarding the Analytical Engine under this point of view will, we think, become more and more obvious, as the reader proceeds with M. Menabrea's clear and masterly article. The calculus of operations is likewise in itself a topic of so much interest, and has of late years been so much more written on and thought on than formerly, that any bearing which that engine, from its mode of constitution, may possess upon the illustraiton of this branch of mathematical science, should not be overlooked. Whether the inventor of this engine had any such views in his mind while working out the invention, or whether he may subsequently ever have regarded it under this phase, we do not know; but it is one that forcibly occurred to ourselves on becoming acquainted with the means through which analytical combinations are actually attained by the mechanism. We cannot forbear suggesting one practical result which it appears to us must be greatly facilitated by the independent manner in which the engine orders and combines its *operations*: we allude to the attainment of those combinations into which *imaginary quantities* enter. This is a branch of its processes into which we have not had the opportunity of inquiring, and our conjecture therefore as to the principle on which we conceive the accomplishment of such results may have been made to depend, is very probably not in accordance with the fact, and less subservient for the purpose than some other principles, or at least requiring the cooperation of others. It seems to us obvious, however, that where operations are so independent in their mode of acting, it must be easy by means of a few simple provisions and additions in arranging the mechanism, to bring out a *double* set of *results*, viz.—1st, the *numerical magnitudes* which are the results of operations performed on *numerical data*. (These results are the *primary* object of the engine). 2ndly, the *symbolical results* to be attached to those numerical results, which symbolical results are not less the necessary and logical consequences of operations performed upon *symbolical data*, than are numerical results when the data are numerical*.

If we compare together the powers and the principles of construction of the Difference and of the Analytical Engines, we shall perceive that the capabilities of the latter are immeasurably more extensive than those of the former, and that they in fact hold to each other the same relationship as that of analysis to arithmetic. The Difference Engine can effect but one particular series of operations, viz. that required for tabulating the integral of the special function

$$\Delta^n u_z = 0;$$

and as it can only do this for values of n up to 7†, it cannot be considered as being the most *general* expression even of *one particular* function, much less as being the

* In fact such an extension as we allude to, would merely constitute a further and more perfected development of any system introduced for making the proper combinations of the signs *plus* and *minus*. How ably M. Menabrea has touched on this restricted case is pointed out in Note B.

† The machine might have been constructed so as to tabulate for a higher value of n than seven. Since, however, every unit added to the value of n increases the extent of the mechanism requisite, there would on this account be a limit beyond which it could not be practically carried. Seven is sufficiently high for the calculation of all ordinary tables.

The fact that, in the Analytical Engine, the same extent of mechanism suffices for the solution of $\Delta^n u_z = 0$, whether $n = 7$, $n = 100,000$, or $n =$ any number whatever, at once suggests how entirely distinct must be the *nature of the principles* through whose application matter has been enabled to become the working agent of abstract mental operations in each of these engines respectively; and it affords an equally obvious presumption, that in the case of the Analytical Engine, not only are those principles in themselves of a higher and more comprehensive description, but also such as must vastly extend the *practical* value of the engine whose basis they constitute.

expression of any and all possible functions of all degrees of generality. The Difference Engine can in reality (as has been already partly explained) do nothing but *add*; and any other processes, not excepting those of simple subtraction, multiplication and division, can be performed by it only just to that extent in which it is possible, by judicious mathematical arrangement and artifices, to reduce them to a *series of additions*. The method of differences is, in fact, a method of additions; and as it includes within its means a larger number of results attainable by *addition* simply, than any other mathematical principle, it was very appropriately selected as the basis on which to construct *an Adding Machine*, so as to give to the powers of such a machine the widest possible range. The Analytical Engine, on the contrary, can either add, subtract, multiply or divide with equal facility; and performs each of these four operations in a direct manner, without the aid of any of the other three. This one fact implies everything; and it is scarcely necessary to point out, for instance, that while the Difference Engine can merely *tabulate*, and is incapable of *developing*, the Analytical Engine can *either tabulate or develope*.

The former engine is in its nature strictly *arithmetical*, and the results it can arrive at lie within a very clearly defined and restricted range, while there is no finite line of demarcation which limits the powers of the Analytical Engine. These powers are co-extensive with our knowledge of the laws of analysis itself, and need be bounded only by our acquaintance with the latter. Indeed we may consider the engine as the *material and mechanical representative* of analysis, and that our actual working powers in this department of human study will be enabled more effectually than heretofore to keep pace with our theoretical knowledge of its principles and laws, through the complete control which the engine gives us over the *executive manipulation* of algebraical and numerical symbols.

Those who view mathematical science not merely as a vast body of abstract and immutable truths, whose intrinsic beauty, symmetry and logical completeness, when regarded in their connexion together as a whole, entitle them to a prominent place in the interest of all profound and logical minds, but as possessing a yet deeper interest for the human race, when it is remembered that this science constitutes the language through which alone we can adequately express the great facts of the natural world, and those unceasing changes of mutual relationships which, visibly or invisibly, consciously or unconsciously to our immediate physical perceptions, are interminably going on in the agencies of the creation we live amidst: those who thus think on mathematical truth as the instrument through which the weak mind of man can most effectually read his Creator's works, will regard with especial interest all that can tend to facilitate the translation of its principles into explicit practical forms.

The distinctive characteristic of the Analytical Engine, and that which has rendered it possible to endow mechanism with such extensive faculties as bid fair to make this engine the executive right-hand of abstract algebra, is the introduction into it of the principle which Jacquard devised for regulating, by means of punched cards, the most complicated patterns in the fabrication of brocaded stuffs. It is in this that the distinction between the two engines lies. Nothing of the sort exists in the Difference Engine. We may say most aptly that the Analytical Engine *weaves algebraical patterns* just as the Jacquard-loom weaves flowers and leaves. Here, it seems to us, resides much more of originality than the Difference Engine can be fairly entitled to claim. We do not wish to deny to this latter all such claims. We believe that it is the only proposal or attempt ever made to construct a calculating machine *founded on the principle of successive orders of differences*, and capable of *printing off its own results*; and that this engine surpasses its predecessors, both in the extent of the calculations which it can perform, in the facility, certainty and accuracy with which it can effect them, and in the absence of all necessity for the intervention of human intelligence *during the performance of its calculations*. Its nature is, however, limited to the strictly arithmetical, and it is far from being the first or only scheme for constructing *arithmetical* calculating machines with more or less of success.

The bounds of *arithmetic* were however outstepped the moment the idea of applying the cards had occurred; and the Analytical Engine does not occupy common ground with mere 'calculating machines.' It holds a position wholly its own; and the considerations it suggests are most interesting in their nature. In enabling mechanism to combine together *general* symbols, in successions of unlimited variety and extent, a uniting link is established between the operations of matter and the abstract mental processes of the *most abstract* branch of mathematical science. A new, a vast, and a powerful language is developed for the future use of analysis, in which to wield its truths so that these may become of more speedy and accurate practical application for the purposes of mankind than the means hitherto in our possession have rendered possible. Thus not only the mental and the material, but the theoretical and the practical in the mathematical world, are brought into more intimate and effective connexion with each other. We are not aware of its being on record that anything partaking in the nature of what is so well designated the *Analytical* Engine has been hitherto proposed, or even thought of, as a practical possibility any more than the idea of a thinking or of a reasoning machine.

We will touch on another point which constitutes an important distinction in the modes of operating of the Difference and Analytical Engines. In order to enable the former to do its business, it is necessary to put into its columns the series of numbers constituting the first terms of the several orders of differences for whatever is the particular table under consideration. The machine then works *upon* these as its data. But these data must themselves have been already computed through a series of calculations by a human head. Therefore that engine can only produce results depending on data which have been arrived at by the explicit and actual working out of processes that are in their nature different from any that come within the sphere of its own powers. In other words an *analysing* process must have been gone through by a human mind in order to obtain the data upon which the engine then *synthetically* builds its results. The Difference Engine is in its character exclusively *synthetical*, while the Analytical Engine is equally capable of analysis or of synthesis.

It is true that the Difference Engine can calculate to a much greater extent with these few preliminary data, than the data themselves required for their own determination. The table of squares, for instance, can be calculated to any extent whatever, when the numbers *one* and *two* are furnished; and a very few differences computed at any part of a table of logarithms would enable the engine to calculate many hundreds or even thousands of logarithms. Still the circumstance of its requiring, as a previous condition, that any function whatever shall have been numerically worked out, makes it very inferior in its nature and advantages to an engine which, like the Analytical Engine, requires merely that we should know the *succession and distribution of the operations* to be performed; without there being any occasion*, in order to obtain data on which it can work, for our ever having gone through either the same particular operations which it is itself to effect, or any others. Numerical data must of course be given it, but they are mere arbitrary ones; not data that could only be arrived at through a systematic and necessary series of previous numerical calculations, which is quite a different thing.

To this it may be replied that an analyzing process must equally have been performed in order to furnish the Analytical Engine with the necessary *operative* data; and that herein may also lie a possible source of error. Granted that the actual mechanism is unerring in its processes, the *cards* may give it wrong orders. This is unquestionably the case; but there is much less chance of error, and likewise far less expenditure of time and labour, where operations only, and the distribution of these operations, have to be made out, than where explicit numerical results are to be attained. In the case of the Analytical Engine we have undoubtedly to lay out a certain capital of analytical labour in one particular line; but this is in order that the engine may bring us in a much larger return in another line. It should be remembered also that the cards when once made out for any formula, have all the generality of algebra, and include an infinite number of particular cases.

We have dwelt considerably on the distinctive peculiarities of each of these engines, because we think it essential to place their respective attributes in strong relief before the apprehension of the public; and to define with clearness and accuracy the wholly different nature of the principles on which each is based, so as to make it self-evident to the reader (the mathematical reader at least) in what manner and degree the powers of the Analytical Engine transcend those of an engine, which, like the Difference Engine, can only work out such results as may be derived from *one restricted and particular series of processes*, such as those included in $\Delta^n u_z = 0$. We think this of importance, because we know that there exists considerable vagueness and inaccuracy in the mind of persons in general on the subject. There is a misty notion amongst most of those who have attended at all to it, that *two* 'calculating machines' have been successively invented by the same person within the last few years; while others again have never heard but of the one original 'calculating

---

* This subject is further noticed in Note F.

machine,' and are not aware of there being any extension upon this. For either of these two classes of persons the above considerations are appropriate. While the latter require a knowledge of the fact that there *are two* such inventions, the former are not less in want of accurate and well-defined information on the subject. No very clear or correct ideas prevail as to the characteristics of each engine, or their respective advantages or disadvantages; and, in meeting with those incidental allusions, of a more or less direct kind, which occur in so many publications of the day, to these machines, it must frequently be matter of doubt *which* 'calculating machine' is referred to, or whether *both* are included in the general allusion.

We are desirous likewise of removing two misapprehensions which we know obtain, to some extent, respecting these engines. In the first place it is very generally supposed that the Difference Engine, after it had been completed up to a certain point, *suggested* the idea of the Analytical Engine; and that the second is in fact the improved offspring of the first, and *grew out* of the existence of its predecessor, through some natural or else accidental combination of ideas suggested by this one. Such a supposition is in this instance contrary to the facts; although it seems to be almost an obvious inference, wherever two inventions, similar in their nature and objects, succeed each other closely in order of *time*, and strikingly in order of *value*; more especially when the same individual is the author of both. Nevertheless the ideas which led to the Analytical Engine occurred in a manner wholly independent of any that were connected with the Difference Engine. These ideas are indeed in their own intrinsic nature independent of the latter engine, and might equally have occurred had it never existed nor been even thought of at all.

The second of the misapprehensions above alluded to, relates to the well-known suspension, during some years past, of all progress in the construction of the Difference Engine. Respecting the circumstances which have interfered with the actual completion of either invention, we offer no opinion; and in fact are not possessed of the data for doing so, had we the inclination. But we know that some persons suppose these obstacles (be they what they may) to have arisen *in consequence* of the subsequent invention of the Analytical Engine while the former was in progress. We have ourselves heard it even *lamented* that an idea should ever have occurred at all, which had turned out to be merely the means of arresting what was already in a course of successful execution, without substituting the superior invention in its stead. This notion we can contradict in the most unqualified manner. The progress of the Difference Engine had long been suspended, before there were even the least crude glimmerings of any invention superior to it. Such glimmerings, therefore, and their subsequent development, were in no way the original *cause* of that suspension; although, where difficulties of some kind or other evidently already existed, it was not perhaps calculated to remove or lessen them that an invention should have been meanwhile thought of, which, while including all that the first was capable of, possesses powers so extended as to eclipse it altogether.

We leave it for the decision of each individual (*after he has possessed himself of* competent information as to the characteristics of each engine), to determine how far it ought to be matter of regret that such an accession has been made to the powers of human science, even if it *has* (which we greatly doubt) increased to a certain limited extent some already existing difficulties that had arisen in the way of completing a valuable but lesser work. We leave it for each to satisfy himself as to the wisdom of desiring the obliteration (were that now possible) of all records of the more perfect invention, in order that the comparatively limited one might be finished. The Difference Engine would doubtless fulfil all those practical objects which it was originally destined for. It would certainly calculate all the tables that are more directly necessary for the physical purposes of life, such as nautical and other computations. Those who incline to very strictly utilitarian views, may perhaps feel that the peculiar powers of the Analytical Engine bear upon questions of abstract and speculative science, rather than upon those involving every-day and ordinary human interest. These persons being likely to possess but little sympathy, or possibly acquaintance, with any branches of science which they do not find to be *useful* (according to *their* definition of that word), may conceive that the undertaking of that engine, now that the other one is already in progress, would be a barren and unproductive laying out of yet more money and labour; in fact, a work of supererogation. Even in the utilitarian aspect, however, we do not doubt that very valuable practical results would be developed by the extended faculties of the Analytical Engine; some of which results we think we could now hint at, had we the space; and others, which it may not yet be possible to foresee, but which would be brought forth by the daily increasing requirements of science, and by a more intimate practical acquaintance with the powers of the engine, were it in actual existence.

On general grounds, both of an *à priori* description as well as those founded on the scientific history and experience of mankind, we see strong presumptions that such would be the case. Nevertheless all will probably concur in feeling that the completion of the Difference Engine would be far preferable to the non-completion of any calculating engine at all. With whomsoever or wheresoever may rest the present causes of difficulty that apparently exist towards either the completion of the old engine, or the commencement of the new one, we trust they will not ultimately result in this generation's being acquainted with these inventions through the medium of pen, ink and paper merely; and still more do we hope, that for the honour of our country's reputation in the future pages of history, these causes will not lead to the completion of the undertaking by some *other* nation or government. This could not but be matter of just regret; and equally so, whether the obstacles may have originated in private interests and feelings, in considerations of a more public description, or in causes combining the nature of both such solutions.

We refer the reader to the *Edinburgh Review* of July 1834, for a very able account of the Difference Engine. The writer of the article we allude to, has selected as his prominent matter for exposition, a wholly different view of the subject from that which M. Menabrea has chosen. The former chiefly treats it under its mechanical aspect, entering but slightly into the mathematical principles of which that engine is the representative, but giving, in considerable length, many details of the mechanism and contrivances by means of which it tabulates the various orders of differences. M. Menabrea, on the contrary, exclusively developes the analytical view, taking it for granted that mechanism is able to perform certain processes, but without attempting to explain *how*; and devoting his whole attention to explanations and illustrations of the manner in which analytical laws can be so arranged and combined as to bring every branch of that vast subject within the grasp of the assumed powers of mechanism. It is obvious that, in the invention of a calculating engine, these two branches of the subject are equally essential fields of investigation, and that on their mutual adjustment, one to the other, must depend all success. They must be made to meet each other, so that the weak points in the powers of either department may be compensated by the strong points in those of the other. They are indissolubly connected, though so different in their intrinsic nature that perhaps the same mind might not be likely to prove equally profound or successful in both. We know those who doubt whether the powers of mechanism will in practice prove adequate in all respects to the demands made upon them in the working of such complicated trains of machinery as those of the above engines, and who apprehend that unforeseen practical difficulties and disturbances will arise in the way of accuracy and of facility of operation. The Difference Engine, however, appears to us to be in a great measure an answer to these doubts. It is complete as far as it goes, and it does work with all the anticipated success. The Analytical Engine, far from being more complicated, will in many respects be of simpler construction; and it is a remarkable circumstance attending it, that with very *simplified* means it is so much more powerful.

The article in the *Edinburgh Review* was written some time previous to the occurrence of any ideas such as afterwards led to the invention of the Analytical Engine; and in the nature of the Difference Engine there is much less that would invite a writer to take exclusively, or even prominently, the mathematical view of it, than in that of the Analytical Engine; although mechanism has undoubtedly gone much further to meet mathematics, in the case of this engine, than of the former one. Some publication embracing the *mechanical* view of the Analytical Engine is a desideratum which we trust will be supplied before long.

Those who may have the patience to study a moderate quantity of rather dry details, will find ample compensation, after perusing the article of 1834, in the clearness with which a succinct view will have been attained of the various practical steps through which mechanism can accomplish certain processes; and they will also find themselves still further capable of appreciating M. Menabrea's more comprehensive and generalized memoir. The very difference in the style and object of these two articles, makes them peculiarly valuable to each other; at least for the purposes of those who really desire something more than a merely superficial and popular comprehension of the subject of calculating engines.

A. A. L.

NOTE B.—PAGES 252–3

That portion of the Analytical Engine here alluded to is called the storehouse. It contains an indefinite number of the columns of discs described by M. Menabrea. The reader may picture to himself a pile of rather large draughtsmen heaped perpendicularly one above another to a considerable height, each counter having the digits from 0 to 9 inscribed on its *edge* at equal intervals; and if he then conceives that the counters do not actually lie one upon another so as to be in contact, but are fixed at small intervals of vertical distance on a common axis which passes perpendicularly through their centres, and around which each disc can *revolve horizontally* so that any required digit amongst those inscribed on its margin can be brought into view, he will have a good idea of one of these columns. The *lowest* of the discs on any column belongs to the units, the next above to the tens, the next above this to the hundreds, and so on. Thus, if we wished to inscribe 1345 on a column of the engine, it would stand thus:—

1
3
4
5

In the Difference Engine there are seven of these columns placed side by side in a row, and the working mechanism extends behind them; the general form of the whole mass of machinery is that of a quadrangular prism (more or less approaching to the cube); the results always appearing on that perpendicular face of the engine which contains the columns of discs, opposite to which face a spectator may place himself. In the Analytical Engine there would be many more of these columns, probably at least two hundred. The precise form and arrangement which the whole mass of its mechanism will assume is not yet finally determined.

We may conveniently represent the columns of discs on paper in a diagram like the following:—

V_1	V_2	V_3	V_4	&c.
O	O	O	O	&c.
0	0	0	0	
0	0	0	0	
0	0	0	0	&c.
0	0	0	0	
□	□	□	□	&c.

The V's are for the purpose of convenient reference to any column, either in writing or speaking, and are consequently numbered. The reason why the letter V is chosen for this purpose in preference to any other letter, is because these columns are designated (as the reader will find in proceeding with the Memoir) the *Variables*, and sometimes the *Variable columns, or the columns of Variables*. The origin of this appellation is, that the values on the columns are destined to change, that is to *vary*, in every conceivable manner. But it is necessary to guard against the natural misapprehension that the columns are only intended to receive the values of the *variables* in an analytical formula, and not of the *constants*. The columns are called Variables on a ground wholly unconnected with the *analytical* distinction between constants and variables. In order to prevent the possibility of confusion, we have, both in the translation and in the notes, written Variable with a capital letter when we use the word to signify a *column of the engine*, and variable with a small letter when we mean the *variable of a formula*. Similarly, *Variable-cards* signify any cards that belong to a column of the engine.

To return to the explanation of the diagram: each circle at the top is intended to contain the algebraic sign + or − , either of which can be substituted* for the other, according as the number represented on the column below is positive or negative. In a similar manner any other purely *symbolical* results of algebraical processes might be made to appear in these circles. In Note A. the practicability of developing *symbolical* with no less ease than *numerical* results has been touched on.

The zeros beneath the *symbolic* circles represent each of them a disc, supposed to have the digit 0 presented in front. Only four tiers of zeros have been figured in the diagram, but these may be considered as representing thirty or forty, or any number of tiers of discs that may be required. Since each disc can present any digit, and each circle any sign, the discs of every column may be so adjusted† as to express any positive or negative number whatever within the limits of the machine; which limits depend on the *perpendicular* extent of the mechanism, that is, on the number of discs to a column.

Each of the squares below the zeros is intended for the inscription of any *general* symbol or combination of symbols we please; it being understood that the number represented on the column immediately above, is the numerical value of that symbol,

---

* A fuller account of the manner in which the *signs* are regulated, is given in M. Menabrea's Memoir, pages 257–9. He himself expresses doubts (in a note of his own at the bottom of the latter page) as to his having been likely to hit on the precise methods really adopted; his explanation being merely a conjectural one. That it *does* accord precisely with the fact is a remarkable circumstance, and affords a convincing proof how completely M. Menabrea has been imbued with the true spirit of the invention. Indeed the whole of the above Memoir is a striking production, when we consider that M. Menabrea had had but very slight means for obtaining any adequate ideas respecting the Analytical Engine. It requires however a considerable acquaintance with the abstruse and complicated nature of such a subject, in order fully to appreciate the penetration of the writer who could take so just and comprehensive a view of it upon such limited opportunity.

† This adjustment is done by hand merely.

or combination of symbols. Let us, for instance, represent the three quantities a, n, x, and let us further suppose that $a = 5$, $n = 7$, $x = 98$. We should have—

V_1	V_2	V_3	V_4	&c.
+*	+	+	+	
0	0	0	0	
0	0	0	0	&c.
0	0	9	0	
5	7	8	0	&c.
$\boxed{a}$	$\boxed{n}$	$\boxed{x}$	$\boxed{}$	

We may now combine these symbols in a variety of ways, so as to form any required function or functions of them, and we may then inscribe each such function below brackets, every bracket uniting together those quantities (and those only) which enter into the function inscribed below it. We must also, when we have decided on the particular function whose numerical value we desire to calculate, assign another column to the right-hand for receiving the *results*, and must inscribe the function in the square below this column. In the above instance we might have any one of the following functions:—

$$ax^n, \quad x^{an}, \quad a \cdot n \cdot x, \quad \frac{a}{n}x, \quad a + n + x, \text{ &c. &c.}$$

Let us select the first. It would stand as follows, previous to calculations:—

V_1	V_2	V_3	V_4	&c.
+	+	+	+	
0	0	0	0	&c.
0	0	0	0	
0	0	9	0	
5	7	8	0	&c.
$\boxed{a}$	$\boxed{n}$	$\boxed{x}$	$\boxed{ax^n}$	&c.

$$\underbrace{\qquad\qquad\qquad\qquad}_{ax_n}$$

The data being given, we must now put into the engine the cards proper for directing the operations in the case of the particular function chosen. These operations would in this instance be, —

Firstly, six multiplications in order to get x^n ($= 98^7$ for the above particular data).

Secondly, one multiplication in order then to get $a \cdot x^n$ ($= 5 \cdot 98^7$).

In all, seven multiplications to complete the whole process. We may thus represent them:—

$$(\times, \times, \times, \times, \times, \times, \times), \text{ or } 7 \, (\times).$$

---

* It is convenient to omit the circles whenever the signs $+$ or $-$ can be actually represented.

The multiplications would, however, at successive stages in the solution of the problem, operate on pairs of numbers, derived from *different* columns. In other words, the *same operation* would be performed on different *subjects of operation*. And here again is an illustration of the remarks made in the preceding Note on the independent manner in which the engine directs its *operations*. In determining the value of ax^n, the *operations are homogeneous*, but are distributed amongst different *subjects of operation*, at successive stages of the computation. It is by means of certain punched cards, belonging to the Variables themselves, that the action of the operations is so *distributed* as to suit each particular function. The *Operation-cards* merely determine the succession of operations in a general manner. They in fact throw all that portion of the mechanism included in the *mill*, into a series of different *states*, which we may call the *adding state*, or the *multiplying state*, &c. respectively. In each of these states the mechanism is ready to act in the way peculiar to that state, on any pair of numbers which may be permitted to come within its sphere of action. Only *one* of these operating states of the mill can exist at a time; and the nature of the mechanism is also such that only *one pair of numbers* can be received and acted on at a time. Now, in order to secure that the mill shall receive a constant supply of the proper pairs of numbers in succession, and that it shall also rightly locate the result of an operation performed upon any pair, each Variable has cards of its own belonging to it. It has, first, a class of cards whose business it is to *allow* the number on the Variable to pass into the mill, there to be operated upon. These cards may be called the *Supplying-cards. They* furnish the mill with its proper food. Each Variable has, secondly, another class of cards, whose office it is to allow the Variable to *receive* a number *from* the mill. These cards may be called the *Receiving-cards. They* regulate the location of results, whether temporary or ultimate results. The Variable-cards in general (including both the preceding classes) might, it appears to us, be even more appropriately designated the Distributive-cards, since it is through their means that the action of the operations, and the results of this action, are rightly *distributed*.

There are *two varieties* of the *Supplying* Variable-cards, respectively adapted for fulfilling two distinct subsidiary purposes: but as these modifications do not bear upon the present subject, we shall notice them in another place.

In the above case of ax^n, the Operation-cards merely order seven multiplications, that is, they order the mill to be in the *multiplying state* seven successive times (without any reference to the particular columns whose numbers are to be acted upon). The proper Distributive Variable-cards step in at each successive multiplication, and cause the distributions requisite for the particular case.

For x^{an} the operations would be   34 ($\times$)

For $a \cdot n \cdot x$ the operations would be ($\times$, $\times$), or 2 ($\times$)

For $\dfrac{a}{n} \cdot x$ the operations would be   ($\div$, $\times$)

For $a + n + x$ the operations would be ($+$, $+$), or 2 ($+$)

The engine might be made to calculate all these in succession. Having completed ax^n, the function x^{an} might be written under the brackets instead of ax^n, and a new calculation commenced (the appropriate Operation and Variable-cards for the new function of course coming into play). The results would then appear on V_5. So on for any number of different functions of the quantities a, n, x. Each *result* might either permanently remain on its column during the succeeding calculations, so that when all the function had been computed, their values would simultaneously exist on V_4, V_5, V_6, &c.; or each result might (after being printed off, or used in any specified manner) be effaced, to make way for its successor. The square under V_4 ought, for the latter arrangement, to have the functions ax^n, x^{an}, anx, &c. successively inscribed in it.

Let us now suppose that we have *two* expressions whose values have been computed by the engine independently of each other (each having its own group of columns for data and results). Let them be ax^n, $b \cdot p \cdot y$. They would then stand as follows on the columns:—

V_1	V_2	V_3	V_4	V_5	V_6	V_7	V_8	V_9
$+$	$+$	$+$	$+$	$+$	$+$	$+$	$+$	$+$
0	0	0	0	0	0	0	0	0
0	0	0	0	0	0	0	0	0
0	0	0	0	0	0	0	0	0
0	0	0	0	0	0	0	0	0
a	n	x	ax^n	b	p	y	bpy	$\dfrac{ax^n}{bpy}$

We may now desire to combine together these two *results*, in any manner we please; in which case it would only be necessary to have an additional card or cards, which should order the requisite operations to be performed with the numbers on the two result-columns, V_4 and V_8, and the *result of these further operations* to appear on a new column, V_9. Say that we wish to divide ax^n by $b \cdot p \cdot y$. The numerical value of this division would then appear on the column V_9, beneath which we have inscribed $\dfrac{ax^n}{bpy}$. The whole series of operations from the beginning would be as follows (n being $= 7$):—

$$\{7(\times),\ 2\ (\times),\ \div \},\ \text{or}\ \{9\ (\times),\ \div \}.$$

This example is introduced merely to show that we may, if we please, retain separately and permanently any *intermediate* results (like ax^n, $b \cdot p \cdot y$), which occur in the course of processes have an ulterior and more complicated result as their chief and final object $\left(\text{like } \dfrac{ax^n}{bpy}\right)$.

Any group of columns may be considered as representing a *general* function, until a *special* one has been implicitly impressed upon them through the introduction into the engine of the Operation and Variable-cards made out for a *particular* function. Thus, in the preceding example, V_1, V_2, V_3, V_5, V_6, V_7 represent the *general* function ϕ (a, n, b, p, x, y) until the function $\dfrac{ax^n}{b \cdot p \cdot y}$ has been determined on, and *implicitly* expressed by the placing of the right cards in the engine. The actual working of the mechanism, as regulated by these cards, then *explicitly* developes the value of the function. The inscription of a function under the brackets, and in the square under the result-column, in no way influences the processes or the results, and is merely a memorandum for the observer, to remind him of what is going on. It is the Operation and Variable-cards only, which in reality determine the function. Indeed it should be distinctly kept in mind that the inscriptions with *any* of the squares, are quite independent of the mechanism or workings of the engine, and are nothing but arbitrary memorandums placed there at pleasure to assist the spectator.

The further we analyse the manner in which such an engine performs its processes and attains its results, the more we perceive how distinctly it places in a true and just light the mutual relations and connexion of the various steps of mathematical analysis, how clearly it separates those things which are in reality distinct and independent, and unites those which are mutually dependent. A. A. L.

NOTE C.——PAGE 254

Those who may desire to study the principles of the Jacquard-loom in the most effectual manner, viz. that of practical observation, have only to step into the Adelaide Gallery or the Polytechnic Institution. In each of these valuable repositories of scientific *illustration*, a weaver is constantly working at a Jacquard-loom, and is ready to give any information that may be desired as to the construction and modes of acting of his apparatus. The volume of the manufacture of silk, in Lardner's *Cyclopaedia*, contains a chapter on the Jacquard-loom, which may also be consulted with advantage.

The mode of application of the cards, as hitherto used in the art of weaving, was not found, however, to be sufficiently powerful for all the simplifications which it was desirable to attain in such varied and complicated processes as those required in order to fulfil the purposes of an Analytical Engine. A method was devised of what was technically designated *backing* the cards in certain groups according to certain laws. The object of this extension is to secure the possibility of bringing any particular card or set of cards into use *any number of times successively* in the solution of one problem. Whether this power shall be taken advantage of or not, in each particular instance, will depend on the nature of the operations which the problem under consideration may require. The process is alluded to by M. Menabrea in page 257, and it is a very important simplification. It has been proposed to use it for the reciprocal benefit of that art, which, while it has itself no apparent connexion with the domains of abstract science, has yet proved so valuable to the latter, in suggesting the principles which, in their new and singular field of appliction, seem likely to place *algebraical* combinations not less completely within the province of mechanism, than are all those varied intricacies of which *intersecting threads* are susceptible. By the introduction of the system of *backing* into the Jacquard-loom itself, patterns which should possess symmetry, and follow regular laws of any extent, might be woven by means of comparatively few cards.

Those who understand the mechanism of this loom will perceive that the above improvement is easily effected in practice, by causing the prism over which the train of pattern-cards is suspended, to revolve *backwards* instead of *forwards*, at pleasure, under the requisite circumstances; until, by so doing, any particular card, or set of cards, that has done duty once, and passed on in the ordinary regular succession, is brought back to the position it occupied just before it was used the preceding time. The prism then resumes its *forward* rotation, and thus brings the card or set of cards in question into play a second time. This process may obviously be repeated any number of times.   A. A. L.

NOTE D.—PAGE 255

We have represented the solution of these two equations, with every detail, in a diagram* similar to those used in Note G.; but additional explanations are requisite, partly in order to make this more complicated case perfectly clear, and partly for the comprehension of certain indications and notations not used in the preceding diagrams. Those who may wish to understand Note B. completely, are recommended to pay particular attention to the contents of the present Note, or they will not otherwise comprehend the similar notation and indications when applied to a much more complicated case.

In all calculations, the columns of Variables used may be divided into three classes:—

1st.   Those on which the data are inscribed:

2ndly.   Those intended to receive the final results:

3dly.   Those intended to receive such intermediate and temporary combinations of the primitive data as are not to be permanently retained, but are merely needed for *working with*, in order to attain the ultimate results. Combinations of this kind might properly be called *secondary data*. They are in fact so many *successive stages* towards the final result. The columns which receive them are rightly named *Working-Variables* for their office is in its nature purely *subsidiary* to other purposes. They develope an intermediate and transient class of results, which unite the original data with the final results.

The Result-Variables sometimes partake of the nature of Working-Variables. It frequently happens that a Variable destined to receive a final result is the recipient of one or more intermediate values successively, in the course of the processes.

---

* See the diagram of page 289.

Similarly, the Variables for data often become Working-Variables, or Result-Variables, or even both in succession. It so happens, however, that in the case of the present equations the three sets of offices remain throughout perfectly separate and independent.

It will be observed, that in the squares below the *Working*-Variables nothing is inscribed. Any one of these Variables is in many cases destined to pass through various values successively during the performance of a calculation (although in these particular equations no instance of this occurs). Consequently no *one fixed* symbol, or combination of symbols, should be considered as properly belonging to a merely *Working*-Variable; and as a general rule their squares are left blank. Of course in this, as in all other cases where we mention a *general* rule, it is understood that many particular exceptions may be expedient.

In order that all the indications contained in the diagram may be completely understood, we shall now explain two or three points, not hitherto touched on. When the value on any Variable is called into use, one of two consequences may be made to result. Either the value may *return* to the Variable after it has been used, in which case it is ready for a second use if needed; or the Variable may be made zero. (We are of course not considering a third case, of not unfrequent occurrence, in which the same Variable is destined to receive the *result* of the very operation which it has just supplied with a number.) Now the ordinary rule is, that the value *returns* to the Variable; unless it has been foreseen that no use for that value can recur, in which case zero is substituted. At the *end* of a calculation, therefore, every column ought as a general rule to be zero, excepting those for results. Thus it will be seen by the diagram, that when m, the value on V_0, is used for the second time by Operation 5, V_0 becomes o, since m is not again needed; that similarly, when $(mn' - m'n)$, on V_{12}, is used for the third time by Operation 11, V_{12} becomes zero, since $(mn' - m'n)$ is not again needed. In order to provide for the one or the other of the courses above indicated, there are *two* varieties of the *Supplying* Variable-cards. One of these varieties has provisions which cause the number given off from any Variable to *return* to that Variable after doing its duty in the mill. The other variety has provisions which cause *zero* to be substituted on the Variable, for the number given off. These two varieties are distinguished, when needful, by the respective appellations of the *Retaining* Supply-cards and the *Zero* Supply-cards. We see that the *primary* office (see Note B.) of both these varieties of cards is the same; they only differ in their *secondary* office.

Every Variable thus has belonging to it *one* class of *Receiving* Variable-cards and *two* classes of *Supplying* Variable-cards. It is plain however that only the *one* or the *other* of these two latter classes can be used by any one Variable for *one* operation; never *both* simultaneously; their respective functions being mutually incompatible.

It should be understood that the Variable-cards are not placed in *immediate contiguity* with the columns. Each card is connected by means of wires with the column it is intended to act upon.

Our diagram ought in reality to be placed side by side with M. Menabrea's corresponding table, so as to be compared with it, line for line belonging to each operation. But it was unfortunately inconvenient to print them in this desirable form. The diagram is, in the main, merely another manner of indicating the various relations denoted in M. Menabrea's table. Each mode has some advantages and some disadvantages. Combined, they form a complete and accurate method of registering every step and sequence in all calculations performed by the engine.

No notice has yet been taken of the *upper* indices which are added to the left of each V in the diagram; an addition which we have also taken the liberty of making to the V's in M. Menabrea's tables of pages 258, 261, since it does not *alter* anything therein represented by him, but merely *adds* something to the previous indications of those tables. The *lower* indices are obviously indices of *locality* only, and are wholly independent of the operations performed or of the results obtained, their value continuing unchanged during the performance of calculations. The *upper* indices, however, are of a different nature. Their office is to indicate any *alteration* in the value which a Variable represents; and they are of course liable to changes during the processes of a calculation. Whenever a Variable has only zeros upon it, it is called ^0V; the moment a value appears on it (whether that value be placed there arbitrarily, or appears in the natural course of a calculation), it becomes ^1V. If this value gives place to another value, the Variable becomes ^2V, and so forth. Whenever a *value* again gives place to *zero*, the Variable again becomes ^0V, even if it have been ^nV the moment before. If a *value* then again be substituted, the Variable becomes ^{n+1}V (as it would have done if it had not passed through the intermediate ^0V); &c. &c. Just before any calculation is commenced, and after the data have been given, and everything adjusted and prepared for setting the mechanism in action, the upper indices of the Variables for data are all unity, and those for the Working and Result-variables are all zero. In this state the diagram represents them.

There are several advantages in having a set of indices of this nature; but these advantages are perhaps hardly of a kind to be immediately perceived, unless by a mind somewhat accustomed to trace the successive steps by means of which the engine accomplishes its purposes. We have only space to mention in a general way, that the whole notation of the tables is made more consistent by these indices, for they are able to mark a *difference* in certain cases, where there would otherwise be an apparent *identity* confusing in its tendency. In such a case as $V_n = V_p + V_n$ there is more clearness and more consistency with the usual laws of algebraical notation, in being able to write $^{m+1}V_n = {}^qV_p + {}^mV_n$. It is also obvious that the indices furnish a powerful means of tracing back the derivation of any result; and of registering various circumstances concerning that *series of successive substitutions*, of which every *result* is in fact merely the final consequence; circumstances that may in certain cases involve relations which it is important to observe, either for purely analytical reasons, or for practically adapting the workings of the engine to their occurrence. The series of substitutions which lead to the equations of the diagram are as follow:—

$$\begin{array}{cccc} (1.) & (2.) & (3.) & (4.) \end{array}$$

$${}^{\mathrm{I}}V^{*}{}_{16}=\frac{{}^{\mathrm{I}}V_{14}}{{}^{\mathrm{I}}V_{12}}=\frac{{}^{\mathrm{I}}V_{10}-{}^{\mathrm{I}}V_{11}}{{}^{\mathrm{I}}V_6-{}^{\mathrm{I}}V_7}=\frac{{}^{\mathrm{I}}V_0\cdot{}^{\mathrm{I}}V_5-{}^{\mathrm{I}}V_2\cdot{}^{\mathrm{I}}V_3}{{}^{\mathrm{I}}V_0\cdot{}^{\mathrm{I}}V_4-{}^{\mathrm{I}}V_3\cdot{}^{\mathrm{I}}V_1}=\frac{d'm-dm'}{mn'-m'n}$$

$$\begin{array}{cccc} (1.) & (2.) & (3.) & (4.) \end{array}$$

$${}^{\mathrm{I}}V_{15}=\frac{{}^{\mathrm{I}}V_{13}}{{}^{\mathrm{I}}V_{12}}=\frac{{}^{\mathrm{I}}V_8-{}^{\mathrm{I}}V_9}{{}^{\mathrm{I}}V_6-{}^{\mathrm{I}}V_7}=\frac{{}^{\mathrm{I}}V_2\cdot{}^{\mathrm{I}}V_4-{}^{\mathrm{I}}V_5\cdot{}^{\mathrm{I}}V_1}{{}^{\mathrm{I}}V_0\cdot{}^{\mathrm{I}}V_4-{}^{\mathrm{I}}V_3\cdot{}^{\mathrm{I}}V_1}=\frac{dn'-d'n}{mn'-m'n}.$$

There are *three* successive substitutions for each of these equations. The formulae (2.), (3.), and (4.) are *implicitly* contained in (1.), which latter we may consider as being in fact the *condensed* expression of any of the former. It will be observed that every succeeding substitution must contain *twice* as many V's as its predecessor. So that if a problem require n substitutions, the successive series of numbers for the V's in the whole of them will be 2, 4, 8, 16 . . . 2^n.

The substitutions in the preceding equations happen to be of little value towards illustrating the power and uses of the upper indices; for owing to the nature of these particular equations the indices are all unity throughout. We wish we had space to enter more fully into the relations which these indices would in many cases enable us to trace.

M. Menabrea incloses the three centre columns of his table under the general title *Variable-cards*. The V's however in reality all represent the actual *Variable-columns* of the engine, and not the cards that belong to them. Still the title is a very just one, since it is through the special action of certain Variable-cards (when *combined* with the more generalized agency of the Operation-cards) that every one of the particular relations he has indicated under that title is brought about.

Suppose we wish to ascertain how often any *one* quantity, or combination of quantities, is brought into use during a calculation. We easily ascertain *this*, from the inspection of any vertical column or columns of the diagram in which that quantity may appear. Thus, in the present case, we see that all the data, and all the intermediate results likewise, are used twice, excepting $(mn'-m'n)$, which is used three times.

The *order* in which it is possible to perform the operations for the present example, enables us to effect all the eleven operations of which it consists, with only *three* *Operation-cards*; because the problem is of such a nature that it admits of each *class* of operations being performed in a group together; all the multiplications one after another, all the subtractions one after another, &c. The operations are $\{6(\times), 3(-), 2(\div)\}$.

Since the very definition of an operation implies that there must be *two* numbers to

---

* We recommend the reader to trace the successive substitutions backwards from (1.) to (4.), in M. Menabrea's Table. This he will easily do by means of the upper and lower indices, and it is interesting to observe how each V successively ramifies (so to speak) into two other V's in some other column of the Table; until at length the V's of the original data are arrived at.

act upon, there are of course *two Supplying* Variable-cards necessarily brought into action for every operation, in order to furnish the two proper numbers. (See Note B.) Also, since every operation must produce a *result*, which must be placed *somewhere*, each operation entails the action of a *Receiving* Variable-card, to indicate the proper locality for the result. Therefore, at least three times as many Variable-cards as there are *operations* (not *Operation-cards*, for these, as we have just seen, are by no means always as numerous as the *operations*) are brought into use in every calculation. Indeed, under certain contingencies, a still larger proportion is requisite; such, for example, would probably be the case when the same result has to appear on more than one Variable simultaneously (which is not unfrequently a provision necessary for subsequent purposes in a calculation), and in some other cases which we shall not here specify. We see therefore that a great disproportion exists between the amount of *Variable* and of *Operation*-cards requisite for the working of even the simplest calculation.

All calculations do not admit, like this one, of the operations of the same nature being performed in groups together. Probably very few do so without exceptions occurring in one or other stage of the progress; and some would not admit it at all. The *order* in which the operations shall be performed in every particular case, is a very interesting and curious question, on which our space does not permit us fully to enter. In almost every computation a great *variety* of arrangements for the succession of the processes is possible, and various considerations must influence the selection amongst them for the purposes of a Calculating Engine. One essential object is to choose that arrangement which shall tend to reduce to a minimum the *time* necessary for completing the calculation.

It must be evident how multifarious and how mutually complicated are the considerations which the workings of such an engine involve. There are frequently several distinct *sets of effects* going on simultaneously; all in a manner independent of each other, and yet to a greater or less degree exercising a mutual influence. To adjust each to every other, and indeed even to perceive and trace them out with perfect correctness and success, entails difficulties whose nature partakes to a certain extent of those involved in every question where *conditions* are very numerous and inter-complicated; such as for instance the estimation of the mutual relations amongst *statistical* phaenomena, and of those involved in many other classes of facts.   A. A. L.

NOTE E.—PAGE 260

This example has evidently been chosen on account of its brevity and simplicity, with a view merely to explain the *manner* in which the engine would proceed in the case of an *analytical calculation containing variables*, rather than to illustrate the *extent of its powers* to solve cases of a difficult and complex nature. The equations of page 255 are in fact a more complicated problem than the present one.

We have not subjoined any diagram of its development for this new example, as we

Diagram belonging to Note D

Number of operations	Nature of operations	$_1V_0$	$_1V_1$	$_1V_2$	$_1V_3$	$_1V_4$	$_1V_5$	$_0V_6$	$_0V_7$	$_0V_8$	$_0V_9$	$_0V_{10}$	$_0V_{11}$	$_0V_{12}$	$_0V_{13}$	$_0V_{14}$	$_0V_{15}$	$_0V_{16}$
																	Variables for results	
		Variables for data						Working Variables										
		$+$	$+$	$+$	$+$	$+$	$+$	$+$	$+$	$+$	$+$	$+$	$+$	$+$	$+$	$+$	$+$	$+$
1	$\times$	o	o	o	o	o	o	o	o	o	o	o	o	o	o	o	o	o
2	$\times$	o	o	o	o	o	o	o	o	o	o	o	o	o	o	o	o	o
3	$\times$	o	o	o	o	o	o	o	o	o	o	o	o	o	o	o	o	o
4	$\times$	o	o	o	o	o	o	o	o	o	o	o	o	o	o	o	o	o
5	$\times$	o	o	o	o	o	o	o	o	o	o	o	o	o	o	o	o	o
6	$\times$	o	o	o	o	o	o	o	o	o	o	o	o	o	o	o	o	o
7	$-$	o	o	o	o	o	o	o	o	o	o	o	o	o	o	o	o	o
8	$-$	o	o	o	o	o	o	o	o	o	o	o	o	o	o	o	o	o
9	$-$	o	o	o	o	o	o	o	o	o	o	o	o	o	o	o	o	o
10	$\div$	o	o	o	o	o	o	o	o	o	o	o	o	o	o	o	o	o
11	$\div$	o	o	o	o	o	o	o	o	o	o	o	o	o	o	o	o	o
Data		m	n	d	m'	n'	d'	mn'	$m'n$	dn'	$d'n$	$d'm$	dm'	$mn'-m'n$	$dn'-d'n$	$d'm-dm'$	$\dfrac{dn'-d'n}{mn'-m'n}=x$	$\dfrac{d'm-dm'}{mn'-m'n}=y$
Results		m	n	0	m'	0	d'	0	0	0	0	0	0	$mn'-m'n$	0	0	$\dfrac{dn'-d'n}{mn'-m'n}=x$	$\dfrac{d'm-dm'}{mn'-m'n}=y$

289

did for the former one, because this is unnecessary after the full application already made of those diagrams to the illustration of M. Menabrea's excellent tables.

It may be remarked that a slight discrepancy exists between the formulae

$$(a + bx')$$
$$(A + B \cos' x)$$

given in the Memoir as the *data* for calculation, and the *results* of the calculation as developed in the last division of the table which accompanies it. To agree perfectly with this latter, the data should have been given as

$$(ax^0 + bx')$$
$$(A \cos^0 x + B \cos' x).$$

The following is a more complicated example of the manner in which the engine would compute a trigonometrical function containing variables. To multiply

$$A + A_1 \cos \theta + A_2 \cos 2 \theta + A_3 \cos 3 \theta + \ldots$$

by $$B + B_1 \cos \theta$$

let the resulting products be represented under the general form

$$C_0 + C_1 \cos \theta + C_2 \cos 2 \theta + C_3 \cos 3 \theta + \ldots \qquad (1.)$$

This trigonometrical series is not only in itself very appropriate for illustrating the processes of the engine, but is likewise of much practical interest from its frequent use in astronomical computations. Before proceeding further with it, we shall point out that there are three very distinct classes of ways in which it may be desired to deduce numerical values from any analytical formula.

First. We may wish to find the collective numerical value of the *whole formula*, without any reference to the quantities of which that formula is a function, or to the particular mode of their combination and distribution, of which the formula is the result and representative. Values of this kind are of a strictly arithmetical nature in the most limited sense of the term, and retain no trace whatever of the processes through which they have been deduced. In fact, any one such numerical value may have been attained from an *infinte variety* of data, or of problems. The values of x and y in the two equations (see Note D.), come under the class of numerical results.

Secondly. We may propose to compute the collective numerical value of *each term* of a formula, or of a series, and to keep these results separate. The engine must in such a case appropriate as many columns to *results* as there are terms to compute.

Thirdly. It may be desired to compute the numerical value of various *subdivisions of each term*, and to keep all these results separate. It may be required, for instance, to compute each coefficient separately from its variable, in which particular case the engine must appropriate *two* result-columns to *every term that contains both a variable and coefficient*.

There are many ways in which it may be desired in special cases to distribute and keep separate the numerical values of different parts of an algebraical formula; and the power of effecting such distributions to any extent is essential to the *algebraical* character of the Analytical Engine. Many persons who are not conversant with mathematical studies, imagine that because the business of the engine is to give its results in *numerical notation*, the *nature of its processes* must consequently be *arithmetical* and *numerical*, rather than *algebraical* and *analytical*. This is an error. The engine can arrange and combine its numerical quantities exactly as if they were *letters* or any other *general* symbols; and in fact it might bring out its results in algebraical *notation*, were provisions made accordingly. It might develope three sets of results simultaneously, viz. *symbolic* results (as already alluded to in Notes A. and B.); *numerical* results (its chief and primary object); and *algebraical* results in *literal* notation. This latter however has not been deemed a necessary or desirable addition to its powers, partly because the necessary arrangements for effecting it would increase the complexity and extent of the mechanism to a degree that would not be commensurate with the advantages, where the main object of the invention is to translate into *numerical* language general formulae of analysis already known to us, or whose laws of formation are known to us. But it would be a mistake to suppose that because its *results* are given in the *notation* of a more restricted science, its *processes* are therefore restricted to those of that science. The object of the engine is in fact to give the *utmost practical efficiency* to the resources of *numerical interpretations* of the higher science of analysis, while it uses the processes and combinations of this latter.

To return to the trigonometrical series. We shall only consider the four first terms of the factor $(A + A_I \cos \theta + \&c.)$, since this will be sufficient to show the method. We propose to obtain separately the numerical value of *each coefficient* C_0, C_I, &c. of (1.). The direct multiplication of the two factors gives

$$\left. \begin{array}{l} BA + BA_I \cos \theta + BA_2 \quad\quad \cos 2\,\theta + BA_3 \quad\quad \cos 3\,\theta + \ldots \\ + B_I A \cos \theta + B_I A_I \cos \theta \cdot \cos \theta + B_I A_2 \cos 2\theta \cdot \cos \theta + B_I A_3 \cos 3\,\theta \cdot \cos \theta \end{array} \right\} (2.)$$

a result which would stand thus on the engine:—

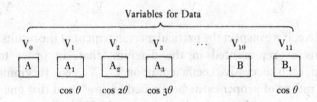

		Variables for Data				
V_0	V_1	V_2	V_3	$\cdots$	V_{10}	V_{11}
A	A_1	A_2	A_3		B	B_1
	$\cos \theta$	$\cos 2\theta$	$\cos 3\theta$			$\cos \theta$

		Variables for results					
V_{20}	V_{21}	V_{22}	V_{23}	$\cdots$ V_{31}	V_{32}	V_{33}	V_{34}
BA	BA_1	BA_2	BA_3	$B_I A$	$B_I A_1$	$B_I A_2$	$B_I A_3$
	$\cos \theta$	$\cos 2\theta$	$\cos 3\theta$	$\cos \theta$	$(\cos \theta \cdot \cos \theta)$	$\vdots$	$(\cos 3\theta \cdot \cos \theta)$
						$(\cos 2\theta \cdot \cos \theta)$	

The variable belonging to each coefficient is written below it, as we have done in the diagram, by way of memorandum. The only further reduction which is at first apparently possible in the preceding result, would be the addition of V_{21} to V_{31} (in which case B_IA should be effaced from V_{31}). The whole operations from the beginning would then be —

First Series of Operations	Second Series of Operations	Third Series, which contains only one (final) operation
$^IV_{10} \times {}^IV_0 = {}^IV_{20}$	$^IV_{11} \times {}^IV_0 = {}^IV_{31}$	$^IV_{21} + {}^IV_{31} = {}^2V_{21}$, and
$^IV_{10} \times {}^IV_1 = {}^IV_{21}$	$^IV_{11} \times {}^IV_1 = {}^IV_{32}$	V_{31} becomes $= 0$.
$^IV_{10} \times {}^IV_2 = {}^IV_{22}$	$^IV_{11} \times {}^IV_2 = {}^IV_{33}$	
$^IV_{10} \times {}^IV_3 = {}^IV_{23}$	$^IV_{11} \times {}^IV_3 = {}^IV_{34}$	

We do not enter into the same detail of *every* step of the processes as in the examples of Notes D. and G., thinking it unnecessary and tedious to do so. The reader will remember the meaning and use of the upper and lower indices, &c., as before explained.

To proceed: we know that

$$\cos n\,\theta \cdot \cos\theta = \tfrac{1}{2}\cos\overline{n+1}\,\theta + \tfrac{1}{2}\overline{n-1}\cdot\theta \ldots \tag{3.}$$

Consequently, a slight examination of the second line of (2.) will show that by making the proper substitutions, (2.) will become

BA	$+BA_I \cdot \cos\theta$	$+BA_2 \cdot \cos 2\theta$	$+BA_3 \cdot \cos 3\theta$	
	$+B_IA \cdot \cos\theta$			
$+\tfrac{1}{2}B_IA_I$		$+\tfrac{1}{2}B_IA_I \cdot \cos 2\theta$		
	$+\tfrac{1}{2}B_IA_2 \cdot \cos\theta$		$+\tfrac{1}{2}B_IA_2 \cdot \cos 3\theta$	
		$+\tfrac{1}{2}B_IA_3 \cdot \cos 2\theta$		$+\tfrac{1}{2}B_IA_3 \cdot \cos 4\theta$
C_0	C_I	C_2	C_3	C_4

These coefficients should respectively appear on

V_{20}	V_{21}	V_{22}	V_{23}	V_{24}

We shall perceive, if we inspect the particular arrangement of the results in (2.) on the Result-columns as represented in the diagram, that, in order to effect this transformation, each successive coefficient upon V_{32}, V_{33}, &c. (beginning with V_{32}), must through means of proper cards be divided by *two**; and that one of the halves thus obtained must be added to the coefficient on the Variable which precedes it by ten columns, and the other half to the coefficient on the Variable which precedes it by twelve columns; V_{32}, V_{33}, &c. themselves becoming zeros during the process.

---

* This division would be managed by ordering the number two to appear on any separate new column which should be conveniently situated for the purpose, and then directing this column (which is in the strictest sense a *Working*-Variable) to divide itself successively with V_{32}, V_{33}, &c.

This series of operations may be thus expressed:—

Fourth Series

$$\begin{cases} {}^IV_{32} \div 2 + {}^IV_{22} = {}^2V_{22} = BA_2 + \tfrac{1}{2}B_IA_I \\ {}^IV_{32} \div 2 + {}^IV_{20} = {}^2V_{20} = BA + \tfrac{1}{2}B_IA_I \;\dots\dots\; = C_0 \end{cases}$$

$$\begin{cases} {}^IV_{33} \div 2 + {}^IV_{23} = {}^2V_{23} = BA_3 + \tfrac{1}{2}B_IA_2 \dots\dots\; = C_3{}^* \\ {}^IV_{33} \div 2 + {}^2V_{21} = {}^3V_{21} = BA_I + B_IA + \tfrac{1}{2}B_IA_2 \;\; = C_I \end{cases}$$

$$\begin{cases} {}^IV_{34} \div 2 + {}^0V_{24} = {}^IV_{24} = \tfrac{1}{2}B_IA_3 \;\dots\dots\dots\; = C_4 \\ {}^IV_{34} \div 2 + {}^2V_{22} = {}^3V_{22} = BA_2 + \tfrac{1}{2}B_IA_I + \tfrac{1}{2}B_IA_3 = C_2 \end{cases}$$

The calculation of the coefficients C_0, C_I, &c. of (1.), would now be completed, and they would stand ranged in order on V_{20}, V_{21}, &c. It will be remarked, that from the moment the fourth series of operations is ordered, the Variables V_{31}, V_{32}, &c. cease to be *Result*-Variables, and become mere *Working*-Variables.

The substitution made by the engine of the processes in the second side of (3.) for those in the first side, is an excellent illustration of the manner in which we may arbitrarily order it to substitute any function, number, or process, at pleasure, for any other function, number or process, on the occurrence of a specified contingency.

We will now suppose that we desire to go a step further, and to obtain the numerical value of each *complete* term of the product (1.), that is of each *coefficient and variable united*, which for the $(n+1)$th term would be $C_n \cdot \cos n\,\theta$.

We must for this purpose place the variables themselves on another set of columns, V_{41}, V_{42}, &c., and then order their successive multiplication by V_{21}, V_{22}, &c., each for each. There would thus be a final series of operations as follows:—

Fifth and Final Series of Operations

$$^2V_{20} \times {}^0V_{40} = {}^IV_{40}$$
$$^3V_{21} \times {}^0V_{41} = {}^IV_{41}$$
$$^3V_{22} \times {}^0V_{42} = {}^IV_{42}$$
$$^2V_{23} \times {}^0V_{43} = {}^IV_{43}$$
$$^IV_{24} \times {}^0V_{44} = {}^IV_{44}$$

(N.B. that V_{40} being intended to receive the coefficient on V_{20} which has *no* variable, will only have cos o θ ($=1$) inscribed on it, preparatory to commencing the fifth series of operations.)

From the moment that the fifth and final series of operations is ordered, the Variables V_{20}, V_{21}, &c. then in their turn cease to be *Result*-Variables and become mere *Working*-Variables; V_{40}, V_{41}, &c. being now the recipients of the ultimate results.

We should observe, that if the variables cos θ, cos 2 θ, cos 3 θ, &c. are furnished, they would be placed directly upon V_{41}, V_{42}, &c., like any other data. If not, a separate computation might be entered upon in a separate part of the engine, in order to calculate them, and place them on V_{41}, &c.

We have now explained how the engine might compute (1.) in the most direct manner, supposing we knew nothing about the *general* term of the resulting series. But the engine would in reality set to work very differently, whenever (as in this case) we *do* know the law for the general term.

---

* It should be observed, that were the rest of the factor $(A + A \cos \theta + \&c.)$ taken into account, instead of *four* terms only, C_3 would have the additional term $\tfrac{1}{2}B_IA_4$; and C_4 the two additional terms, BA_4, $\tfrac{1}{2}B_IA_5$. This would indeed have been the case had even *six* terms been multiplied.

The two first terms of (1.) are

$$(BA + \tfrac{1}{2}B_1 A_1) + \overline{(BA_1 + B_1 A + \tfrac{1}{2}B_1 A_2 \cdot \cos \theta)} \tag{4.}$$

and the general term for all after these is

$$(BA_n + B_1 \cdot \overline{A_{n-1} + A_{n+2}}) \cos n\,\theta \tag{5.}$$

which is the coefficient of the $(n+1)$th term. The engine would calculate the two first terms by means of a separate set of suitable Operation-cards, and would then need another set for the third term; which last set of Operation-cards would calculate all the succeeding terms *ad infinitum*; merely requiring certain new Variable-cards for each term to direct the operations to act on the proper columns. The following would be the successive sets of operations for computing the coefficients of $n + 2$ terms:—

$$(\times, \times, \div, +), (\times, \times, \times, \div, +, +), n\,(\times, +, \times, \div, +).$$

Or we might represent them as follows, according to the numerical order of the operations:—

$$(1, 2 \ldots 4), (5, 6 \ldots 10), n\,(11, 12 \ldots 15).$$

The brackets, it should be understood, point out the relation in which the operations may be *grouped*, while the comma marks *succession*. The symbol $+$ might be used for this latter purpose, but this would be liable to produce confusion, as $+$ is also necessarily used to represent one class of the actual operations which are the subject of that succession. In accordance with this meaning attached to the comma, care must be taken when any one group of operations recurs more than once, as is represented above by $n\,(11 \ldots 15)$, not to insert a comma after the number or letter prefixed to that group. $n, (11 \ldots 15)$ would stand for *an operation n, followed by the group of operations* $(11 \ldots 15)$; instead of denoting *the number of groups which are to follow each other*.

Wherever a *general term* exists, there will be a *recurring group* of operations, as in the above example. Both for brevity and for distinctness, a *recurring group* is called a *cycle*. A *cycle* of operations, then, must be understood to signify any *set of operations* which is repeated *more than once*. It is equally a *cycle*, whether it be repeated *twice* only, or an indefinite number of times; for it is the fact of a *repetition occurring at all* that constitutes it such. In many cases of analysis there is a *recurring group* of one or more *cycles*; that is, a *cycle of a cycle*, or a *cycle of cycles*. For instance: suppose we wish to divide a series by a series,

$$(1.) \qquad \frac{a + bx + cx^2 + \ldots}{a' + b'x + c'x^2 + \ldots},$$

it being required that the result shall be developed, like the dividend and the divisor, in successive powers of x. A little consideration of (1.), and of the steps through which algebraical division is effected, will show that (if the denominator be supposed to consist of p terms) the first partial quotient will be completed by the following operations:—

(2.) $\{(\div), p(\times, -)\}$ or $\{(1), p(2, 3)\}$,

that the second partial quotient will be completed by an exactly similar set of operations, which acts on the remainder obtained by the first set, instead of on the original dividend. The whole of the processes therefore that have been gone through, by the time the *second* partial quotient has been obtained, will be,—

(3.) $2\{(\div), p(\times, -)\}$ or $2\{(1), p(2, 3)\}$,

which is a cycle that includes a cycle, or a cycle of the second order. The operations for the *complete* division, supposing we propose to obtain n terms of the series constituting the quotient, will be,—

(4.) $n\{(\div), p(\times, -)\}$ or $n\{(1), p(2, 3)\}$.

It is of course to be remembered that the process of algebraical division in reality continues *ad infinitum*, except in the few exceptional cases which admit of an exact quotient being obtained. The number n in the formula (4.), is always that of the number of terms we propose to ourselves to obtain; and the nth partial quotient is the coefficient of the $(n-1)$th power of x.

There are some cases which entail *cycles of cycles of cycles*, to an indefinite extent. Such cases are usually very complicated, and they are of extreme interest when considered with reference to the engine. The algebraical development in a series, of the nth function of any given function, is of this nature. Let it be proposed to obtain the nth function of

(5.) $\phi(a, b, c \ldots x)$, x being the variable.

We should premise that we suppose the reader to understand what is meant by an nth function. We suppose him likewise to comprehend distinctly the difference between developing an nth *function algebraically*, and merely *calculating an nth function arithmetically*. If he does not, the following will be by no means very intelligible; but we have not space to give any preliminary explanations. To proceed: the law, according to which the successive functions of (5.) are to be developed, must first be fixed on. This law may be of very various kinds. We may propose to obtain our results in successive *powers* of x, in which case the general form would be

$$C + C_1 x + C_2 x^2 + \&c.,$$

or, in successive powers of n itself, the index of the function we are ultimately to obtain, in which case the general form would be

$$C + C_1 n + C_2 n^2 + \&c.,$$

and x would only enter in the coefficients. Again, other functions of x or of n instead of *powers*, might be selected. It might be in addition proposed, that the coefficients themselves should be arranged according to given functions of a certain quantity.

Another mode would be to make equations arbitrarily amongst the coefficients only, in which case the several functions, according to either of which it might be possible to develop the *n*th function of (5.), would have to be determined from the combined consideration of these equations and of (5.) itself.

The *algebraical* nature of the engine (so strongly insisted on in a previous part of this Note) would enable it to follow out any of these various modes indifferently; just as we recently showed that it can distribute and separate the numerical results of any one prescribed series of processes, in a perfectly aribtrary manner. Were it otherwise, the engine could merely *compute the arithmetical nth functions*, a result which, like any other purely arithmetical results, would be simply a collective number, bearing no traces of the data or the processes which had led to it.

Secondly, the *law* of development for the *n*th function being selected, the next step would obviously be to develope (5.) itself, according to this law. This result would be the first function, and would be obtained by a determinate series of processes. These in most cases would include amongst them one or more *cycles* of operations.

The third step (which would consist of the various processes necessary for effecting the actual substitution of the series constituting the *first function*, for the *variable* itself) might proceed in either of two ways. It might make the substitution either wherever *x* occurs in the original (5.), or it might similarly make it wherever *x* occurs in the first function itself which is the equivalent of (5.). In some cases the former mode might be best, and in others the latter.

Whichever is adopted it must be understood that the result is to appear arranged in a series following the law originally prescribed for the development of the *n*th function. This result constitutes the second function; with which we are to proceed exactly as we did with the first function, in order to obtain the third function; and so on, $n-1$ times, to obtain the *n*th function. We easily perceive that since every successive function is arranged in a series *following the same law* these would (after the *first* function is obtained) be a *cycle, of a cycle, of a cycle*, &c. of operations*, one, two, three, up to $n-1$ times, in order to get the *n*th function. We say, *after the first function is obtained*, because (for reasons on which we cannot here enter) the *first* function might in many cases be developed through a set of processes peculiar to itself, and not recurring for the remaining functions.

We have given but a very slight sketch, of the principal *general* steps which would be requisite for obtaining an *n*th function of such a formula as (5.). The question is so exceedingly complicated, that perhaps few persons can be expected to follow, to their own satisfaction, so brief and general a statement as we are here restricted to on this

---

* A cycle that includes *n* other cycles, successively *contained one within another*, is called a cycle of the $(n+1)$th order. A cycle may simply *include* many other cycles, and yet only be of the second order. If a series follows a certain law for a certain number of terms, and then another law for another number of terms, there will be a cycle of operations for every new law; but these cycles will not be *contained one within another*,—they merely *follow each other*. Therefore their number may be infinite without influencing the *order* of a cycle that includes a repetition of such a series.

subject. Still it is a very important case as regards the engine, and suggests ideas peculiar to itself, which we should regret to pass wholly without allusion. Nothing could be more interesting than to follow out, in every detail, the solution by the engine of such a case as the above; but the time, space and labour this would necessitate, could only suit a very extensive work.

To return to the subject of *cycles* of operations: some of the notation of the integral calculus lends itself very aptly to express them: (2.) might be thus written:—

$$(6.) \qquad (\div), \ \Sigma \ (+1)^p \ (\times, \ -) \text{ or } (1), \ \Sigma \ (+1)^p \ (2, 3),$$

where p stands for the variable; $(+1)^p$ for the function of the variable, that is, for ϕp; and the limits are from 1 to p, or from 0 to $p-1$, each increment being equal to unity. Similarly, (4.) would be,—

$$(7.) \qquad \Sigma \ (+1)^n \ \{(\div), \ \Sigma \ (+1)^p \ (\times, \ -)\}$$

the limits of n being from 1 to n, or from 0 to $n-1$,

$$(8.) \qquad \text{or } \Sigma \ (+1)^n \ \{(1), \ \Sigma \ (+1)^p \ (2, 3)\}.$$

Perhaps it may be thought that this notation is merely a circuitous way of expressing what was more simply and as effectually expressed before; and, in the above example, there may be some truth in this. But there is another description of cycles which *can* only effectually be expressed, in a condensed form, by the preceding notation. We shall call them *varying cycles*. They are of frequent occurrence, and include successive cycles of operations of the following nature:—

$$(9.) \ p \ (1, 2, \ldots m), \ \overline{p-1} \ (1, 2 \ldots m), \ \overline{p-2} \ (1, 2 \ldots m) \ldots \overline{p-n} \ (1, 2 \ldots m),$$

where each cycle contains the same group of operations, but in which the number of repetitions of the group varies according to a fixed rate, with every cycle. (9.) can be well expressed as follows:—

$$(10.) \qquad \Sigma \ p \ (1, 2, \ldots m), \text{ the limits of } p \text{ being from } p-n \text{ to } p.$$

Independent of the intrinsic advantages which we thus perceive to result in certain cases from this use of the notation of the integral calculus, there are likewise considerations which make it interesting, from the connections and relations involved in this new application. It has been observed in some of the former Notes, that the processes used in analysis form a logical system of much higher generality than the applications to number merely. Thus, when we read over any algebraical formula, considering it exclusively with reference to the processes of the engine, and putting aside for the moment its abstract signification as to the relations of quantity, the symbols $+$, $\times$, &c., in reality represent (as their immediate and proximate effect, when the formula is applied to the engine) that a certain prism which is a part of the mechanism (see Note C.), turns a new face, and thus presents a new card to act on the bundles of the levers of the engine; the new card being perforated with holes, which are arranged according to the peculiarities of the operation of addition, or of multiplication, &c. Again, the *numbers* in the preceding formula (8.), each of them really represents one of these very pieces of card that are hung over the prism.

Now in the use made in the formulae (7.), (8.) and (10.), of the notation of the integral calculus, we have glimpses of a similar new application of the language of the *higher* mathematics. Σ, in reality, here indicates that when a certain number of cards have acted in succession, the prism over which they revolve must *rotate backwards*, so as to bring those cards into their former position; and the limits 1 to n, 1 to p, &c., regulate how often this backward rotation is to be repeated.   A. A. L.

NOTE F.—PAGE 264

There is in existence a beautiful woven portrait of Jacquard, in the fabrication of which 24,000 cards were required.

The power of *repeating* the cards, alluded to by M. Menabrea in page 256, and more fully explained in Note C., reduces to an immense extent the number of cards required. It is obvious that this mechanical improvement is especially applicable wherever *cycles* occur in the mathematical operations, and that, in preparing data for calculations by the engine, it is desirable to arrange the order and combination of the processes with a view to obtain them as much as possible *symmetrically* and in cycles, in order that the mechanical advantages of the *backing* system may be applied to the utmost. It is here interesting to observe the manner in which the value of an *analytical* resource is *met* and *enhanced* by an ingenious *mechanical* contrivance. We see in it an instance of one of those mutual *adjustments* between the purely mathematical and the mechanical departments, mentioned in Note A. as being a main and essential condition of success in the invention of a calculating engine. The nature of the resources afforded by such adjustments would be of two principal kinds. In some cases, a difficulty (perhaps in itself insurmountable) in the one department, would be overcome by facilities in the other; and sometimes (as in the present case) a strong point in the one, would be rendered still stronger and more available, by combination with a corresponding strong point in the other.

As a mere example of the degree to which the combined systems of cycles and of backing can diminish the *number* of cards requisite, we shall choose a case which places it in strong evidence, and which has likewise the advantage of being a perfectly different *kind* of problem from those that are mentioned in any of the other Notes. Suppose it be required to eliminate nine variables from ten simple equations of the form—

$$ax_0 + bx_1 + cx_2 + dx_3 + \ldots = p \qquad (1.)$$
$$a^1x_0 + b^1x_1 + c^1x_2 + d^1x_3 + \ldots = p^1 \qquad (2.)$$
$$\text{\&c.} \qquad \text{\&c.} \qquad \text{\&c.} \qquad \text{\&c.}$$

We should explain, before proceeding, that it is not our object to consider this problem with reference to the actual arrangement of the data on the Variables of the engine, but simply as an abstract question of the *nature* and *number* of the *operations* required to be performed during its complete solution.

The first step would be the elimination of the first unknown quantity x_0 between the two first equations. This would be obtained by the form—

$$(a^1a - aa^1)\, x_0 + (a^1b - ab^1)\, x_1 + (a^1c - ac^1)\, x_2$$
$$+ (a^1d - ad^1)\, x_3 + \ldots = a^1p - ap^1,$$

for which the operations $10\,(\times, \times, -)$ would be needed. The second step would be the elimination of x_0 between the second and third equations, for which the operations would be precisely the same. We should then have had altogether the following operations:—

$$10\,(\times, \times, -),\ 10\,(\times, \times, -). = 20\,(\times, \times, -).$$

Continuing in the same manner, the total number of operations for the complete elimination of x^0 between all the successive pairs of equations, would be—

$$9 \cdot 10\,(\times, \times, -) = 90\,(\times, \times, -).$$

We should then be left with nine simple equations of nine variables from which to eliminate the next variable x_1; for which the total of the processes would be—

$$8 \cdot 9\,(\times, \times, -) = 72\,(\times, \times, -).$$

We should then be left with eight simple equations of eight variables from which to eliminate x_2, for which the processes would be—

$$7 \cdot 8\,(\times, \times, -) = 56\,(\times, \times, -),$$

and so on. The total operations for the elimination of all the variables would thus be—

$$9 \cdot 10 + 8 \cdot 9 + 7 \cdot 8 + 6 \cdot 7 + 5 \cdot 6 + 4 \cdot 5 + 3 \cdot 4 + 2 \cdot 3 + 1 \cdot 2 = 330.$$

So that *three* Operation-cards would perform the office of 330 such cards.

If we take n simple equations containing $n-1$ variables, n being a number unlimited in magnitude, the case becomes still more obvious, as the same three cards might then take the place of thousands or millions of cards.

We shall now draw further attention to the fact, already noticed, of its being by no means necessary that a formula proposed for solution should ever have been actually worked out, as a condition for enabling the engine to solve it. Provided we know the *series of operations* to be gone through, that is sufficient. In the foregoing instance this will be obvious enough on a slight consideration. And it is a circumstance which deserves particular notice, since herein may reside a latent value of such an engine almost incalculable in its possible ultimate results. We already know that there are functions whose numerical value it is of importance for the purposes both of abstract and of practical science to ascertain, but whose determination requires processes so lengthy and so complicated, that, although it is possible to arrive at them through great expenditure of time, labour and money, it is yet on these accounts practically almost unattainable; and we can conceive [of] there being some results which it may be *absolutely impossible* in practice to attain with any accuracy, and whose precise determination it may prove highly important for some of the future wants of science in its manifold, complicated and rapidly-developing fields of inquiry, to arrive at.

Without, however, stepping into the region of conjecture we will mention a particular problem which occurs to us at this moment as being an apt illustration of the use to which such an engine may be turned for determining that which human brains find it difficult or impossible to work out unerringly. In the solution of the famous problem of the Three Bodies, there are, out of about 295 coefficients of lunar perturbations given by M. Calusen (*Astroe. Nachrichten*, No. 406) as the result of the calculations by Burg, of two by Damoiseau, and of one by Burckhardt, fourteen coefficients that differ in the nature of their algebraic sign; and out of the remainder there are only 101 (or about one-third) that agree precisely both in signs and in amount. These discordances, which are generally small in individual magnitude, may arise either from an erroneous determination of the abstract coefficients in the development of the problem, or from discrepancies in the data deduced from observation, or from both causes combined. The former is the most ordinary source of error in astronomical computations, and this the engine would entirely obviate.

We might even invent laws for series or formulae in an arbitrary manner, and set the engine to work upon them, and thus deduce numerical results which we might not otherwise have thought of obtaining. But this would hardly perhaps in any instance be productive of any great practical utility, or calculated to rank higher than as a kind of philosophical amusement.   A. A. L.

NOTE G.——PAGE 265

It is desirable to guard against the possibility of exaggerated ideas that might arise as to the powers of the Analytical Engine. In considering any new subject, there is frequently a tendency, first, to *overrate* what we find to be already interesting or remarkable; and, secondly, by a sort of natural reaction, to *undervalue* the true state of the case, when we do discover that our notions have surpassed those that were really tenable.

The Analytical Engine has no pretentions whatever to *originate* any thing. It can do whatever we *know how to order it* to perform. It can *follow* analysis; but it has no power of *anticipating* any analytical relations or truths. Its province is to assist us in making *available* what we are already acquainted with. This it is calculated to effect primarily and chiefly of course, through its executive faculties; but it is likely to exert an *indirect* and reciprocal influence on science itself in another manner. For, in so distributing and combining the truth and the formulae of analysis, that they may become most easily and rapidly amenable to the mechanical combinations of the engine, the relations and the nature of many subjects in that science are necessarily thrown into new lights, and more profoundly investigated. This is a decidedly indirect, and a somewhat *speculative*, consequence of such an invention. It is however pretty evident, on general principles, that in devising for mathematical truths a new form in which to record and throw themselves out for actual use, views are likely to be induced, which should again react on the more theoretical phase of the subject. There are in all extensions of human power, or additions to human knowledge, various *collateral* influences, besides the main and primary object attained.

To return to the executive faculties of this engine: the question must arise in every mind, are they *really* even able to *follow* analysis in its whole extent? No reply, entirely satisfactory to all minds, can be given to this query, excepting the actual existence of the engine, and actual experience of its practical results. We will however sum up for each reader's consideration the chief elements with which the engine works:—

1. It performs the four operations of simple arithmetic upon any numbers whatever.
2. By means of certain artifices and arrangements (upon which we cannot enter within the restricted space which such a publication as the present may admit of), there is no limit either to the *magnitude* of the *numbers* used, or to the *number of quantities* (either variables or constants) that may be employed.
3. It can combine these numbers and these quantities either algebraically or arithmetically, in relations unlimited as to variety, extent, or complexity.
4. It uses algebraic *signs* according to their proper laws, and developes the logical consequences of these laws.
5. It can arbitrarily substitute any formula for any other; effacing the first from the columns on which it is represented, and making the second appear in its stead.
6. It can provide for singular values. Its power of doing this is referred to in M. Menabrea's memoir, page 262, where he mentions the passage of values through zero and infinity. The practicability of causing it arbitrarily to change its processes at any moment, on the occurrence of any specified contingency (of which its substitution of $(\frac{1}{2} \cos \overline{n+1}\, \theta + \frac{1}{2} \cos \overline{n-1}\, \theta)$ for $(\cos n\, \theta \cdot \cos \theta)$ explained in Note E., is in some degree an illustration), at once secures this point.

The subject of integration and of differentiation demands some notice. The engine can effect these processes in either of two ways:—

First. We may order it, by means of the Operation and of the Variable-cards, to go through the various steps by which the required *limit* can be worked out for whatever function is under consideration.

Secondly. It may (if we know the form of the limit for the function in question) effect the integration or differentiation by direct* substitution. We remarked in Note

---

* The engine cannot of course compute limits for perfectly *simple* and *uncompounded* functions, except in this manner. It is obvious that it has no power of representing or of manipulating with any but *finite* increments or decrements; and consequently that wherever the computation of limits (or of any other functions) depends upon the *direct* introduction of quantities which either increase or decrease *indefinitely*, we are absolutely beyond the sphere of its powers. Its nature and arrangements

B., that any *set* of columns on which numbers are inscribed, represents merely a *general* function of the several quantities, until the special function have been impressed by means of the Operation and Variable-cards. Consequently, if instead of requiring the value of the function, we require that of its integral, or of its differential coefficient, we have merely to order whatever particular combination of the ingredient quantities may constitute that integral or that coefficient. In ax^n, for instance, instead of the quantities

$$V_0 \qquad V_1 \qquad V_2 \qquad V_3$$

$$\boxed{a} \qquad \boxed{n} \qquad \boxed{x} \qquad \boxed{ax^n}$$

$$\underbrace{\qquad\qquad\qquad\qquad\qquad}_{ax^n}$$

being ordered to appear on V_3 in the combination ax^n, they would be ordered to appear in that of

$$anx^{n-1}.$$

They would then stand thus:—

$$V_0 \qquad V_1 \qquad V_2 \qquad V_3$$

$$\boxed{a} \qquad \boxed{n} \qquad \boxed{x} \qquad \boxed{anx^{n-1}}$$

$$\underbrace{\qquad\qquad\qquad\qquad\qquad}_{anx^{n-1}}$$

Similarly, we might have $\frac{a}{n}x^{(n+1)}$, the integral of ax^n.

An interesting example for following out the processes of that engine would be such a form as

$$\int \frac{x^n dx}{\sqrt{a^2 - x^2}},$$

or any other cases of integration by successive reductions, where an integral which contains an operation repeated n times can be made to depend upon another which contains the same $n-1$ or $n-2$ times, and so on until by continued reduction we arrive at a certain *ultimate* form, whose value has then to be determined.

The methods in Arbogast's *Calcul des Dérivations* are peculiarly fitted for the

are remarkably adapted for taking into account all *finite* increments or decrements (however small or large), and for developing the true and logical modifications of form or value dependent upon differences of this nature. The engine may indeed be considered as including the whole Calculus of Finite Differences; many of whose theorems would be especially and beautifully fitted for development by its processes, and would offer peculiarly interesting considerations. We may mention, as an example, the calculation of the Numbers of Bernoulli by means of the *Differences of Nothing*.

notation and the processes of the engine. Likewise the whole of the Combinatorial Analysis, which consists first in a purely numerical calculation of indices, and secondly in the distribution and combination of the quantities according to laws prescribed by these indices.

We will terminate these Notes by following up in detail the steps through which the engine could compute the Numbers of Bernoulli, this being (in the form in which we shall deduce it) a rather complicated example of its powers. The simplest manner of computing these numbers would be from the direct expansion of

$$\frac{x}{\epsilon^x - 1} = \frac{1}{1 + \frac{x}{2} + \frac{x^2}{2 \cdot 3} + \frac{x^3}{2 \cdot 3 \cdot 4} + \&c.} \tag{1.}$$

which is in fact a particular case of the development of

$$\frac{a + bx + cx^2 + \&c.}{a' + b'x + c'x^2 + \&c.}$$

mentioned in Note E. Or again, we might compute them from the well-known form

$$B_{2n-1} = 2 \cdot \frac{1 \cdot 2 \cdot 3 \dots 2n}{(2\pi)^{2n}} \cdot \{1 + \frac{1}{2^{2n}} + \frac{1}{3^{2n}} + \dots\} \tag{2.}$$

or from the form

$$B_{2n-1} = \frac{\pm 2n}{(2^{2n} - 1) 2^{n-1}} \left\{ \begin{array}{l} \frac{1}{2} n^{2n-1} \\[2mm] -(n-1)^{2n-1} \left\{ 1 + \frac{1}{2} \cdot \frac{2n}{1} \right\} \\[2mm] +(n-2)^{2n-1} \left\{ 1 + \frac{2n}{1} + \frac{1}{2} \cdot \frac{2n \cdot (2n-1)}{1 \cdot 2} \right\} \\[2mm] -(n-3)^{2n-1} \left\{ 1 + \frac{2n}{1} + \frac{2n \cdot (2n-1)}{1 \cdot 2} + \\[2mm] + \frac{1}{2} \cdot \frac{2n \cdot (2n-1) \cdot (2n-2)}{1 \cdot 2 \cdot 3} \right\} \\[2mm] + \dots \qquad \dots \qquad \dots \qquad \dots \end{array} \right\} \tag{3.}$$

or from many others. As however our object is not simplicity or facility of computation, but the illustration of the powers of the engine, we prefer selecting the formula below, marked (8.). This is derived in the following manner:—

If in the equation

$$\frac{x}{\epsilon^x - 1} = 1 - \frac{x}{2} + B_1 \frac{x^2}{2} + B_3 \frac{x^4}{2 \cdot 3 \cdot 4} + B_5 \frac{x^6}{2 \cdot 3 \cdot 4 \cdot 5 \cdot 6} + \dots \tag{4.}$$

(in which $B_1, B_3 \dots$, &c. are the Numbers of Bernoulli), we expand the denominator of the first side in powers of x, and then divide both numerator and denominator by x, we shall derive

$$1 = \left(1 - \frac{x}{2} + B_1 \frac{x^2}{2} + B_3 \frac{x^4}{2 \cdot 3 \cdot 4} + \ldots \right) \left(1 + \frac{x}{2} + \frac{x^2}{2 \cdot 3} + \frac{x^3}{2 \cdot 3 \cdot 4} + \ldots \right) \qquad (5.)$$

If this latter multiplication be actually performed, we shall have a series of the general form

$$1 + D_1 x + D_2 x^2 + D_3 x^3 + \ldots \qquad (6.)$$

in which we see, first, that all the coefficients of the powers of x are severally equal to zero; and secondly, that the general form for D_{2n} the co-efficient of the $(2n + 1)$th *term*, (that is of x^{2n} any *even* power of x), is the following:—

$$\frac{1}{2 \cdot 3 \ldots 2n + 1} - \frac{1}{2} \cdot \frac{1}{2 \cdot 3 \ldots 2n} + \frac{B_1}{2} \cdot \frac{1}{2 \cdot 3 \ldots 2n - 1} +$$

$$\frac{B_3}{2 \cdot 3 \cdot 4} \cdot \frac{1}{2 \cdot 3 \ldots 2n - 3} + \frac{B_5}{2 \cdot 3 \cdot 4 \cdot 5 \cdot 6} \cdot \frac{1}{2 \cdot 3 \ldots 2n - 5} + \qquad (7.)$$

$$\ldots + \frac{B_{2n - 1}}{2 \cdot 3 \ldots 2n} \cdot 1 = 0$$

Multiplying every term by $(2 \cdot 3 \ldots 2n)$, we have

$$0 = -\frac{1}{2} \cdot \frac{2n - 1}{2n + 1} + B_1 \left(\frac{2n}{2}\right) + B_3 \left(\frac{2n \cdot 2n - 1 \cdot 2n - 2}{2 \cdot 3 \cdot 4}\right) + \\ + B_5 \left(\frac{2n \cdot 2n - 1 \ldots 2n - 4}{2 \cdot 3 \cdot 4 \cdot 5 \cdot 6}\right) + \ldots + B_{2n - 1} \Bigg\} \qquad (8.)$$

which it may be convenient to write under the general form:—

$$0 = A_0 + A_1 B_1 + A_3 B_3 + A_5 B_5 + \ldots + B_{2n - 1} \qquad (9.)$$

A_1, A_3, &c. being those functions of n which respectively belong to B_1, B_3, &c.

We might have derived a form nearly similar to (8.), from D_{2n-1} the coefficient of any *odd* power of x in (6.); but the general form is a little different for the coefficients of the *odd* powers, and not quite so convenient.

On examining (7.) and (8.), we perceive that, when these formulae are isolated from (6.) whence they are derived, and considered in themselves separately and independently, n may be any whole number whatever; although when (7.) occurs *as one of the* D's in (6.), it is obvious that n is then not arbitrary, but is always a certain function of the *distance of that* D *from the beginning*. If that distance be $= d$, then

$$2n + 1 = d, \text{ and } n = \frac{d - 1}{2} \text{ (for any } even \text{ power of } x.)$$

$$2n = d, \text{ and } n = \frac{d}{2} \text{ (for any } odd \text{ power of } x.)$$

It is with the *independent* formula (8.) that we have to do. Therefore it must be remembered that the conditions for the value of n are now modified, and that n is a perfectly *arbitrary* whole number. This circumstance, combined with the fact (which we may easily perceive) that whatever n is, every term of (8.) after the $(n+1)$th is $= 0$, and that the $(n+1)$th term itself is always $= B_{2n-1} \cdot \dfrac{1}{1} = B_{2n-1}$, enables us to find the value (either numerical or algebraical) of any nth Number of Bernoulli B_{2n-1}, *in terms of all the preceding ones*, if we but know the values of B_1, B_3 . . . B_{2n-3}. We append to this Note a Diagram and Table, containing the details of the computation for B_7, (B_1, B_3, B_5 being supposed given).

On attentively considering (8.), we shall likewise perceive that we may derive from it the numerical value of *every* Number of Bernoulli in succession, from the very beginning, *ad infinitum*, by the following series of computations:—

1st Series.— Let $n = 1$, and calculate (8.) for this value of n. The result is B_1,

2nd Series.—Let $n = 2$. Calculate (8.) for this value of n, substituting the value of B_1 just obtained. The result is B_3.

3rd Series.— Let $n = 3$. Calculate (8.) for this value of n, substituting the values of B_1, B_3 before obtained. The result is B_5. And so on, to any extent.

The diagram* represents the columns of the engine when just prepared for computing B_{2n-1} (in the case of $n = 4$); while the table beneath them presents a complete simultaneous view of all the successive changes which these columns then severally pass through in order to perform the computation. (The reader is referred to Note D, for explanations respecting the nature and notation of such tables.)

Six numerical *data* are in this case necessary for making the requisite combinations. These data are 1, 2, n ($=4$), B_1, B_3, B_5. Were $n = 5$, the additional datum B_7 would be needed. Were $n = 6$, the datum B_9 would be needed; and so on. Thus the actual *number of data* needed will always be $n + 2$, for $n = n$; and out of these $n + 2$ data, $(n + 2 - 3)$ of them are successive Numbers of Bernoulli. The reason why the Bernoulli Numbers used as data, are nevertheless placed on *Result*-columns in the diagram, is because they may properly be supposed to have been previously computed in succession by the *engine* itself; under which circumstances each B will appear as a *result*, previous to being used as a *datum* for computing the succeeding B. Here then is an instance (of the kind alluded to in Note D.) of the same Variables filling more than one office in turn. It is true that if we consider our computation of B_7 as a perfectly *isolated* calculation, we may conclude B_1, B_3, B_5 to have been arbitrarily placed on the columns; and it would then perhaps be more consistent to put them on V_4, V_5, V_6 as data and not results. But we are not taking this view. On the contrary, we suppose the engine to be *in the course of* computing the Numbers to an indefinite extent, from the very beginning; and that we merely single out, by way of example, *one amongst* the successive but distinct series of computations it is thus performing. Where the B's are fractions, it must be understood that they are computed and appear in the notation of *decimal* fractions. Indeed this is a circumstance that should be noticed with reference to all calculations. In any of the examples already given in the

---

* See the diagram at the end of these Notes.

translation and in the Notes, some of the *data*, or of the temporary or permanent results, might be fractional, quite as probably as whole numbers. But the arrangements are so made, that the nature of the processes would be the same as for whole numbers.

In the above table and diagram we are not considering the *signs* of any of the B's, merely their numerical magnitude. The engine would bring out the sign for each of them correctly of course, but we cannot enter on *every* additional detail of this kind, as we might wish to do. The circles for the sign are therefore intentionally left blank in the diagram.

Operation-cards 1, 2, 3, 4, 5, 6 prepare $-\dfrac{1}{2}\cdot\dfrac{2n-1}{2n+1}$. Thus, Card 1 multiplies *two* into n, and the three *Receiving* Variable-cards belonging respectively to V_4, V_5, V_6, allow the result $2n$ to be placed on each of these latter columns (this being a case in which a triple receipt of the result is needed for subsequent purposes); we see that the upper indices of the two Variables used, during Operation 1, remain unaltered.

We shall not go through the details of every operation singly, since the table and diagram sufficiently indicate them; we shall merely notice some few peculiar cases.

By Operation 6, a *positive* quantity is turned into a *negative* quantity, by simply subtracting the quantity from a column which has only zero upon it. (The sign at the top of V_8 would become $-$ during this process.)

Operation 7 will be unintelligible, unless it be remembered that if we were calculating for $n=1$ instead of $n=4$, Operation 6 would have completed the computation of B_1 itself; in which case the engine, instead of continuing its processes, would have to put B_1 on V_{21}; and then either to stop altogether, or to begin Operations 1, 2 . . . 7 all over again for value of $n(=2)$, in order to enter on the computation of B_3; (having however taken care, previous to the recommencement, to make the number of V_3 equal to *two*, by the addition of unity to the former $n=1$ on that column). Now Operation 7 must either bring out a result equal to zero (if $n=1$); or a result *greater* than *zero*, as in the present case; and the engine follows the one or the other of the two courses just explained, contingently on the one or the other result of Operation 7. In order fully to perceive the necessity of this *experimental* operation, it is important to keep in mind what was pointed out, that we are not treating a perfectly isolated and independent computation, but one out of a series of antecedent and prospective computations.

Cards 8, 9, 10 produce $-\dfrac{1}{2}\cdot\dfrac{2n-1}{2n+1}+B_1\dfrac{2n}{2}$. In Operation 9 we see an example of an upper index which again becomes a value after having passed from preceding values to zero. V_{11} has successively been ${}^0V_{11}$, ${}^1V_{11}$, ${}^2V_{11}$, ${}^0V_{11}$, ${}^3V_{11}$; and, from the nature of the office which V_{11} performs in the calculation, its index will continue to go through further changes of the same description, which, if examined, will be found to be regular and periodic.

Card 12 has to perform the same office as Card 7 did in the preceding section; since, if n had been $=2$, the 11th operation would have completed the computation of B_3.

Cards 13 to 20 make A_3. Since A_{2n-1}, always consists of $2n-1$ factors, A_3 has three factors; and it will be seen that Cards 13, 14, 15, 16 make the second of these factors, and then multiply it with the first; and that 17, 18, 19, 20 make the third factor, and then multiply this with the product of the two former factors.

Card 23 has the office of Cards 11 and 7 to perform, since if n were $= 3$, the 21st and 22nd operations would complete the computation of B_5. As our case is B_7, the computation will continue one more stage; and we must now direct attention to the fact, that in order to compute A, it is merely necessary precisely to repeat the group of Operations 13 to 20; and then, in order to complete the computation of B_7, to repeat Operations 21, 22.

It will be perceived that every unit added to n in B_{2n-1}, entails an additional repetition of operations $(13 \ldots 23)$ for the computation of B_{2n-1}. Not only are all the *operations* precisely the same however for every such repetition, but they require to be respectively supplied with numbers from the very *same pairs of columns*; with only the one exception of Operation 21, which will of course, need B_5 (from V_{23}) instead of B_3 (from V_{22}). This identity in the *columns* which supply the requisite numbers, must not be confounded with identity in the *values* those columns have upon them and give out to the mill. Most of those values undergo alterations during a performance of the operations $(13 \ldots 23)$, and consequently the columns present a new set of values for the *next* performance of $(13 \ldots 23)$ to work on.

At the termination of the *repetition* of operations $(13 \ldots 23)$ in computing B_7, the alterations in the values on the Variables are, that

$V_6 = 2n-4$ instead of $2n-2$.
$V_7 = 6$ instead of 4.
$V_{10} = 0$ instead of 1.
$V_{13} = A_0 + A_1 B_1 + A_3 B_3 + A_5 B_5$ instead of $A_0 + A_1 B_1 + A_3 B_3$

In this state the only remaining processes are first: to transfer the value which is on V_{15}, to V_{24}; and secondly to reduce V_6, V_7, V_{13} to zero, and to add* *one* to V_3, in order that the engine may be ready to commence computing B_9. Operations 24 and 25 accomplish these purposes. It may be thought anomalous that Operation 25 is represented as leaving the upper index of V_3 still $=$ unity. But it must be remembered that these indices always begin anew for a separate calculation, and that Operation 25 places upon V_3 the *first* value *for the new calculation*.

It should be remarked, that when the group $(13 \ldots 23)$ is *repeated*, changes occur in some of the *upper* indices during the course of the repetition: for example, $^3V_6$ would become $^4V_6$ and $^5V_6$.

We thus see that when $n = 1$, nine Operation-cards are used; that when $n = 2$,

---

* It is interesting to observe, that so complicated a case as this calculation of the Bernoullian Numbers, nevertheless, presents a remarkable simplicity in one respect; viz., that during the processes for the computation of *millions* of these Numbers, no other arbitrary modification would be requisite in the arrangements, excepting the above simple and uniform provision for causing one of the data periodically to receive the finite increment unity.

fourteen Operation-cards are used; and that when $n > 2$, twenty-five Operation-cards are used; but that no *more* are needed, however great n may be; and not only this, but that these same twenty-five cards suffice for the successive computation of all the Numbers from B_1 to B_{2n-1} inclusive. With respect to the number of *Variable*-cards, it will be remembered, from the explanations in previous Notes, that an average of three such cards to each *operation* (not however to each Operation-*card*) is the estimate. According to this the computation of B_1 will require twenty-seven Variable-cards; B_3 forty-two such cards; B_5 seventy-five; and for every succeeding B after B_5, there would be thirty-three additional Variable-cards (since each repetition of the group $(13 \ldots 23)$ adds eleven to the number of operations required for computing the previous B). But we must now explain, that whenever there is a *cycle of operations*, and if these merely require to be supplied with numbers from the *same pairs of columns* and likewise each operation to place its *result* on the *same* column for every repetition of the whole group, the process then admits of a *cycle of Variable-cards* for effecting its purposes. There is obviously much more symmetry and simplicity in the arrangements, when cases do admit of repeating the Variable as well as the Operation-cards. Our present example is of this nature. The only exception to a *perfect identity* in *all* the processes and columns used, for every repetition of Operations $(13 \ldots 23)$ is, that Operation 21 always requires one of its factors from a new column, and Operation 24 always puts its result on a new column. But as these variations follow the same law at each repetition, (Operation 21 always requiring its factor from a column *one* in advance of that which it used the previous time, and Operation 24, always putting its result on the column *one* in advance of that which received the previous result), they are easily provided for in arranging the recurring group (or cycle) of variable-cards.

We may here remark that the average estimate of three Variable-cards coming into use to each operation, is not to be taken as an absolutely and literally correct amount for all cases and circumstances. Many special circumstances, either in the nature of a problem, or in the arrangements of the engine under certain contingencies, influence and modify this average to a greater or less extent. But it is a very safe and correct *general* rule to go upon. In the preceding case it will give us seventy-five Variable-cards as the total number which will be necessary for computing any B after B_3. This is very nearly the precise amount really used, but we cannot here enter into the minutae of the few particular circumstances which occur in this example (as indeed at some one stage or other of probably most computations) to modify slightly this number.

It will be obvious that the very *same* seventy-five Variable-cards may be repeated for the computation of every succeeding Number, just on the same principle as admits of the repetition of the thirty-three Variable-cards of Operations $(13 \ldots 23)$ in the computation of any *one* Number. Thus there will be a *cycle of a cycle* of Variable-cards.

If we now apply the notation for cycles, as explained in Note E, we may express the operations for computing the Numbers of Bernoulli in the following manner:—

$(1 \ldots 7), (24, 25)$ gives B_1 $=$ 1st number; (n being $= 1$).

$(1 \ldots 7), (8 \ldots 12), (24, 25)$ gives B_3 $=$ 2nd number; (n being $= 2$).

$(1 \ldots 7), (8 \ldots 12), (13 \ldots 23), (24, 25)$ gives B_5 $=$ 3rd number; (n being $= 3$).

$(1 \ldots 7), (8 \ldots 12), 2 (13 \ldots 23), (24, 25)$ gives B_7 $=$ 4th number; (n being $= 4$).

$$\vdots \qquad \vdots \qquad \vdots \qquad \vdots \qquad \vdots \qquad \vdots$$

$(1 \ldots 7), (8 \ldots 12), \Sigma (+1)^{n-2} (13 \ldots 23), (24, 25)$ gives $B_{2n-1} = n$th number; (n being $= n$).

Again,

$$(1 \ldots 7), (24, 25), \underset{\text{limits 1 to } n}{\Sigma (+1)^n} \{(1 \ldots 7), (8 \ldots 12), \underset{\text{limits 0 to } (n+2)}{\Sigma (n+2)} (13 \ldots 23), (24, 25)\}$$

represents the total operations for computing every number in succession, from B_1 to B_{2n-1} inclusive.

In this formula we see a *varying cycle* of the *first* order, and an ordinary cycle of the *second* order. The latter cycle in this case includes in it the varying cycle.

On inspecting the ten Working-Variables of the diagram, it will be perceived, that although the *value* on any one of them (excepting V_4 and V_5) goes through a series of changes, the *office* which each performs is in this calculation *fixed* and *invariable*. Thus V_6 always prepares the *numerators* of the factors of any A; V_7 the *denominators*. V_8 always receives the $(2n-3)$th factor of A_{2n-1}, and V_9 the $(2n-1)$th. V_{10} always decides which of two courses the succeeding processes are to follow, by feeling for the value of n through means of a subtraction; and so on; but we shall not enumerate further. It is desirable in all calculations, so to arrange the processes, that the *offices* performed by the Variables may be as uniform and fixed as possible.

Supposing that it was desired not only to tabulate B_1, B_3, &c., but A_0, A_1, &c.; we have only then to appoint another series of variables, V_{41}, V_{42}, &c., for receiving these latter results as they are successively produced upon V_{11}. Or again, we may, instead of this, or in addition to this second series of results, wish to tabulate the value of each successive *total* term of the series (8), viz: $A_0, A_1 B_1, A_3 B_3$, &c. We have then merely to multiply each B with each corresponding A, as produced, and to place these successive products on Result-columns appointed for the purpose.

The formula (8.) is interesting in another point of view. It is one particular case of the general Integral of the following Equation of Mixed Differences:—

$$\frac{d^2}{dx^2} \left(z_{n+1} x^{2n+2} \right) = (2n+1)(2n+2) z^n x^{2n}$$

for certain special suppositions respecting z, x and n.

The *general* integral itself is of the form,

$$z_n = f(n) \cdot x + f_1(n) + f_2(n) \cdot x^{-1} + f_3(n) \cdot x^{-3} + \ldots$$

and it is worthy of remark, that the engine might (in a manner more or less similar to the preceding) calculate the value of this formula upon most *other* hypotheses for the functions in the integral, with as much, or (in many cases) with more, ease than it can formula (8.).   A. A. L.

Diagram for the computation by the Engine of the Numbers of Bernoulli.

Number of Operation.	Nature of Operation.	Variables acted upon.	Variables receiving results.	Indication of change in the value on any Variable.	Statement of Results.	Data. ${}^1V_1$ $\begin{smallmatrix}0\\0\\0\\1\end{smallmatrix}$ $\boxed{1}$	${}^1V_2$ $\begin{smallmatrix}0\\0\\0\\2\end{smallmatrix}$ $\boxed{2}$	${}^1V_3$ $\begin{smallmatrix}0\\0\\0\\4\end{smallmatrix}$ $\boxed{n}$	${}^0V_4$ $\begin{smallmatrix}0\\0\\0\\0\end{smallmatrix}$ $\boxed{\ }$	${}^0V_5$ $\begin{smallmatrix}0\\0\\0\\0\end{smallmatrix}$ $\boxed{\ }$	${}^0V_6$ $\begin{smallmatrix}0\\0\\0\\0\end{smallmatrix}$ $\boxed{\ }$
1	×	${}^1V_2 \times {}^1V_3$	${}^1V_4, {}^1V_5, {}^1V_6$	$\left\{\begin{smallmatrix}{}^1V_2={}^1V_2\\{}^1V_3={}^1V_3\end{smallmatrix}\right\}$	$= 2n$	...	2	n	$2n$	$2n$	$2n$
2	−	${}^1V_4 - {}^1V_1$	${}^2V_4$	$\left\{\begin{smallmatrix}{}^1V_4={}^2V_4\\{}^1V_1={}^1V_1\end{smallmatrix}\right\}$	$= 2n-1$	1	...	...	$2n-1$		
3	+	${}^1V_5 + {}^1V_1$	${}^2V_5$	$\left\{\begin{smallmatrix}{}^1V_5={}^2V_5\\{}^1V_1={}^1V_1\end{smallmatrix}\right\}$	$= 2n+1$	1	...	...	...	$2n+1$	
4	÷	${}^2V_5 \div {}^2V_4$	${}^1V_{11}$	$\left\{\begin{smallmatrix}{}^2V_5={}^0V_5\\{}^2V_4={}^0V_4\end{smallmatrix}\right\}$	$= \dfrac{2n-1}{2n+1}$	...	...	...	0	0	...
5	÷	${}^1V_{11} \div {}^1V_2$	${}^2V_{11}$	$\left\{\begin{smallmatrix}{}^1V_{11}={}^2V_{11}\\{}^1V_2={}^1V_2\end{smallmatrix}\right\}$	$= \dfrac{1}{2} \cdot \dfrac{2n-1}{2n+1}$	...	2	...	...	...	...
6	−	${}^0V_{13} - {}^2V_{11}$	${}^1V_{13}$	$\left\{\begin{smallmatrix}{}^2V_{11}={}^0V_{11}\\{}^0V_{13}={}^1V_{13}\end{smallmatrix}\right\}$	$= -\dfrac{1}{2} \cdot \dfrac{2n-1}{2n+1} = A_0$	...	...	...	...	...	...
7	−	${}^1V_3 - {}^1V_1$	${}^1V_{10}$	$\left\{\begin{smallmatrix}{}^1V_3={}^1V_3\\{}^1V_1={}^1V_1\end{smallmatrix}\right\}$	$= n-1 \;(= 3)$	1	...	n	...	...	
8	+	${}^1V_2 + {}^0V_7$	${}^1V_7$	$\left\{\begin{smallmatrix}{}^1V_2={}^1V_2\\{}^0V_7={}^1V_7\end{smallmatrix}\right\}$	$= 2+0 = 2$	...	2	...	...	...	...
9	÷	${}^1V_6 \div {}^1V_7$	${}^3V_{11}$	$\left\{\begin{smallmatrix}{}^1V_6={}^1V_6\\{}^0V_{11}={}^3V_{11}\end{smallmatrix}\right\}$	$= \dfrac{2n}{2} = A_1$	...	...	...	...	...	$2n$
10	×	${}^1V_{21} \times {}^3V_{11}$	${}^1V_{12}$	$\left\{\begin{smallmatrix}{}^1V_{21}={}^1V_{21}\\{}^3V_{11}={}^3V_{11}\end{smallmatrix}\right\}$	$= B_1 \cdot \dfrac{2n}{2} = B_1 A_1$	...	...	...	...	...	...
11	+	${}^1V_{12} + {}^1V_{13}$	${}^2V_{13}$	$\left\{\begin{smallmatrix}{}^1V_{12}={}^0V_{12}\\{}^1V_{13}={}^2V_{13}\end{smallmatrix}\right\}$	$= -\dfrac{1}{2}\dfrac{2n-1}{2n+1} + B_1 \cdot \dfrac{2n}{2}$	...	...	...	...	...	...
12	−	${}^1V_{10} - {}^1V_1$	${}^2V_{10}$	$\left\{\begin{smallmatrix}{}^1V_{10}={}^2V_{10}\\{}^1V_1={}^1V_1\end{smallmatrix}\right\}$	$= n-2 \;(= 2)$	1	...	...	...	...	...
13	−	${}^1V_6 - {}^1V_1$	${}^2V_6$	$\left\{\begin{smallmatrix}{}^1V_6={}^2V_6\\{}^1V_1={}^1V_1\end{smallmatrix}\right\}$	$= 2n-1$	1	...	...	...	...	$2n-1$
14	+	${}^1V_1 + {}^1V_7$	${}^2V_7$	$\left\{\begin{smallmatrix}{}^1V_1={}^1V_1\\{}^1V_7={}^2V_7\end{smallmatrix}\right\}$	$= 2+1 = 3$	1	...	...	...	...	...
15	÷	${}^2V_6 \div {}^2V_7$	${}^1V_8$	$\left\{\begin{smallmatrix}{}^2V_6={}^2V_6\\{}^2V_7={}^2V_7\end{smallmatrix}\right\}$	$= \dfrac{2n-1}{3}$	...	...	...	...	...	$2n-1$
16	×	${}^1V_8 \times {}^3V_{11}$	${}^4V_{11}$	$\left\{\begin{smallmatrix}{}^1V_8={}^0V_8\\{}^3V_{11}={}^4V_{11}\end{smallmatrix}\right\}$	$= \dfrac{2n}{2} \cdot \dfrac{2n-1}{3}$	...	...	...	...	...	...
17	−	${}^2V_6 - {}^1V_1$	${}^3V_6$	$\left\{\begin{smallmatrix}{}^2V_6={}^3V_6\\{}^1V_1={}^1V_1\end{smallmatrix}\right\}$	$= 2n-2$	1	...	...	...	...	$2n-2$
18	+	${}^1V_1 + {}^2V_7$	${}^3V_7$	$\left\{\begin{smallmatrix}{}^2V_7={}^3V_7\\{}^1V_1={}^1V_1\end{smallmatrix}\right\}$	$= 3+1 = 4$	1	...	...	...	...	...
19	÷	${}^3V_6 \div {}^3V_7$	${}^1V_9$	$\left\{\begin{smallmatrix}{}^3V_6={}^3V_6\\{}^3V_7={}^3V_7\end{smallmatrix}\right\}$	$= \dfrac{2n-2}{4}$	...	...	...	...	...	$2n-2$
20	×	${}^1V_9 \times {}^4V_{11}$	${}^5V_{11}$	$\left\{\begin{smallmatrix}{}^1V_9={}^0V_9\\{}^4V_{11}={}^5V_{11}\end{smallmatrix}\right\}$	$= \dfrac{2n}{2} \cdot \dfrac{2n-1}{3} \cdot \dfrac{2n-2}{4} = A_3$	...	...	...	...	...	...
21	×	${}^1V_{22} \times {}^5V_{11}$	${}^0V_{12}$	$\left\{\begin{smallmatrix}{}^1V_{22}={}^1V_{22}\\{}^0V_{12}={}^2V_{12}\end{smallmatrix}\right\}$	$= B_3 \cdot \dfrac{2n}{2} \cdot \dfrac{2n-1}{3} \cdot \dfrac{2n-2}{3} = B_3 A_3$	...	...	...	...	...	...
22	+	${}^2V_{12} + {}^2V_{13}$	${}^3V_{13}$	$\left\{\begin{smallmatrix}{}^2V_{12}={}^0V_{12}\\{}^2V_{13}={}^3V_{13}\end{smallmatrix}\right\}$	$= A_0 + B_1 A_1 + B_3 A_3$	...	...	...	...	...	...
23	−	${}^2V_{10} - {}^1V_1$	${}^3V_{10}$	$\left\{\begin{smallmatrix}{}^2V_{10}={}^3V_{10}\\{}^1V_1={}^1V_1\end{smallmatrix}\right\}$	$= n-3 \;(= 1)$	1	...	...	...	...	...

Here follows a repetition of

24	+	${}^4V_{13} + {}^0V_{24}$	${}^1V_{24}$	$\left\{\begin{smallmatrix}{}^4V_{13}={}^0V_{13}\\{}^0V_{24}={}^1V_{24}\end{smallmatrix}\right\}$	$= B_7$	...	...	...	...	...	...
25	+	${}^1V_1 + {}^1V_3$	${}^1V_3$	$\left\{\begin{smallmatrix}{}^1V_1={}^1V_1\\{}^1V_3={}^1V_3\\{}^5V_6={}^0V_6\\{}^5V_7={}^0V_7\end{smallmatrix}\right\}$	$= n+1 = 4+1 = 5$ .... by a Variable-card. by a Variable-card.	1	...	$n+1$	...	...	0

			Working Variables.						Result Variables.		
$^0V_7$ ◯ 0 0 0 ▢	$^0V_8$ ◯ 0 0 0 ▢	$^0V_9$ ◯ 0 0 0 ▢	$^0V_{10}$ ◯ 0 0 0 ▢	$^0V_{11}$ ◯ 0 0 0 ▢	$^0V_{12}$ ◯ 0 0 0 ▢	$^0V_{13}$ ...... ◯ 0 0 0	$^1V_{21}$ ◯ B₁ in a decimal fraction. ▢ B₁	$^1V_{22}$ ◯ B₃ in a decimal fraction. ▢ B₃	$^1V_{23}$ ◯ B₅ in a decimal fraction. ▢ B₅	$^0V_{24}$... ◯ 0 0 0 ▢ B₇
...	...	...	...	$\dfrac{2n-1}{2n+1}$						
...	...	...	...	$\dfrac{1}{2}\cdot\dfrac{2n-1}{2n+1}$						
...	...	...	...	0		.... $-\dfrac{1}{2}\cdot\dfrac{2n-1}{2n+1}=A_0$				
...	...	$n-1$	...							
2										
2	...	...	...	$\dfrac{2n}{2}=A_1$						
...	...	...	...	$\dfrac{2n}{2}=A_1$	$B_1\cdot\dfrac{2n}{2}=B_1A_1$	....	B₁			
...	...	...	...	....	0	$\left\{-\dfrac{1}{2}\cdot\dfrac{2n-1}{2n+1}+B_1\cdot\dfrac{2n}{2}\right\}$				
...	...	$n-2$	...							
3										
3	$\dfrac{2n-1}{3}$									
...	0	...	...	$\dfrac{2n}{2}\cdot\dfrac{2n-1}{3}$						
4										
4	...	$\dfrac{2n-2}{4}$	...	$\left\{\dfrac{2n}{2}\cdot\dfrac{2n-1}{3}\cdot\dfrac{2n-2}{3}=A_3\right\}$						
...	...	0								
...	...	...	...	0						
...	...	...	...	0	B_3A_3	....	......	B₃		
...	...	...	...	....	0	$\left\{A_3+B_1A_1+B_3A_3\right\}$				
...	...	...	$n-3$							

Operations thirteen to twenty-three.

...	...	...	...	....	....	....	......	......	......	B₇
	0									

In 1851, in connection with the Great Exhibition, Babbage published a version of his mechanical notation without which development of the Analytical Engines would have been impossible.

Laws of Mechanical Notation, pp. 1–6

ON LETTERING DRAWINGS

All machinery consists of—

	Framing.			Parts, or Pieces.	
Fixed.		Moveable.		Moveable as axes, springs, &c.	

Every *Piece* possesses one or more *Working Points*. These are divided into two classes, those by which the Piece acts on others, and those by which it receives action from them: these are called *Driving* and *Driven Points*. A *Working Point* may fulfil both these offices, as for example, the same teeth which are driven by one wheel, may in another part of their course drive other wheels.

The following alphabets of large letters are used in Drawings:—

Etruscan.		Roman.		Writing.	
A	**A**	A	*A*	*A*	*A*
B	**B**	B	*B*	*B*	*B*
C	**C**	C	*C*	*C*	*C*
...		...		...	

The following alphabets of small letters are used:—

a   b   c   d ...
a   b   c   d ...

It is most convenient, and generally sufficient, to use only the letters *a, c, e, i, m, n, o, r, s, u, v, w, x, z*, of both these latter alphabets.

Rule 1.—Every separate portion of *Frame-work* must be indicated by a large *upright* letter.

Rule 2.—Every *Working Point* of Frame-work must be indicated by a *small* printed letter.

Rule 3.—*Frame-work* which is itself moveable, must be represented by a

large upright letter, with the sign of motion in its proper place below it (*see
Signs of Motion*), as

G   H̤

Rule 4.—In lettering Drawings, commence with the axes. These must be
lettered with *large inclined* letters, of either of the three alphabets. Whenever
the wheels or arms of any two or more adjacent axes cross each other on the
plan, avoid denoting those axes by letters of the *same* alphabet.

Rule 5.—In lettering *Pieces* as wheels, arms, &c. belonging to any axis,
whether they are fixed to it or moveable upon it, always use *inclined capitals* of
the *same* alphabet as that of the letter representing the Axis.

Rule 6.—Beginning with the lowest *Piece* upon an Axis, assign to it any
capital letter of the *same* alphabet. To the Piece next above, assign any other
capital letter which occurs *later* in the *same* alphabet. Continue this process for
each *Piece*.

Thus, although the succession of the letters of the *same* alphabet need not be
continuous, yet their occurrence in *alphabetic* order will never be violated.

Rule 7.—In lettering *Pieces* upon axes perpendicular to the elevation, or to
the end views, looking from the left side, the earliest letters of the alphabet
must be placed on the pieces most remote from the eye.

Rule 8.—No axis, which has a *Piece* crossing any other *Piece* belonging to an
adjacent axis, must have the same identity as that axis.

If there are many *Pieces* on the same axis, it may be necessary to commence
with one of the earlier letters of the alphabet.

Rule 9.—In placing letters representing any *Piece* on which portions of
other *Pieces* are projected, it is always desirable to select such a situation that
no doubt can be entertained as to which of those *Pieces* the letter is intended to
indicate. This can often be accomplished by placing the letter upon some
portion of its own *Piece* which extends beyond the projected parts of the other
Piece.

Rule 10.—When *Pieces* are very small, or when they are crossed by many
other lines, it is convenient to place the letter representing them outside the
Piece itself, and to connect it with the *Piece* it indicates by an arrow. This arrow
should be a short fine line terminated by a head, abutting on, or perhaps
projecting into, the piece represented by the letter.

Rule 11.—When upon any drawing, a letter having a dot beneath it occurs,
it marks the existence of a *Piece* below.

Rule 12.—In case another *Piece*, exactly similar to one already represented
and lettered, exists below it, it cannot be expressed by any visible line. It may,

however, be indicated by placing its proper letter outside, and connecting that letter with a *dotted* arrow abutting on the upper *Piece*.

Rule 13.—The permanent connexion of two pieces of matter, or the permanent gearing of two wheels, is indicated by a short line, crossing, at right angles, the point of contact. The sign | indicates, in a certain sense, fixed connexion. This sign will be found very useful for indicating the boundaries of various pieces of framing.

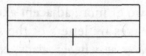

OF THE INDICES OF LETTERS

Rule 14.—Various indices and signs may be affixed to letters. Their position and use are indicated in the subjoined letter:—

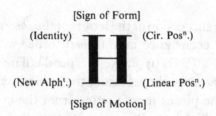

Rule 15.—The index on the left hand upper corner is used to mark the identity of two or more parts of a *Piece* which are permanently united; each being denoted by a letter with the *same* index.

Rule 16.—It is used also to connect any *Piece* itself with its various working points. Thus all the small letters which indicate the working points, must have the *same* index of identity as the letter expressing the *Piece* itself.

Rule 17.—Every *Working Point* must be marked by the *same small letter* as the *Working Point* of the *Piece* upon which it acts.

Rule 18.—The bearings in which axes work, as well as the working surface of the axes themselves, and also the working surfaces of slides, are *Working Points*, and must be lettered as such.

OF THE INDEX OF LINEAR POSITION

The successive order in which the various *Pieces* upon one axis succeed each other is indicated by the alphabetic succession.

It may, however, in some cases be convenient to distinguish between the relative heights of the various arms or wheels which constitute one *Piece*.

This may be easily accomplished by means of the index of *linear position*.

Every Piece may be represented as a whole, by one letter, with its proper index of identity. If, however, it is necessary to distinguish the different arms or parts of which it is composed, so as to indicate their relative position above the plane of projection, this may be accomplished by means of the indices of linear position.

Rule 19.—If ^3P represents the whole of any *Piece*, $^3P_1, ^3P_2, ^3P_3, ^3P_4$, &c. will represent in succession the several arms or parts of which ^3P is composed: $^3P_1$ indicating that which is most distant from the eye.

OF THE INDEX OF CIRCULAR POSITION

It may occasionally be desirable to indicate the order of succession in angular position of the various arms belonging to the same *Piece*, when projected on a plane. The index on the right hand upper corner is applied to this purpose.

Rule 20. ^6R representing any *Piece*,

$^6R_1$ will represent any arm as the origin,

$^6R_2$ the next arm in angular position in the direction '*screw*' that is, from left to right.

$^6R_3$ the next, &c.

Thus,

$$^6R_1^1, \; ^6R_2^2, \; ^6R_3^3 \ldots \, ^6R_n^n$$

would represent n arms placed spirally round an axis at various heights above it.

OF THE INDEX OF NEW ALPHABET

In case the three alphabets given above are found insufficient, the index on the left lower side is reserved to mark new alphabets. In the most complicated drawing I have scarcely ever had occasion to use it. It might in some cases be desirable to have a fourth alphabet, differing in form from those already given.

MR BABBAGE *will feel obliged by any criticisms, or additions to these Rules of Drawing, and to the Mechanical Alphabet, and request they may be addressed to him by post, at No. 1, Dorset Street, Manchester Square.*

JULY *1851.*

ALPHABET of FORM 7ᵀᴴ EDⁿ

#	Description	Symbols	#	Description	Symbols
1	Centres Solid. Hollow		31	Wheel Teeth	
2	Boss		32	Levers Straight. Cranked	
3	Arms Driving or Driver		33	Bar or Link	
4	Arms Driving or Driver		34	Links fixed centre in left	
5	Points D° D°		35	No fixed centre	
6	Pins Solid. Hollow		36	Fixed centre in right	
7	Arms Driving with solid pin Driven		37	Both centres fixed	
8	D° with hollow pin		38	Three centres in line	
9	Stops Stopping Stopped		39	Three centres not in line	
9	Stopping Arms		40		
10	Gearing Single Tooth		41		
11	Arm		42	Slide	
12	Wheel Teeth		43	Slides V's	
13	Pinion		44	Slot	
14	Sector		45	Slot with Arms	
15	Wanting Tooth Wheel		46		
16	Rack		47	Crank	
17	Cam Single Tooth		48	Eccentric Hooks Joint	
18	Arms		49	Bolt	
19	Wheel Teeth		50	Key & Counter Key	
20	Rachet Single Tooth. Right		51	Clutch & Counter Clutch	
21	Wheel Teeth		52	Fork & Fork Collar	
22	Arm		53	Fork & Fork Pin	
23	Single Tooth Left		54	Inclined Planes Driving Driven	
24	Wheel		55	Escapement	
25	Arm		56	Screws Right handed Left	
26	Returning & Correcting Tooth		57	Drums Conical	
27	Arm		58	Pulley	
28	Wheel Teeth		59	Arms Driving with Pulley Driven	
29	Locking Single Tooth		60	Barrels Cam. Jacquard Stud	
30	Arm		61	Piston	

316

No.	Item				No.	Description		
62	Cylinder	⊓			1	Circular in Plan	○	
63	Valve				2	Reciprocating in Plan	⌣	
64	Hammer	→			3	Linear in Plan	—	
65	Bell	⋀			4	Linear in plat at right angles	⋅	
66	Knife	∕			5	Circular in Elevation right side left side	↺	↻
67	Grind Stone	⊕			6	Reciprocating in elevation	⌄	⌢
68	Handle	⌐			7	Linear in Elevation			
69		↺			8	Curvilinear	∼	
70	Weight	⊡			9	Reciprocating in End View		
71	As Counterpoise				10	Circular in End View	⊙	
72	Pendulum				11			
73	Fly Wheel Fly Vane	◉			12			
74	Spring	◎			13			
75	Frame. Fixed	□			14			
76					15			
77	Plummer Blocks Lower Upper Side Double				16			
78	Punch Matrix				17			
79	Churn	∞			18			
80	Strap	=			19			
81					20			
82					21			
83					22			
84					23			
85					24			
86					25			
87	Handle with Ratchet				26			
88	Handle with locking Tooth				27			
89	Handle with both				28			
90					29			
91					30			
92					31			

In 1852 Ada Lovelace died in distressing circumstances. There has been a long-standing story that Ada was heavily in debt caused by gambling on the races. In my biography of Babbage I accepted this story, but I have since come to doubt its truth. All these stories have come from the entirely unreliable Lady Noel Byron, and are probably part of her disinformation programme. The remaining betting slips are only for a few pounds, and the dates when Ada is supposed to have been at the races do not tally with other information about her movements. That there was some murky business involving John Cross is clear, but it may have been, for example, straight blackmail based on compromising letters. On these matters it is better to suspend judgment and wait for a much more thorough study of Ada's life than has yet been published.

Lighthouse signalling

BETWEEN 1848 and 1857 Babbage took a break from detailed work on the Engines. During this period, which included the Great Exhibition, Babbage was involved in a wide range of activities. One of his most interesting developments was a system of signalling using occulting lights. The signalling method used by lighthouses all around the world was developed from this.

Passages from the Life of a Philospher, pp. 452–7

Occulting lights

The great object of all my inquiries has ever been to endeavour to ascertain those laws of thought by which man makes discoveries. It was by following out one of the principles which I had arrived at that I was led to the system of occulting numerical lights for distinguishing lighthouses and for night signals at sea, which I published about twelve years ago. The principle I allude to is this:—

Whenever we meet with any defect in the means we are contriving for the accomplishing a given object, that defect should be noted and reserved for future consideration, and inquiry should be made—

Whether that which is a defect as regards the object in view may not become a source of advantage in some totally different subject.

I had for a long series of years been watching the progress of electric, magnetic, and other lights of that order, with the view of using them for domestic purposes; but their want of uniformity seemed to render them hopeless for that object. Returning from a brilliant exhibition of voltaic light, I thought of applying the above rule. The accidental interruptions might, by breaking the circuit, be made to recur at any required intervals. This remark suggested their adaptation to a system of signals. But it was immediately followed by another, namely: that the interruptions were equally applicable to all lights, and might be effected by simple mechanism.

319

I then, by means of a small piece of clock-work and an argand lamp, made a *numerical* system of occultation, by which any number might be transmitted to all those within sight of the source of light. Having placed this in a window of my house, I walked down the street to the distance of about 250 yards. On turning round I perceived the number 32 clearly indicated by its occultations. There was, however, a small defect in the apparatus. After each occultation there was a kind of semi-occultation. This arose from the arm which carried the shade rebounding from the stop on which it fell. Aware that this defect could be easily remedied, I continued my onward course for about 250 yards more, with my back towards the light. On turning round I was much surprised to observe that the signal 32 was repeated distinctly without the slightest trace of any semi-occultation or blink.

I was very much astonished at this change; and on returning towards my house had the light constantly in view. After advancing a short distance I thought I perceived a very faint trace of the blink. At thirty or forty paces nearer it was clearly visible, and at the half-way point it was again perfectly distinct. I knew that the remedy was easy, but I was puzzled as to the cause.

After a little reflection I concluded that it arose from the circumstance that the small hole through which the light passed was just large enough to be visible at five hundred yards, yet that when the same hole was partially covered by the rebound there did not remain sufficient light to be seen at the full distance of five hundred yards.

Thus prepared, I again applied the principle I had commenced with and proceeded to examine whether this defect might not be converted into an advantage.

I soon perceived that a lighthouse, whose number was continually repeated with a blink, obscuring just half its light, would be seen *without any blink* at all distances beyond half its range; but that at all distances within its half range that fact would be indicated by a blink. Thus with two blinks, properly adjusted the distance of a vessel from a first-class light would be distinguished at from twenty to thirty miles by occultations indicating its number without any blink; between ten and twenty miles by an occultation with one blink, and within ten miles by an occultation with two blinks.

But another advantage was also suggested by this defect. If the opaque cylinder which intercepts the light consists of two cylinders, A and B, connected together by rods: thus—

If the compound cylinder descend to *a*, and then rise again, there will be a single occultation.
If the compound cylinder descend to *b*, and then rise again, there will be a double occultation.
If the compound cylinder descend to *c*, and then rise again, there will be a triple occultation.

Such occultations are very distinct, and are specially applicable to lighthouses.

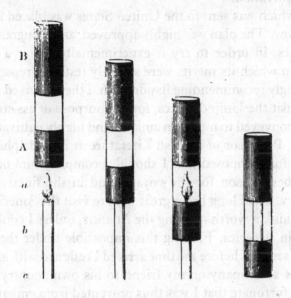

In the year 1851, during the Great Exhibition, the light I have described was exhibited from an upper window of my house in Dorset Street during many weeks. It had not passed unnoticed by foreigners, who frequently reminded me that they had passed my door when I was asleep by writing upon their card the number exhibited by the occulting light and dropping it into my letter-box.

About five or six weeks after its first appearance I received a letter from a friend of mine in the United States, expressing great interest about it, and inquiring whether its construction was a secret. My answer was, that I made no secret of it, and would prepare and send him a short description of it.

I then prepared a description, of which I had a very few copies printed. I sent twelve of these to the proper authorities of the great maritime countries. Most of them were accompanied by a private note of my own to some person of influence with whom I happened to be acquainted.

One of these was addressed to the present Emperor of the French, then a member of their Representative Chamber. It was dated the 30th November, 1852. Three days after I read in the newspapers the account of the *coup* of December 2, and smiled at the inopportune time at which my letter had accidentally been forwarded. However, three days after I received from M. Mocquard the prettiest note, saying that he was commanded by the Prince President to thank me for the communication, and to assure me that the prince was as much attached as ever to science, and should always continue to promote its cultivation.

The letter which was sent to the United States was placed in the hands of the Coast Survey. The plan was highly approved, and Congress made a grant of 5,000 dollars, in order to try it experimentally. After a long series of experiments, in which its merits were severely tested, a report was made to Congress strongly recommending its adoption. I then received a very pressing invitation to visit the United States, for the purpose of assisting to put it in action. It was conveyed to me by an amiable and highly cultivated person, the late Mr. Reed, Professor of English Literature at Philadelphia, who, on his arrival in London, proposed that I should accompany him on his return in October, the best season for the voyage, and in the finest vessel of their mercantile navy. I had long had a great wish to visit the American continent, but I did not think it worth crossing the Atlantic, unless I could have spent a twelvemonth in America. Finding this impossible under the then circumstances, about a month before the time arrived I resigned with great reluctance the pleasure of accompanying my friend to his own country.

It was most fortunate that I was thus prevented from embarking on board the *Arctic*, a steamer of the largest class.

Steaming at the rate of thirteen knots an hour over the banks of Newfoundland during a dense fog, the *Arctic* was run into by a steamer of about half its size, moving at the rate of seven knots. The concussion was in this instance fatal to the larger vessel.

Passages from the Life of a Philosopher, pp. 458–64

In 1853 I spent some weeks at Brussels. During my residence in that city a Congress of naval officers from all the maritime nations assembled to discuss and agree upon certain rules and observations to be arranged for the common benefit of all. One evening I had the great pleasure of receiving the whole party at my house for the purpose of witnessing my occulting lights.

The portable occulting light which I had brought with me was placed in the verandah on the first floor, and we then went along the Boulevards to see its effect at different distances and with various numerical signals. On our return several papers relating to the subject were lying upon the table. The Russian representative, M. —, took up one of the original printed descriptions and was much interested in it. On taking leave he asked, with some hesitation, whether I would lend it to him for a few hours. I told him at once that if I possessed another copy I would willingly give it to him; but that not being the case I could only offer to lend it. M. — therefore took it home with him, and when I sat down to breakfast the next morning I found it upon my table. In the course of the day I met my Russian friend in the Park. I expressed my hope that he had been interested by the little tract he had so speedily returned. He replied that it had interested him so much that he had sat up all night, had copied the whole of it, and that his transcript and a despatch upon the the subject was now on its way by the post to his own Government.

Several years after I was informed that *occulting solar lights* were used by the Russians during the siege of Sebastopol.

Night signals

The system of occulting light applies with remarkable facility to night signals, either on shore or at sea. If it is used numerically, it applies to all the great dictionaries of the various maritime nations. I may here remark, that there exist means by which all such signals may, if necessary, be communicated in cipher.

Sun signals

The distance at which such signals can be rendered visible exceeds that of any other class of signals by means of light. During the Irish Trigonometrical Survey, a mountain in Scotland was observed, with an angular instrument from a station in Ireland, at the distance of 108 miles. This was accomplished by stationing a party on the summit of the mountain in Scotland with a looking-glass of about a foot square, directing the sun's image to the opposite station. No occultations were used; but if the mirror had been larger, and occultation employed, messages might have been sent, and the time of residence upon the mountain considerably diminished. When I was occupied with occulting signals, I made this widely known. I afterwards communicated the plan, during a visit to Paris, to many of my friends in that capital, and, by request, to the Minister of Marine.

I have observed in the *Comptes Rendus* that the system has to a certain extent been since used in the south of Algeria, where, during eight months of the year, the sun is generally unobscured by clouds as long as it is above the horizon. I have not, however, noticed in those communications to the Institute any reference to my own previous publication.

Zenith-light signals

Another form of signal, although not capable of use at very great distance, may, however, be employed with considerable advantage, under certain circumstances. Universality and economy are its great advantages. It consists of a looking-glass, making an angle of 45° with the horizon, placed just behind an opening in a vertical board. This being stuck into the earth, the light of the sky in the zenith, which is usually the brightest, will be projected horizontally through the opening, in whatever direction the person to be communicated with may be placed. The person who makes the signals must stand on one side in front of the instrument; and, by passing his hat slowly before the aperture any number of times, may thus express each unit's figure of his signal.

He must then, leaving the light visible, pause whilst he deliberately counts to himself ten.

He must then with his hat make a number of occultations equal to the tens figure he wishes to express.

This must be continued for each figure in the number of the signal, always pausing between each during the time of counting ten.

When the end of the signal is terminated, he must count sixty in the same manner; and if the signal he gave has not been acknowledged, he should repeat it until it has been observed.

The same simple telegraph may be used in a dark night, by substituting a lantern for the looking-glass. The whole apparatus is simple and cheap, and can be easily carried even by a small boy.

I was led to this contrivance many years ago by reading an account of a vessel stranded within thirty yards of the shore. Its crew consisted of thirteen people, ten of whom got into the boat, leaving the master, who thought himself safer in the ship, with two others of the crew.

The boat put off from the ship, keeping as much out of the breakers as it could, and looking out for a favourable place for landing. The people on shore followed the boat for several miles, urging them not to attempt landing. But not a single word was audible by the boat's crew, who, after rowing several miles, resolved to take advantage of the first favourable lull. They did so—the boat was knocked to pieces, and the whole crew were drowned. If the people on the shore could at that moment have communicated with the boat's crew, they could have informed them that, by continuing their course for half a mile further, they might turn into a cove, and land almost dry.

I was much impressed by the want of easy communication between stranded vessels and those on shore who might rescue them.

I can even now scarcely believe it credible that the very simple means I am about to mention has not been adopted years ago. A list of about a hundred questions, relating to directions and inquiries required to be communicated between the crew of a stranded ship and those on shore who wish to aid it, would, I am told, be amply sufficient for such purposes. Now, if such a list of inquiries were prepared and printed by competent authority, any system of signals by which a number of two places of figures can be expressed might be used. This list of inquiries and answers ought to be printed on cards, and nailed up on several parts of every vessel. It would be still better, by conference with other maritime nations, to adopt the same system of signs, and to have them printed in each language. A looking-glass, a board with a hole in it, and a lantern would be all the apparatus required. The lantern might be used for night, and the looking-glass for day signals.

These simple and inexpensive signals might be occasionally found useful for various social purposes.

Two neighbours in the country whose houses, though reciprocally visible, are separated by an interval of several miles, might occasionally telegraph to each other.

If the looking-glass were of large size, its light and its occultation might be seen perhaps from six to ten miles, and thus become by daylight a cheap guiding light through channels and into harbours.

It may also become a question whether it might not in some cases save the expense of buoying certain channels.

For railway signals during daylight it might in some cases be of great advantage, by saving the erection of very lofty poles carrying dark frames through which the light of the sky is admitted.

Amongst my early experiments, I made an occulting hand-lantern, with a shade for occulting by the pressure of the thumb, and with two other shades of red and of green glass. This might be made available for military purposes, or for the police.

Greenwich time signals

It has been thought very desirable that a signal to indicate Greenwich time should be placed on the Start Point, the last spot which ships going down the Channel on distant voyages usually sight.

The advantage of such an arrangement arises from this—that chronometers having had their rates ascertained on shore, may have them somewhat altered by the motions to which they are submitted at sea. If, therefore, after a run of about two hundred miles, they can be informed of the exact Greenwich time, the sea rate of their chronometers will be obtained.

Of course no other difficulty than that of expense occurs in transmitting Greenwich time by electricity to any points on our coast. The real difficulty is to convey it to the passing vessels. The firing of a cannon at certain fixed hours has been proposed, but this plan is encumbered by requiring the knowledge of the distance of the vessel from the gun, and also from the variation of the velocity of the transmission of sound under various circumstances.

During the night the flash arising from ignited gunpowder might be employed. But this, in case of rain or other atmospheric circumstances, might be impeded. The best plan for night-signals would be to have an occulting light, which might be that of the lighthouse itself, or another specially reserved for the purpose.

During the day, and when the sun is shining, the time might be transmitted by the occultations of reflected solar light, which would be seen at any distance the curvature of the earth admitted.

The application of my Zenith Light might perhaps fulfil all the required conditions during daylight.

I have found that, even in the atmosphere of London, an opening only five inches square can be distinctly seen, and its occultations counted by the naked eye at the distance of a quarter of a mile. If the side of the opening were double the former, then the light transmitted to the eye would be four times as great, and the occultations might be observed at the distance of one mile.

The looking-glass employed must have its side nearly in the proportion of three to two, so that one of five feet by seven and a half ought to be seen at the distance of about eight or nine miles.

Postcript

IN 1856 Babbage returned to work on the Analytical Engines. He attempted to develop constructional methods and to simplify the logical design to the point at which he could construct a minimal Engine with his private resources. In this he was not successful and no working Analytical Engine was made. During this late period the work did not show the systematic application of Babbage's work in his earlier years. Rather it was a hobby for his old age. However he did develop interesting concepts in computing, treating computer design with much of the freedom of a modern logic designer, and the theoretical ideas developed during this period brought him closest to the modern concept of a computer. Details of Babbage's Engines have only begun to become generally known in recent years. The unfamiliarity of Babbage's mechanical notation has hitherto made detail accessible to few, and translation of Babbage's designs into modern terminology has only recently begun.

After Babbage's death his work was carried on by his youngest son, Major-General Henry Prevost Babbage. After his death, detailed knowledge may have been lost, but Babbage's more immediately obvious ideas were absorbed into the general stream of knowledge of mechanical computation. Amongst the mathematical élite of Cambridge, Babbage's ideas were never forgotten and remained almost synonymous with the idea of mechanizing computation.

Curiously commentators seem to have missed the main source of knowledge about Babbage's Engines: the *Encyclopaedia Britannica*. In the eighth edition of the *Britannica* (1853–1860) the introductory dissertation, by James David Forbes, Professor of Natural Philosophy in Edinburgh, section 377 (pp. 880–1) is on Babbage's Calculating Engines, and gives references to the principal published works on the Engines. The ninth edition has a separate entry on 'Calculating Machines' by P. L. Scott, professor of mathematics at St Andrews University. Scott observes that other calculating machines were 'Completely cast in the shade by the wonderful inventions of the late Charles Babbage'. The entry gives a careful discussion of the mathematical principles of the Difference Engine, and discusses the Analytical Engine, explaining crucially the use of 'operation cards' and 'variable cards'. It also gives the reference to the Menabrea paper and Ada's notes in Taylor's *Scientific Memoirs*. Volume 4 of the ninth edition, which includes the entry on Calculating Machines, was published in 1876.

In 1882 Herman Hollerith became an instructor of mechanical engineering at the Massachusetts Institute for Technology, and such was the position of the *Britannica* at the time that it would be remarkable if the entry had failed to come to his notice. One should not look for single-factor theories in invention – the Jacquard loom and the ticket punch doubtless played their parts – but Babbage's Engines must be considered a likely contributing factor. It was simply that the *Britannica* was such a commonplace reference that nobody bothered to mention it.

After the ninth edition, interest in Babbage's Engines waned. The eleventh edition gives no details. It does mention punched-card control, but that says little more than was well known for the Jacquard loom. The fourteenth edition mentions *Babbage's Calculating Engines* by Henry Prevost Babbage, which included all the important published work on the Engines, as well as other valuable material, but the book was privately published and generally available in only a handful of libraries. Thus by the time we come to modern work the fact that Babbage's Engines were described in earlier editions of the *Britannica* was largely forgotten.

In his biography of Alan Turing, *The Enigma of Intelligence* (Burnett, 1983), Andrew Hodges suggests that Turing might have seen the fragment of Babbage's minimal Analytical Engine in the Science Museum in South Kensington, London. If so it would not have told him much. However, Turing certainly knew about Babbage's Engines during the war and it would be interesting to know just when he learned about them. The decisive stimulus on Turing in this respect was a reference to mechanization of computation by M. H. A. Newman in a course of part III lectures in Cambridge on the foundations of mathematics. Whether Newman discussed Babbage's Engines seems not to be known, but for Cambridge mathematicians mechanization of computation and Babbage's work are inextricably woven together.

Since the development of modern computing, Babbage's role as the great pioneer of the computer has become generally recognized. Tribute after tribute has been paid to his genius, leading to the conference on 'Computing, Past, Present, and Future', at Imperial College, London in the summer of 1991, to celebrate the bicentenary of Charles Babbage's birth.

Works represented

Two of these excerpts, on the Difference and on the Analytical Engines, were not written by Babbage himself but by friends of his, with his close cooperation, and are authoritative statements of his views of the Engines.

1813   *Memoirs of the Analytical Society*, Preface, Cambridge; in collaboration with John Herschel.

1822   *A Letter to Sir Humphry Davy, Bart. PRS, on the Application of machinery to the Purpose of Calculating and Printing Mathematical Tables*, J. Booth, London.

1826   *A Comparative View of the Various Institutions for the Assurance of Lives*, J. Mawman, London.

1830   *Reflections on the Decline of Science in England, and Some of its Causes*, B. Fellowes, London.

1833   *A Word to the Wise*, John Murray, London; reprinted in 1856 and subtitled 'Observations on Peerage for Life'.

1834   'Babbage's Calculating Engine', by Dionysius Lardner. *Edinburgh Review*, **CXX**, July.

1835   *On the Economy of Machinery and Manufactures*, 4th edn. Charles Knight, London.

1838   *The Ninth Bridgewater Treatise*, 2nd edn. John Murray, London.

1843   'Sketch of the Analytical Engine', by General Menabrea, translated with notes by Ada Lovelace, *Scientific Memoirs*, iii, 666–731.

1848   *Thoughts on the Principles of Taxation with Reference to a Property Tax, and its Exceptions*, John Murray, London; (3rd edn, 1852).

1851   *Laws of Mechanical Notation*, privately printed, London.

1864   *Passages from the Life of a Philosopher*, Longman, Green, London.

1865   *Thoughts upon an Extension of the Franchise*, Longman, Green, London.

Index

Plates

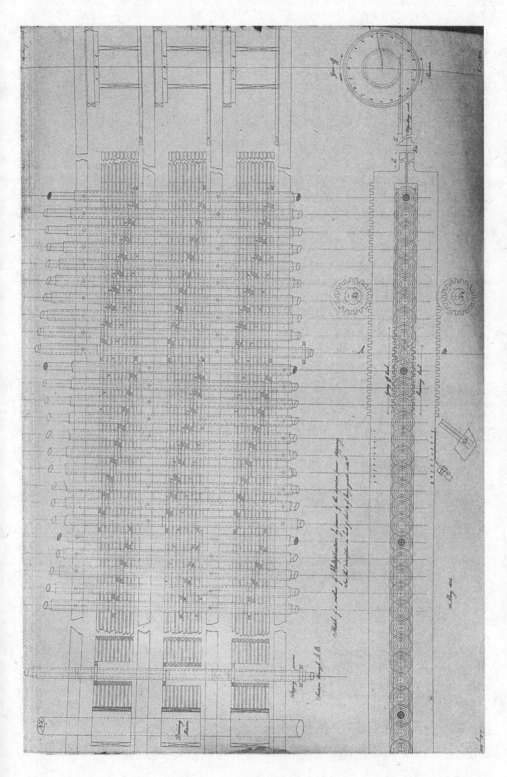

Plate 1 Multiplication by stepping the directive powers; 16 May 1835. An early plan for multiplication.

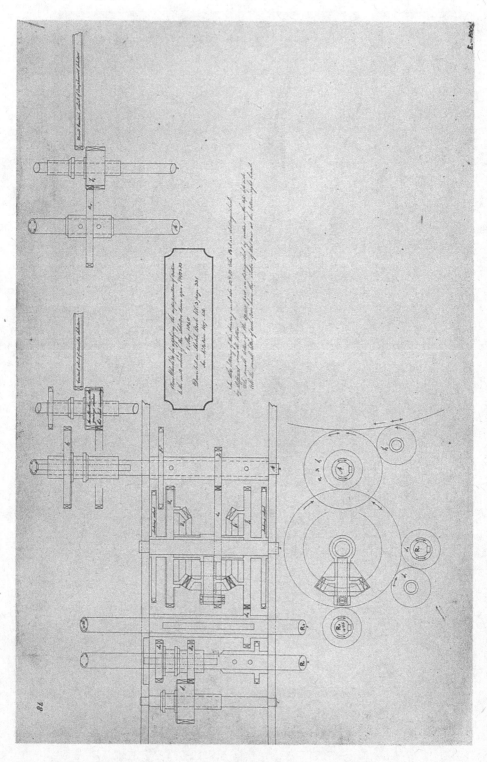

Plate 2 Use of superposition of motion in microprogram (to use a modern term) selection; 1 May 1840. For the detailed logic of the superposition of motion, see: Anthony Hyman, *Charles Babbage, Pioneer of the Computer* (OUP/Princeton, 1982), plate 7.

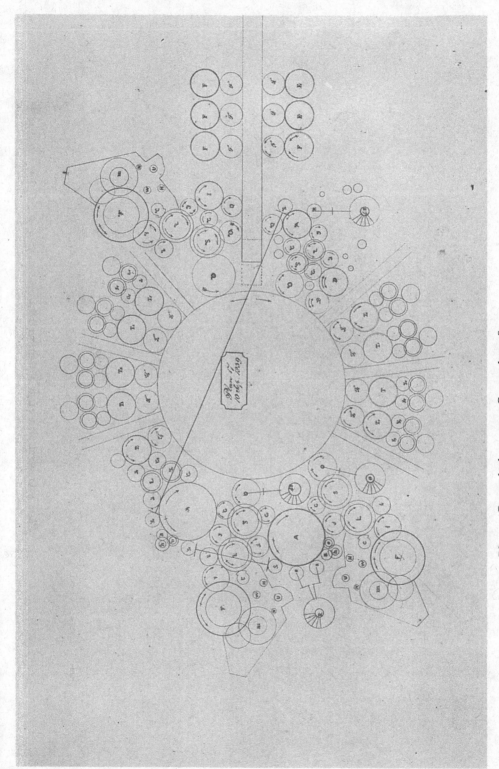

Plate 3 General plan 17; 10 September 1839.

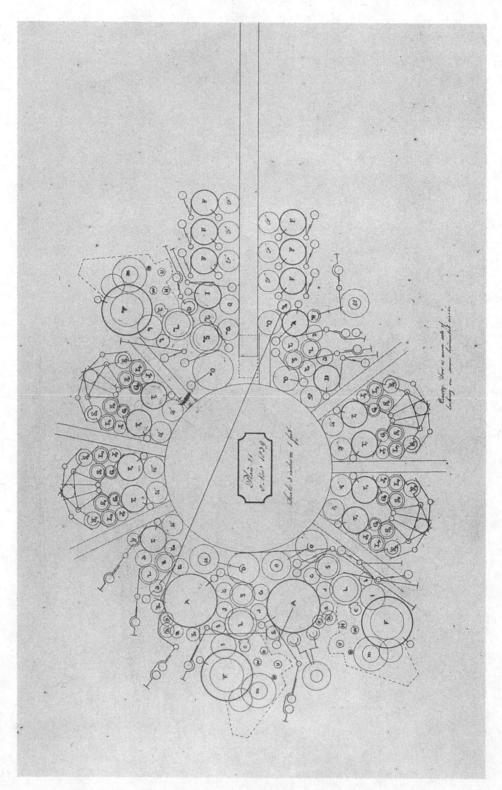

Plate 4 General plan 21; 8 November 1839.

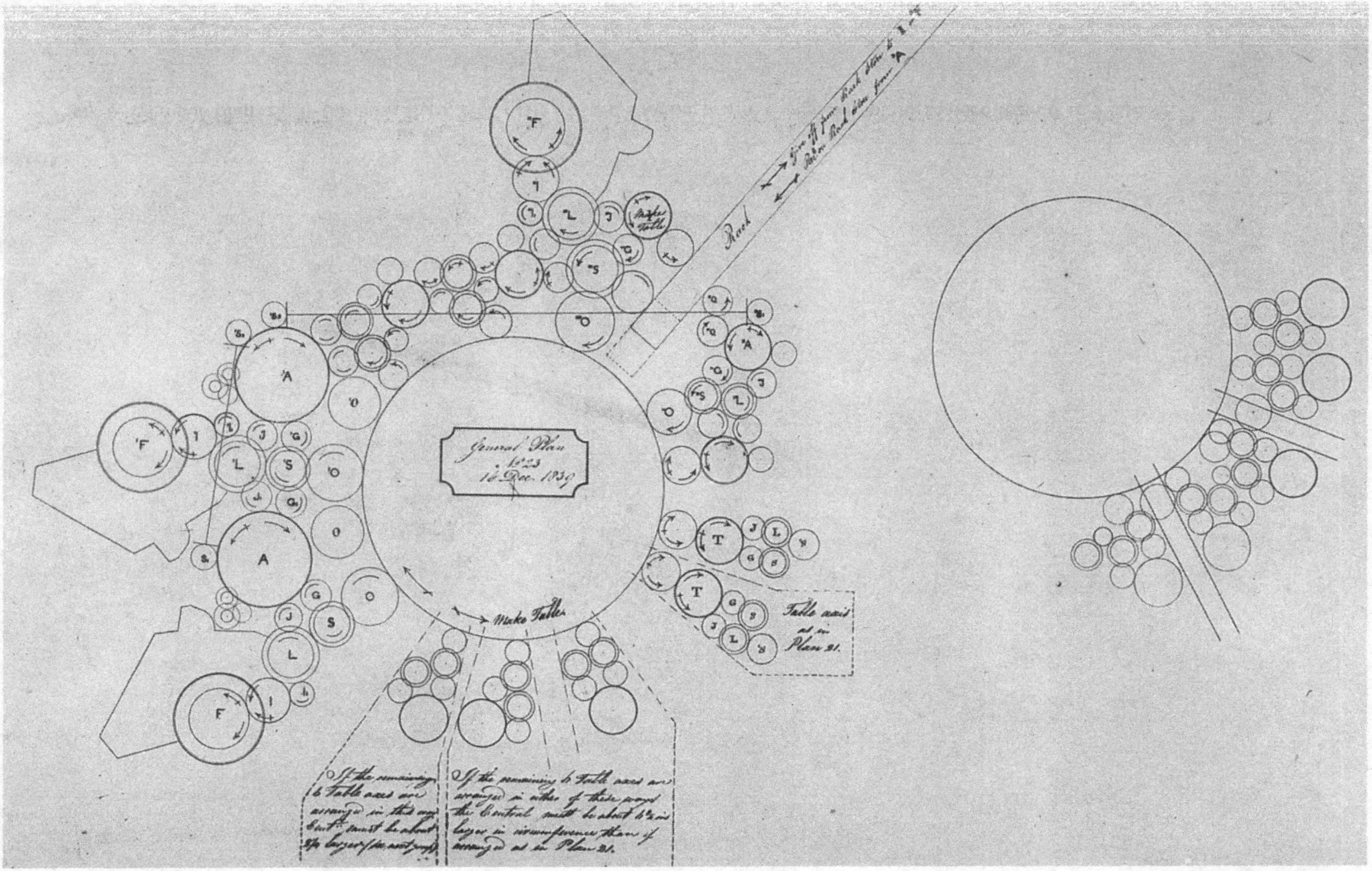

Plate 5 General plan 23; 16 December 1839.

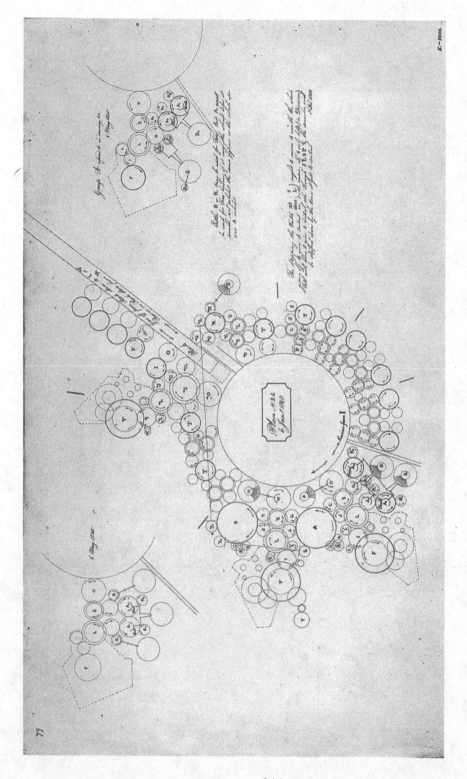

Plate 6 General plan 24; 6 January 1840. For plan 25, see *Pioneer of the Computer*, plate 11 (see legend to Plate 2).

344

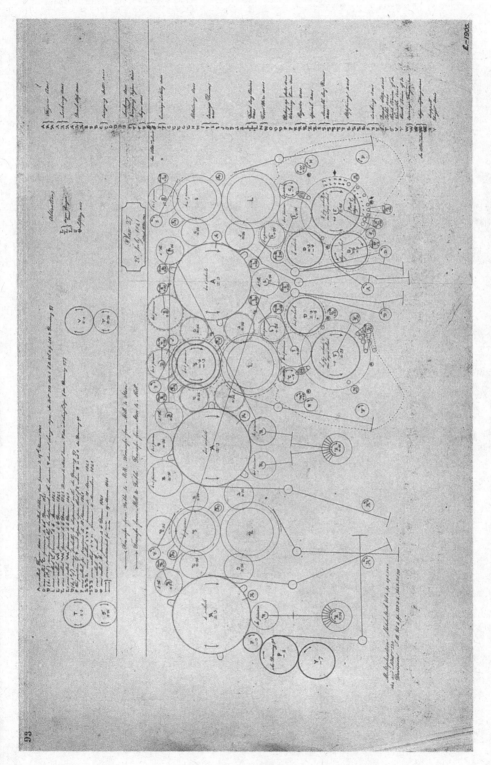

Plate 7 General plan 27, using a rack interconnection system in the mill; 27 July 1841.

345

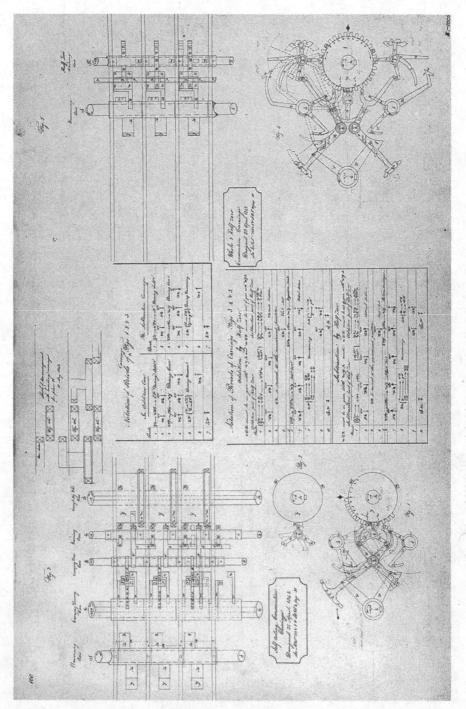

Plate 8 Whole and half zero consecutive carriage; 22 April 1842.

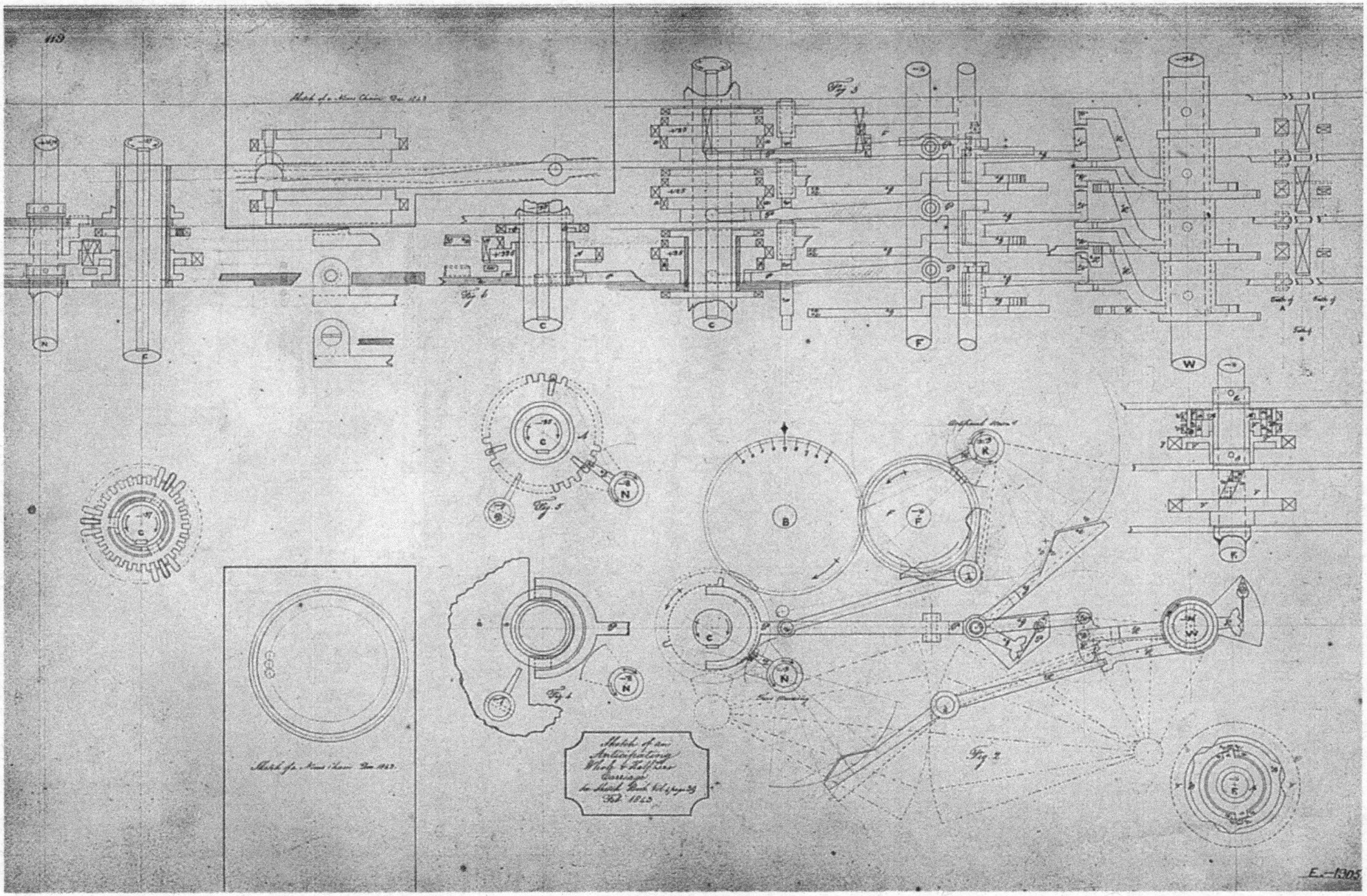

Plate 9 Whole and half zero anticipating carriage; February 1843.

347

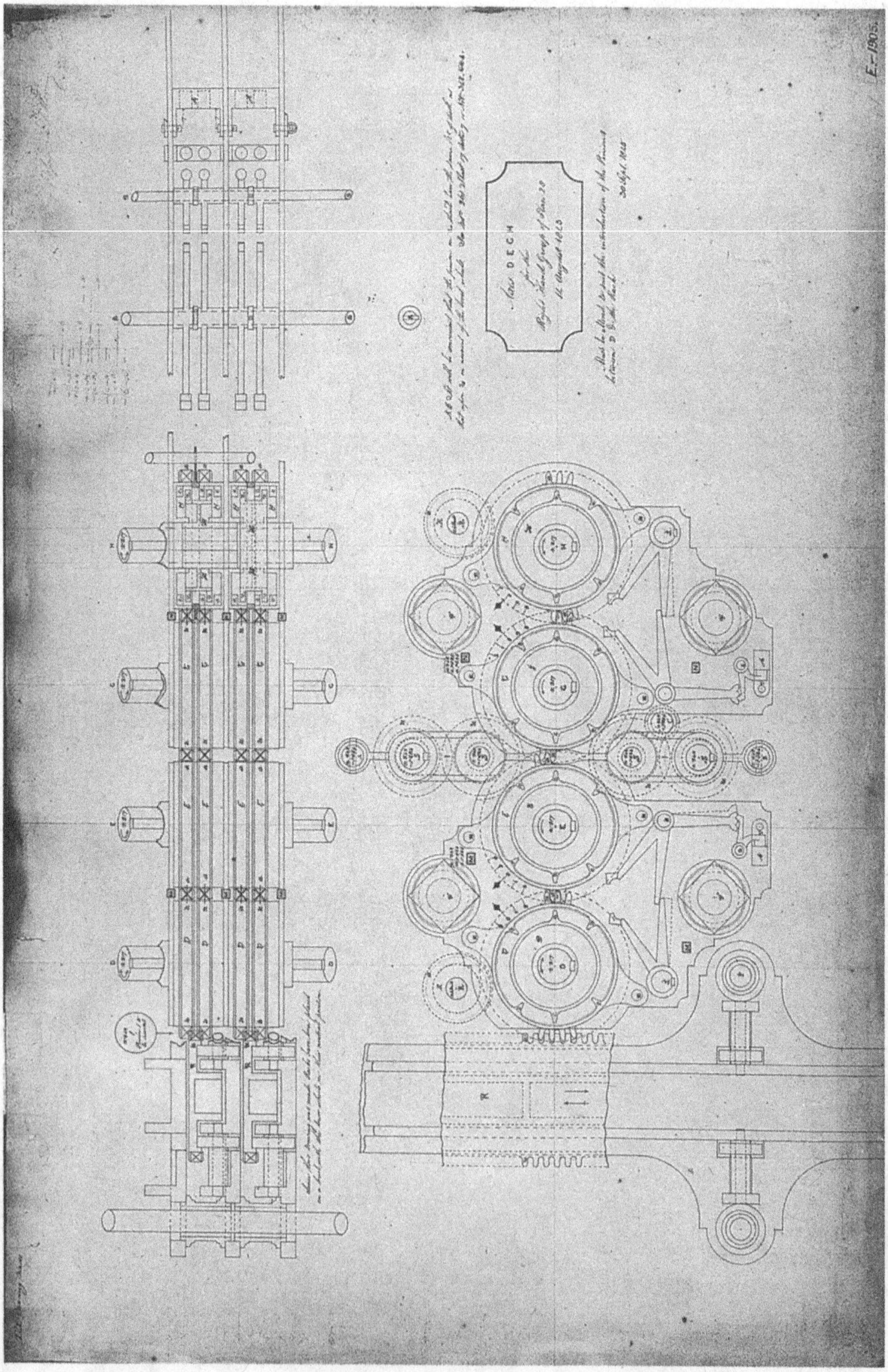

Plate 10 Axes D E G H for right-hand group of plan 28; 14 August 1843.

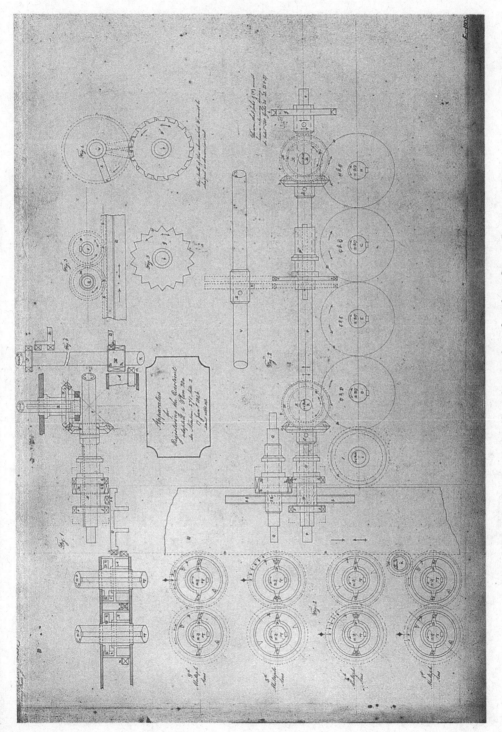

Plate 11 Apparatus for registering the quotient for plan 28a; 17 January 1846.

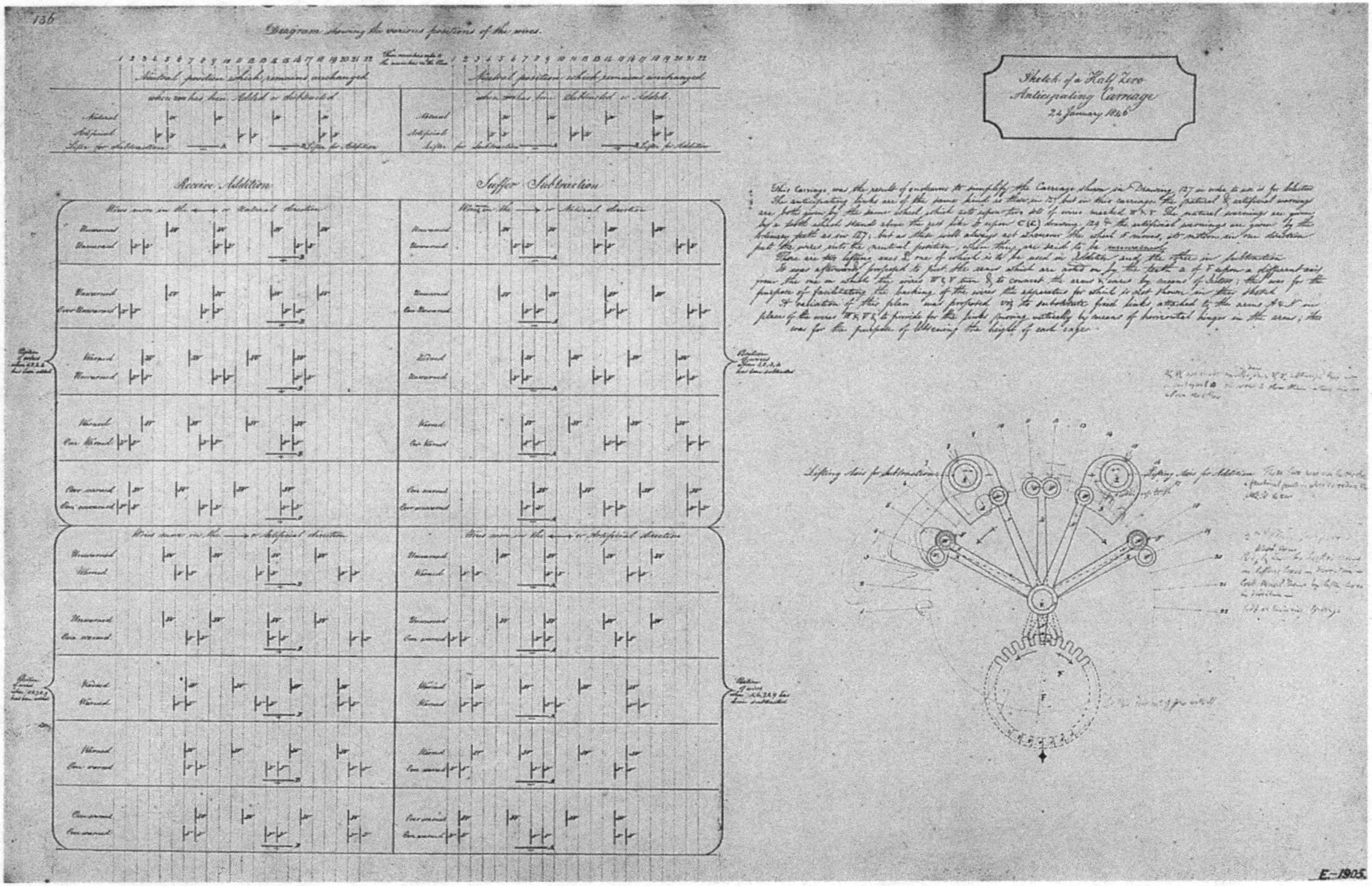

Plate 12 Half zero anticipating carriage; 24 January 1846.

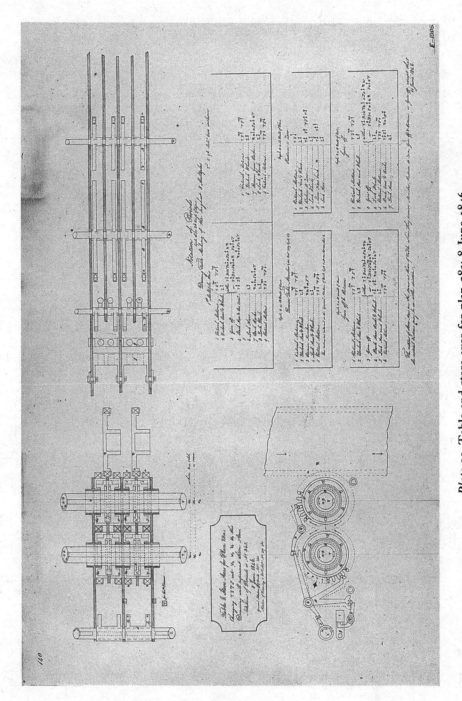

Plate 13 Table and store axes for plan 28a; 8 June 1846.

In the main plan when he returned to detailed work on the Analytical Engines after 1856, Babbage used a rack interconnection system in the mill, as in plans 27, 28, and 28a, but he made the important advance of using an X–Y grid for the store (see *Pioneer of the Computer*, fig. 2: see legend to Plate 2). Plan 28a will be a strong candidate when someone finally settles down to construct one of Babbage's Analytical Engines.

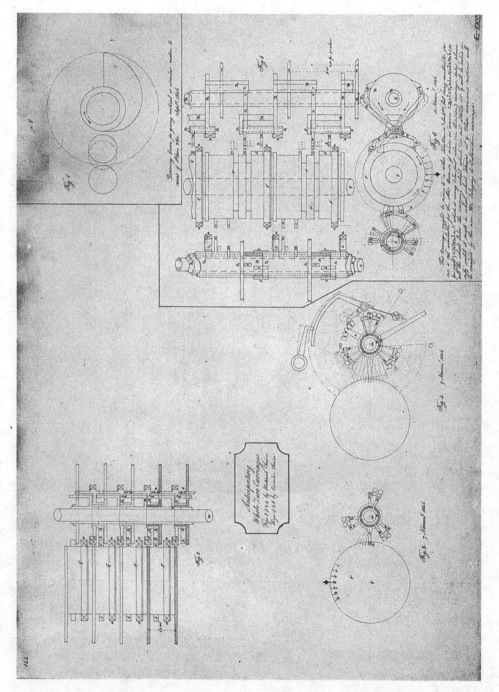

Plate 14 Anticipating whole zero carriage: vertical chain, circular chain; 7 November 1846.

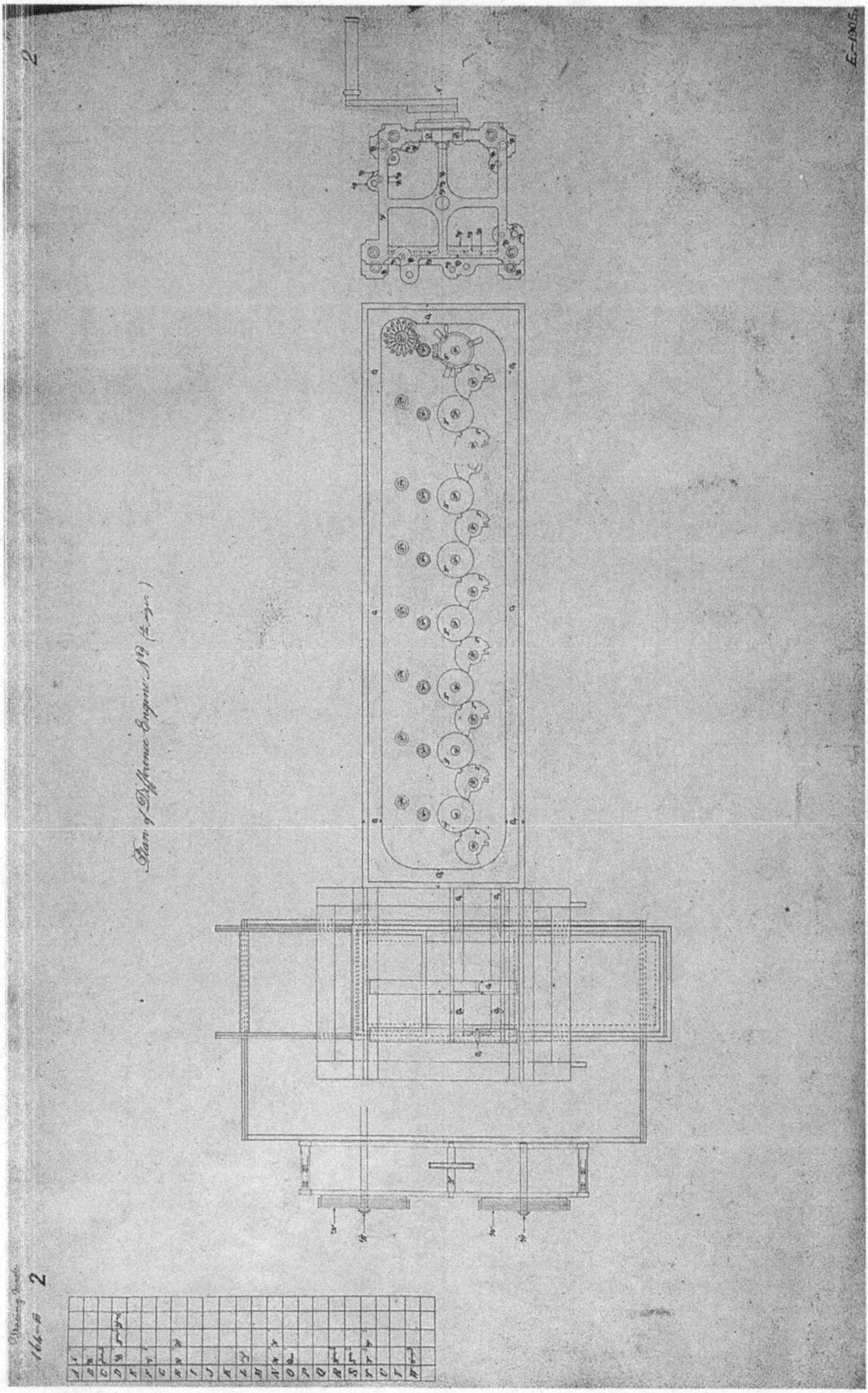

Plate 15 Second Difference Engine: plan.

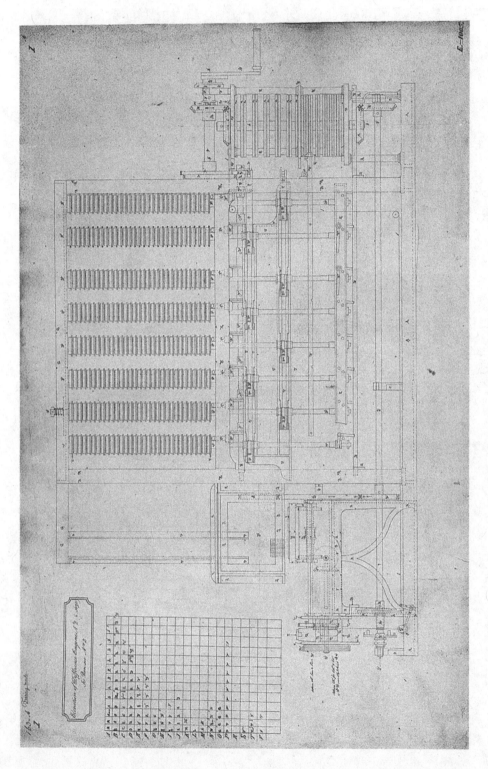

Plate 16 Second Difference Engine: elevation.

354

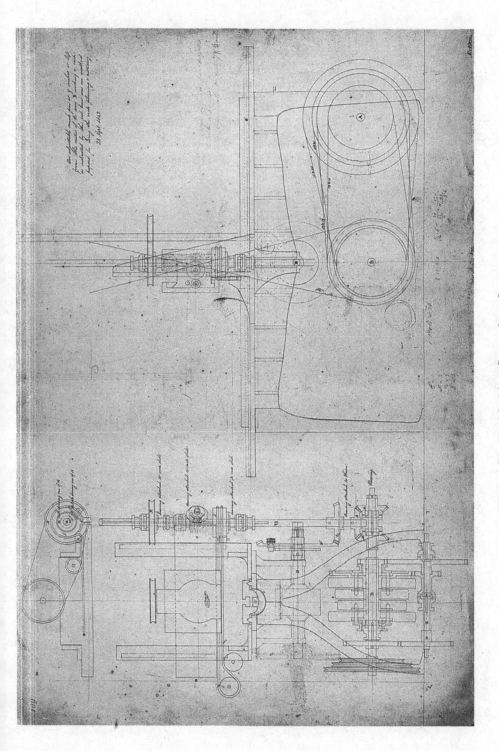

Plate 17 Planing Machine; September 1842.

355

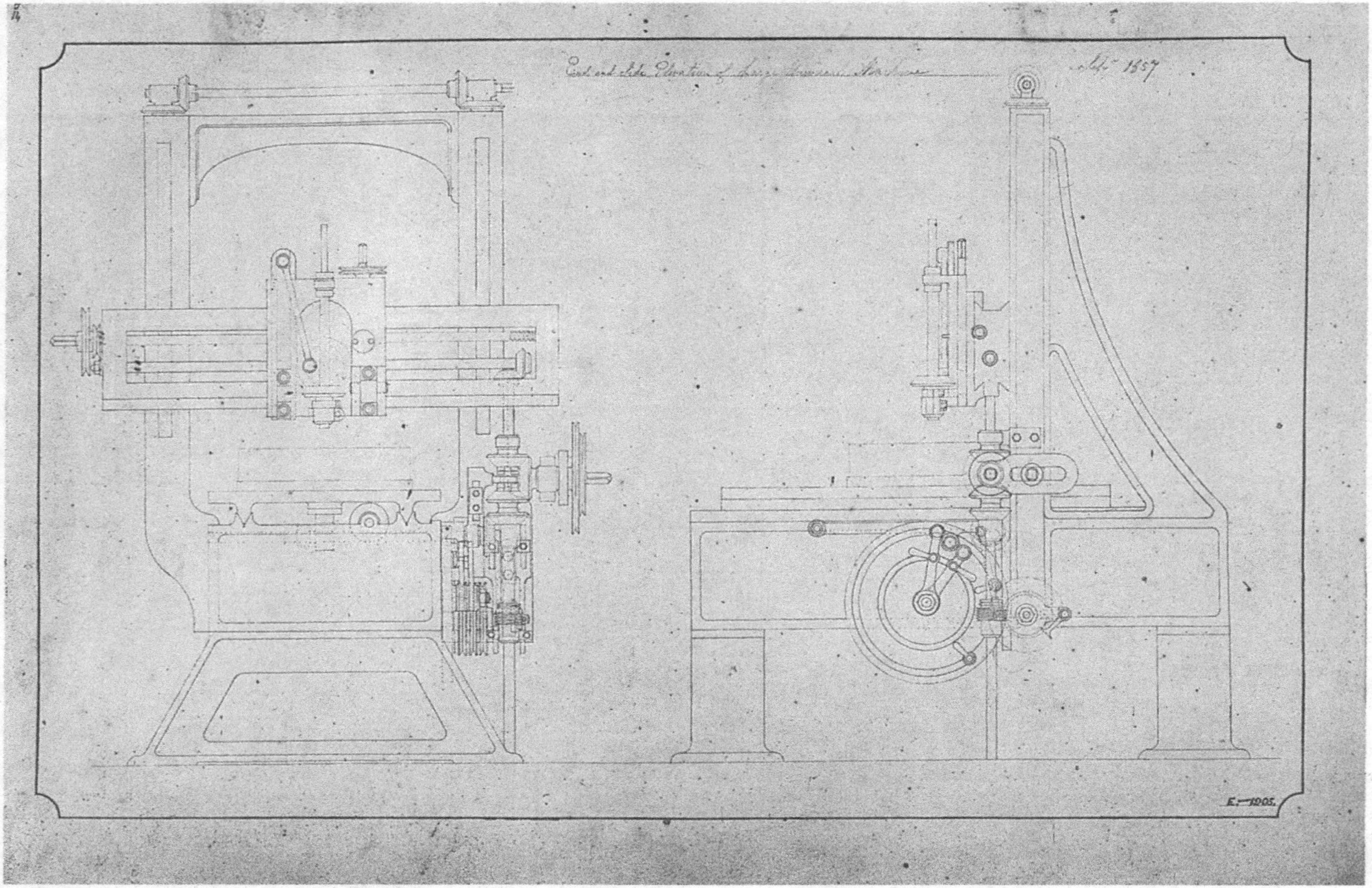

Plate 18 Large Universal Machine-Tool; September 1857.